Einstein, Relativity and Absolute Simultaneity

AF736589

2005 marked the centenary of one of the most remarkable publications in the history of science, Albert Einstein's "On the Electrodynamics of Moving Bodies," in which he presented a theory that later came to be known as the Special Theory of Relativity. This 1905 paper is widely regarded as having destroyed the classical conceptions of absolute time and space, along with absolute simultaneity and absolute length, which had reigned in physics from the times of Galileo and Newton to the dawn of the twentieth century. As we embark upon a new century, the Special Theory is now 100 years old, and a great deal has transpired in both philosophy and physics since its first publication. This volume is a timely reappraisal the theory's central claims, especially concerning the elimination of absolute time and absolute simultaneity.

This collection draws together essays by both philosophers and physicists, and reflects the cutting edge of research and thought on the question of absolute simultaneity. The issues discussed in the book include Aspect's confirm of Bell's theorem, De Broglie-Bohm's quantum mechanics, the privileged cosmic time series in a Friedman universe, Lorentz's ideas and neo-Lorentzian theory and other relevant issues. Almost all the contributors are convinced that the received view that simultaneity is not an absolute relation is not only unwarranted but false, and it is hoped that this collection will stimulate discussion among both philosophers and physicists concerning the warrant for and problems with assertions of the relativity of simultaneity on the basis of Einstein's theory.

William Lane Craig is Research Professor of Philosophy at Talbot School of Theology, USA.

Quentin Smith is Professor of Philosophy at Western Michigan University, USA.

Einstein, Relativity and Absolute Simultaneity

Edited by William Lane Craig
and Quentin Smith

Routledge
Taylor & Francis Group
LONDON AND NEW YORK

First published 2008
by Routledge
2 Park Square, Milton Park, Abingdon, OX14 4RN

Simultaneously published in the USA and Canada
by Routledge
270 Madison Avenue, New York, NY 10016

Routledge is an imprint of the Taylor & Francis Group, an informa business

First issued in paperback 2011

© 2008 William Lane Craig and Quentin Smith, editorial matter and selection; the contributors, their contributions

Typeset in Times New Roman by
Taylor & Francis Books

All rights reserved. No part of this book may be reprinted or reproduced or utilized in any form or by any electronic, mechanical, or other means, now known or hereafter invented, including photocopying and recording, or in any information storage or retrieval system, without permission in writing from the publishers.

British Library Cataloging in Publication Data
A catalogue record for this book is available from the British Library

Library of Congress Cataloging in Publication Data
Einstein, relativity, and absolute simultaneity / edited by William Lane Craig and Quentin Smith.
p. cm.
1. Relativity (Physics) 2. Special relativity (Physics) 3. Absolute simultaneity. 4. Space and time. 5. Bell's theorem. 6. Lorenz transformation. I. Craig, William Lane. II. Smith, Quentin, 1952-
QC173.585.E35 2006
530.11–dc22
2006015456

ISBN13: 978-0-415-70174-7 (hbk)
ISBN13: 978-0-415-59166-9 (pbk)
ISBN13: 978-0-203-70005-1 (ebk)

Contents

Illustrations

Figures

Tables

Introduction

2005 marked the centenary of one of the most remarkable publications in the history of science, Albert Einstein's "On the Electrodynamics of Moving Bodies," in which he presented a theory that later came to be known as the Special Theory of Relativity (STR).

This 1905 paper is widely regarded as having destroyed the classical conceptions of absolute time and space, along with absolute simultaneity and absolute length, which had reigned in physics from the times of Galileo and Newton to the dawn of the twentieth century.

As we embark upon a new century, the Special Theory is now 100 years old, and a great deal has transpired in both philosophy and physics since its first publication. It therefore seems appropriate at this time to seek a fresh appraisal of the theory's central claims, especially concerning the elimination of absolute time and absolute simultaneity.

Part I

The classical concepts of time and space were codified by Isaac Newton in his epochal *Philosophiae Naturalis Principia Mathematica* (1687). In the Scholium to his definitions in the Principia Newton, in order to overcome certain prejudices of "the common people," who conceive of quantities such as time, space, place, and motion only in terms of "the relation they bear to sensible objects," drew a distinction with respect to these quantities between "absolute and relative, true and apparent, mathematical and common".[1] The most striking feature of this distinction is the independence of absolute time and space from the relative measures thereof. Absolute time or simple duration exists regardless of the sensible and external measurements which we try, more or less successfully, to make of it. Similarly, Newtonian space is absolute in the sense that it is distinct from the relatively moving spaces associated with inertial frames and, hence, also independent of the physical measures we apply to it. But Newtonian time and space are also absolute in an important and relevant second sense. Newtonian time is absolute in the sense that the simultaneity of two events e_1 and e_2 requires only a two-place relation of *simultaneous with* between e_1 and e_2, rather than a three-place relation among e_1, e_2, and a reference frame (or a three-dimensional hypersurface of space-time), namely, e_1's being *simultaneous with* e_2 *relative to* a reference frame *F.* Newtonian space is similarly absolute in the sense that length is a monadic property of physical objects, rather than a relational property of physical objects that includes a relation to a reference frame or hypersurface.

Relativity already governed Newtonian mechanics, that is to say, it was impossible for a hypothetical observer associated with an inertial frame to perform mechanical experiments which would disclose to him whether he was in motion or at rest. Newton's three laws of motion and his universal law of gravity applied strictly only to the frame of absolute space or to inertial frames which are at rest with respect to absolute space, but they could be transformed and expressed in any sensible and apparent inertial frame. This principle of relativity had already been enunciated by Galileo and comes to expression in the Galilean transformation equations relating the coordinates of an event in one reference frame to the coordinates assigned to that event relative to some other reference frame. Newton's mechanics assumed that the Galilean transformation equations were the appropriate means of transforming Newton's physics from one frame to another.

There was a loophole, however, in Galilean relativity that became apparent during the nineteenth century. Although mechanics had been relativised, electrodynamics had not. Maxwell's electromagnetic theory (1865) implied that electromagnetic waves propagate through a medium, an "aether" or field, at a constant velocity *c*, which is independent of the motion of the source of the waves. Consequently, by measuring the speed at which light waves in the luminiferous aether pass through one's measuring apparatus, one could determine what mechanics alone could not discover, namely, whether the inertial system in which the experiment was performed was at rest or in motion with respect to the aether. Since the aether was conceived to be at rest with respect to absolute space, any motion through it would be absolute, rather than merely relative. Detection of such motion on the part of any hypothetical observer would therefore provide him with what Newtonian mechanics alone could not, namely, knowledge of his situation with respect to absolute space and time.

The failure of nineteenth century attempts to detect the earth's motion through the aether prompted a crisis in physics which compelled men like FitzGerald, Lorentz, Larmor, and Poincaré to reject the Newtonian assumption that the Galilean transformation equations were sufficient to obtain invariant laws about observed phenomena in electrodynamics and mechanics and to adopt instead the relativistic Lorentz transformations. In so doing, they had already sounded the death knell of Newtonian physics, for they had relativised the sensible measures of absolute time and space in a way undreamt of by Newton. Measures of simultaneity and of length will vary from frame to frame and so become relative to reference frames. But since absolute time and space are independent of their sensible measures, the relativity of these measures did not lead these theorists to abandon the notion that there really exists absolute simultaneity and absolute length, even if these quantities remain undetectable to us.

Einstein inaugurated a "scientific revolution" in his 1905 paper by interrupting the research program of Lorentz and others with a radically different approach. Foundational to this approach was the denial of absolute space and the consequent redefinition of time and simultaneity so as to deny their absolute status as well. What Einstein did, in effect, was to eliminate Newton's absolute time and space on the grounds that they were unobservable in principle, and along with them the aether, thus leaving behind only their sensible measures, so that sensible time became the only time there is and sensible space the only space there is. Since these had by now been relativised to inertial frames, one ends up with the relativity of simultaneity and of length.

Part II

How could Einstein "know" that absolute time and space do not exist? Most historians of science now recognize that Einstein's rejection of Newtonian absolute time and space was predicated upon a positivist philosophy of science. It is now widely acknowledged that at the philosophical roots of Einstein's theory lay an epistemological positivism of Machian provenance which issued in a verificationist analysis of the concepts of time and space. The introductory sections of Einstein's 1905 paper are predicated squarely upon verificationist assumptions. Einstein takes it for granted that *all* our judgments in which time plays a role must have a "physical meaning,"[2] where physical meaning is given by operational definitions. Operationalism is already a strong form of verificationism, but there is more. Einstein goes on to say that "a mathematical description of this kind has no physical meaning unless we are quite clear as to what we will understand by 'time'." The meaning of "time" is made to depend upon the meaning of "simultaneity," which is defined locally in terms of occurrence at the same local clock reading. When it comes to judgments concerning the simultaneity of distant events, the concern is to find a "practical arrangement" to compare clock times. In order to "define" a common time for spatially separated clocks, we adopt the convention that the time which light takes to travel between two relatively stationary observers *A* and *B* is the same from *A* to *B* as from *B* to *A* in a round trip journey – a definition which *presupposes* that absolute space does not exist. For that definition presupposes that *A* and *B* are not at relative rest but both moving in tandem absolutely, or in other words that neither absolute space nor a privileged rest frame exists. The only justification for that assumption is that it is observationally or sensibly impossible to distinguish uniform motion from rest relative to such a frame, and if absolute space and absolute motion or rest cannot be sensibly observed, they therefore do not exist (and may even be said to be meaningless). Such an inference is clearly verificationist and amounts to a variation on Berkeley: "To be is to be perceivable." This sort of inference unjustifiably draws an ontological conclusion from a mere epistemological assumption.

Thus, in Einstein's formulation time and space are reduced to sensible measures based on periodic motions (such as natural or constructed clock pointer movements) and rigid measuring rods, measures that are relative to local inertial frames. Simultaneity is defined in terms of clock synchronization via light signals. All of this is done by mere stipulation. Newton's absolute time and space, which transcend operational definitions, are assumed to be mere figments of our imagination. Through Einstein's operational definitions of time and space, Mach's positivism triumphs in the STR. Reality is reduced to what our physical instruments say.

Unfortunately for Einstein's Special Theory, however, its epistemological and ontological assumptions are now seen to be questionable, unjustified, false, perhaps even illogical. In a recent review, Tyler Burge remarked that "the central event" in philosophy during the second half of the twentieth century has been "the downfall of positivism and the re-opening of discussion of virtually all the traditional problems in philosophy".[3] In light of the collapse of positivism and in light of its essential role in the epistemological foundations of the STR, a reappraisal of the time concept in Relativity Theory is long overdue. For apart from a verificationist critique of some sort, it is no longer obvious why the received view is to be preferred to the views of Lorentz and Poincaré, which are consistent with the Newtonian ideas of absolute time and absolute simultaneity.

Certain contemporary theorists have attempted to free STR of its verificationist assumptions, but it is questionable whether such efforts have been successful. For example, the philosopher of physics Graham Nerlich has been most critical of Einstein's original verificationism and develops a non-reductionistic "realist" theory of relativistic space-time. Such relativistic substantivalism remains verificationist, however, to the extent that it assumes that time and space, although not identical with observable clock movements and rigid rods, must possess only those properties that are measurable by observable clocks and rigid rods. This assumption underlies the views of all those who hold that the "essence" of Special Relativity is Lorentz invariance, since Lorentz invariance ultimately requires that law-like behavior is the behavior that would be measurable by light-clocks and rigid rods if such clocks and rods were present. Similarly, Eli Zahar, who attempts to provide a non-verificationist foundation for Special Relativity, in fact ends up relying on an implicit verificationism after all.[4]

Part III

It is remarkable that the Special Theory has thus far managed to survive largely unscathed the collapse of its essential epistemological underpinnings.

One wonders how this can be so. Undoubtedly a major part of the answer is the understandable one that physicists are not epistemologists; physicists typically know no more about epistemology, the philosophy of language (*e.g.* problems with the verificationist criterion of semantic meaning), and ontology than philosophers typically know about physics. The precise philosophical arguments for the illogicality, falsity, or unjustifiably of the epistemological, semantic, and ontological presuppositions of the Special Theory remain, with a few exceptions, unknown among physicists.

The price paid for the growth of knowledge is increased specialization, which, paradoxically, also prevents or reverses the growth of knowledge, since specialists in one field often base their work on premises that (unbeknownst to them) have been refuted or disconfirmed in another field. The only solution we can see for this problem is that the training or schooling of physicists ought to include schooling in philosophy (and, as we shall see, the converse should hold for philosophers). Perhaps this is most practicable in the form of there being thinkers who take as their specialization the intersection of physics and philosophy and the works of these thinkers, at least in "introductory formats", being a part of the education of both physicists and philosophers. If this proves unfeasible and the situation remains as it presently stands, the unpalatable situation may result that neither physicists nor philosophers are in a position to have adequately justified beliefs about space and time but only philosophers of physics (or the few thinkers who are both philosophers and physicists, such as David Albert and Bas Van Fraassen, and, from the side of physics, Niels Bohr and David Bohm, who developed philosophical theories in addition to physically interpreted equations).

Apart from leaving unaddressed the epistemological and semantic presuppositions of STR, there is an even stronger factor behind physicists' unwillingness to abandon the Special Theory. The Special Theory is a part of orthodox quantum field theory (QFT) (quantum electrodynamics and quantum chromodynamics), which aims to unify the Special Theory with quantum mechanics. Physicists would be at a loss as to how to proceed if they rejected the Special Theory as unjustified, since they (for the most part) believe that this would require them to reject QFT.

In the light of this dependence on Special Relativity, physicists are not likely to abandon it unless it is observationally disconfirmed and there is an observationally adequate theory available to replace it. In fact, there is a theory that is not merely observationally equivalent to the Special Theory, but also observationally superior to it, namely Lorentzian or neo-Lorentzian theory. Lorentz's theory is regarded by many physicists who have studied Lorentzian theory, such as J.S. Bell, to be observationally equivalent to the Special Theory. However a Lorentzian or neo-Lorentzian theory is, in fact, observationally superior to the Special Theory (a fact that Bell, surprisingly, did not point out), since a Lorentzian theory, in contrast to the Special Theory, is consistent with the relations of absolute, instantaneous simultaneity implied by the EPR correlations that were observed in Aspect's confirmations of Bell's theorems. Special Relativistic QFT is no less disconfirmed by Aspect, *et al.* than is Special Relativity. Physicists do not appreciate the fact that a (non-orthodox) QFT can be formulated in terms of a Lorentzian theory, in which case it would avoid both disconfirmation by the Bell-Aspect observational data and the criticism of the untenability of the epistemological basis of the Special Theory.

One reason the Bell-Aspect experiments do not cause worry among most relativistic quantum field theorists is that the discussion of the Bell-Aspect experiments is isolated from the discussions of relativistic theories and is confined (for the most part) to discussions of non-relativistic quantum mechanics. There may be good reasons for this. But it may also reflect the worry that physicists do not know how to solve the problem of reconciling the Bell-Aspect experiments with relativistic QFT.

An even more "glaring" problem that a Lorentzian theory of absolute time can solve is that orthodox (special relativistic) QFT has not been fully or successfully "relativised". The absolute simultaneity implied by either collapse or non-collapse interpretations of QM, shows up in orthodox QFT as instantaneous relations (*e.g.* causal relations or correlations) among space-like separated regions associated with commuting or anti-commuting operators. Since the "special relativistic" ingredients in orthodox QFT are the same as those implied by a space-time formulation of Lorentzian theory, one can formulate a Lorentzian QFT that both explains and predicts the apparent Lorentz invariance and other "relativistic effects" that are observed, and yet is also theoretically and observationally consistent with instantaneous causal relations and correlations (see Chapter 3 for a discussion of the required Lorentzian formulation of QFT).

Physicists who work in the area of General Relativity know that the Special Theory has been surplanted by the General Theory of Relativity (GTR) and that simultaneity does not have the relativity attributed to it by the Special Theory but instead is absolute in the "cosmic time" which emerges through a cosmological application of GTR. However, QFT does not involve GTR and specialists in QFT do not address or take into account the absolute simultaneity relation that belongs to GTR-based cosmogenies (which we discuss in the next section).

It may be concluded that the main reasons many physicists still hold to Special Relativity are (1) an insufficient awareness of the epistemological and other philosophical problems with Special Relativity; (2) the isolation of the discussion of the Bell-Aspect observations from relativistic QFT; (3) the neglect of GTR by specialists in QFT, since GTR is not a part of QFT; and (4) the lack of familiarity with Lorentzian theories and consequent unawareness of how a Lorentzian QFT would not be faced with the manifest problems of orthodox, relativistic QFT. We hope that the discussions in this volume of the epistemological problems with Special Relativity, the discussions of the disconfirmation of the Special Theory by Aspect's confirmation of Bell's theorems, the discussion of the absolute simultaneity relation in GTR, and the discussion of Lorentzian theories, will help physicists who still adhere to the Special Theory to become more aware of these issues and to enable them to formulate more theoretically and observationally adequate theories.

Part IV

An explanation of why many philosophers of time remain wedded to the relativity of simultaneity could in part be that there has developed in twentieth century philosophy of time a habitual or even "orthodox" belief that absolute simultaneity and a tensed or (to borrow McTaggart's convenient nomenclature) A-Theory of time go hand in hand, and that relative simultaneity and the B-Theory go hand in hand. But this "habit of belief" originated with the lay-physicist's belief that the basic contrast in the theory of time is between "the flowing Newtonian time that contains relations of absolute simultaneity" and "the tenseless four-dimensional spacetime of Minkowski" and that physics presented one with a choice between one or the other. But while A-Theorists are arguably committed to a rejection of Einsteinian or Minkowskian relativity,[5] the B-Theory does not in fact require relativity, since a space-time formulation of Lorentzian theory and even of Newtonian absolutism can be given, and there is nothing in positing a preferred foliation of space-time that commits one to objective tenses or temporal becoming. A-Theorists' commitment to tense and temporal becoming may make them more open to absolute simultaneity, but the B-Theory is consistent with either relative or absolute simultaneity. If this "habit of belief" remains operative among B-Theorists who reject absolute simultaneity, then the problem is the lack of a sufficient knowledge of contemporary physics.

The problem here is, in a sense, the reverse of the first problem we mentioned above, namely, physicists' lack of awareness of the many arguments that philosophers have provided against the epistemology, philosophy of language, and ontology presupposed by the Special Theory. The most widespread belief among philosophers is that contemporary physics has proven that simultaneity is relative either in Einstein's sense that there exists no preferred reference frame or in Minkowski's sense that there is no preferred foliation of space-time.

In fact, by the early 1920s this claim had been undermined due to the cosmological application of Einstein's 1915 General Theory. Einstein's cosmological model of 1917, based on his GTR, and Alexander Friedman's cosmological model of 1922 (which is now recognized to be the sort of general relativistic cosmological model that is pertinent to our universe) imply a privileged or preferred foliation of space-time. Although the general relativistic space-time manifold can in theory be sliced up arbitrarily, there obtain *de facto* boundary conditions which determine a preferred foliation.

The hypersurface of homogeneity and isotropy is the preferred hypersurface for the formulation of the laws of physics and the measurement of space and time. For example, the statement that the Big Bang occurred 13.7 billion years ago is not based on a measurement relative to any arbitrary reference frame, but to the privileged reference frame, the frame of homogeneity and isotropy, and is therefore absolute. It is regularly noted that a special relativistic or Minkowski metric can be defined infinitesimally close to any point in a universe with a Freidman metric, even though only a Friedman metric applies to the universe itself. The metric that describes the universe determines its spatio-temporal structure, and Friedman's spatio-temporal structure includes an absolute reference frame, the frame of homogeneity and isotropy. It has often been commented by physicists that Einstein's GTR thus reintroduced the relations of absolute simultaneity that his STR had denied. In fact, the idea that the hypersurface of homogeneity is the privileged frame, which determines absolute temporal and spatial measurements, has been a part of graduate textbooks on General Relativity or cosmology going back to the 1920s.

Thus, philosophers who embrace relativity because they are under the impression that the evidence of physics implies it are quite mistaken; in fact the opposite is true. Similarly, philosophers who hold to a tenseless or B-Theory of time for the reason that it is implied by Einstein's 1905 theory or Minkowski's 1908 theory are unjustified in so doing because the evidence of physics does not in fact imply that those theories are true. Einstein's 1905 theory and Minkowski's 1908 theory are "vacuum solutions" to Einstein's GTR, that is, solutions that hold only if there is no matter in the universe.

The notion that a Friedman universe has an absolute hypersurface of simultaneity remains debatable only in the sense that our universe is not perfectly homogeneous, so that the concept of the hypersurface of simultaneity is the concept of an abstract hypersurface that results from averaging out the variations in the homogeneity of matter on very large scales. Whether or how one can derive a surface of absolute simultaneity in the strict (rather than "averaged out") sense is a matter that is up to debate. Some physicists adopt a "York slicing," but it is not evident that this resolves the problem of the surface of absolute simultaneity being a concept that relies on the "averaging out" of inhomogeneities.

General Relativity is not the only physical theory developed after 1905 or 1908 that falsifies or casts doubt on the relativistic concepts of simultaneity. Louis De Broglie in 1928 and David Bohm in 1952 developed an interpretation of quantum mechanics that implies that there is a strict plane of absolute simultaneity, in the same sense as Lorentz's and Newton's in respect of it being a unique, universal, instantaneous plane of simultaneity that is non-local and space-like. De Broglie's and Bohm's theories of absolute simultaneity were largely ignored until J. S. Bell wrote a series of essays about them in the 1960s, which were collected into his 1987 book *Speakable and Unspeakable in Quantum Mechanics.* Since 1987, the number of adherents to a de Broglie-Bohm interpretation of quantum mechanics has been increasing steadily, as one finds the number of articles about the de Broglie-Bohm theory in physics journals growing from year to year. Several contributors to this volume, Valentini, Maudlin, Smith, and Callender are adherents to the de Broglie-Bohm interpretation of quantum mechanics. Even some competing interpretations of quantum mechanics, namely, those that involve a collapse of the wave function, imply absolute simultaneity, since the collapse of the wave function is instantaneous. The experimental confirmation of Bell's theorem by Aspect and others implies that there is an instantaneous, non-local, space-like relation of simultaneity that coincides with EPR causal correlations (where "cause" is defined in a counterfactual sense, along lines such as suggested, *e.g.* by David Lewis). In this volume Maudlin's essay discusses the most recent experimental findings to date, which give us some idea of the price that must be paid by those who insist upon the relativity of simultaneity in quantum mechanics.

Accordingly, philosophers who believe that simultaneity has the relativistic nature described in Einstein's 1905 paper or Minkowski's 1908 paper make manifest the fact that more education is needed about contemporary physics. If physicists need more education in philosophy, philosophers, especially philosophers of time, need more education in contemporary physics.

Part V

The fact that many philosophers need to become more acquainted with the theories of current physics, theories that assert or imply that simultaneity is not relative in the sense of Einstein's 1905 paper or Minkowski's 1908 paper, makes a volume of essays such as this one useful for philosophers of time and philosophers in general. The same holds for physicists, even if for different reasons. It is appropriate (if coincidental) that the publication of a work that subjects the thesis of the relativity of simultaneity to a critical analysis should fall on the centennial anniversary of Einstein's 1905 essay.

This is the first collection of essays (of which we are aware) that is devoted, for the most part, to arguing that simultaneity is absolute. It is also a collection of essays that include writings in both philosophy and physics, perhaps erasing the distinction between them, at least in some cases. All of the chapters are original creations for this volume, some reflecting the cutting edge of research and reflection on the question of absolute simultaneity. Almost all are convinced that the received view that simultaneity is not an absolute relation is not only unwarranted but false. It is our hope, therefore, that this collection will lead to a better understanding among both philosophers and physicists of the problems inherent in making certain assertions about Einstein's theory of the relativity of simultaneity.

William Lane Craig
Quentin Smith

Notes

1 Isaac Newton (1966) *Sir Isaac Newton's "Mathematical Principles of Natural Philosophy" and his "System of the World,"* trans. Andrew Motte, rev. with an Appendix by Florian Cajori, 2 vols., Los Angeles: University of California Press, 1, p. 6.
2 A. Einstein (1981) 'On the Electrodynamics of Moving Bodies', trans. Arthur I. Miller, Appendix to Arthur I. Miller, *Albert Einstein's Special Theory of Relativity*, Reading, Mass.: Addison-Wesley, p. 393.
3 Tyler Burge (1992) 'Philosophy of Language and Mind', Philosophical Review 101, 49.
4 See Quentin Smith (1998) 'Absolute Simultaneity and the Infinity of the Past', in Robin Le Poidevin (ed.), *Questions of Time and Tense*, Oxford: Oxford University Press.
5 See William Lane Craig's chapter in this volume. Also see Quentin Smith (2000) 'The Inconsistency of STR and the Tensed Theory of Time', in L. Nathan Oaklander (ed.), *The Importance of Time*, Dordrecht, The Netherlands: Kluwer Academic Publishers, and Quentin Smith (1993) *Language and Time*, Oxford: Oxford University Press.

1 The metaphysics of special relativity: three views

William Lane Craig

Introduction

A physical theory is comprised of two components: a mathematical formalism and a physical interpretation of that formalism. Competing theories which differ only in virtue of their divergent physical interpretations can be extremely difficult to assess if they are empirically equivalent in their testable predictions. Considerations which are metaphysical in nature may then become paramount.

The Special Theory of Relativity (hereafter SR) provides a case in point. Herman Bondi has remarked that "there is perhaps no other part of physics that has been checked and tested and cross-checked quite as much as the Theory of Relativity" (Bondi 1964: 168). Indeed, muses J. G. Taylor, "as far as special relativity is concerned all has been worked out and tested;" the theory has enjoyed "remarkable successes, and absolutely no failures" (Taylor 1975, preface). The empirical success of SR's testable predictions can, however, be misleading, dulling us to the truly controversial nature of the correct physical interpretation of the theory's formalism. The fact is that the only version of SR which is experimentally verifiable, as Geoffrey Builder points out, "is the theory that the spatial and temporal coordinates of events, measured in any one inertial reference system, are related to the spatial and temporal coordinates of the same events, as measured in any other inertial reference system, by the Lorentz transformations" (Builder 1971: 422). But this verifiable statement is underdeterminative with regard to the radically different physical interpretations of the Lorentz transformations given, respectively, by Einstein, Minkowski, and Lorentz.

During the decades in which positivism dominated the philosophy of science these differences tended to be glossed over, since empirically equivalent physical interpretations of the same mathematical formalism were regarded as but different linguistic expressions of the same theory. But with the collapse of positivism – arguably the most important event in philosophy in the second half of the twentieth century (Burge 1992: 49) – such indifference toward the fundamentally different ontological structures of space, time, and space-time which appear in these three interpretations can no longer be ignored. Unfortunately the articulation of a post-positivist philosophy of space and time has only scarcely begun. Minkowskians have issued critiques of Einsteinians in the effort to justify the former's space-time realism, and the largely marginalized neo-Lorentzians have criticized what we might call the received interpretation of SR (an Einsteinian-Minkowskian amalgam which fails to differentiate these viewpoints) for its denial of relations of absolute simultaneity; but I know of no critical appraisal in the literature which lays these three interpretations side by side and attempts to come to some adjudication of them. In this paper I propose to do just that.

Three relativistic interpretations

The Einsteinian interpretation

SR, as Einstein originally formulated it, postulates a 3+1-dimensional ontology, not a 4-dimensional ontology (Einstein 1981).[1] That is to say, it is a theory of familiar physical objects enduring through time. Space and time are relativised to reference frames, which serve to define distant simultaneity and along with it notions like rest, motion, speed, and velocity. Light is postulated to have the constant velocity *c* in every reference frame. Because physics is relativised to reference frames, clocks run at different rates and measuring rods have different lengths relative to different frames. Such an interpretation of SR implies an anti-realist or instrumentalist understanding of Minkowski space-time.[2] There is no tenselessly subsisting manifold of events; space-time is a theoretical construct only, a geometrical representation of a theory which is really about physical objects enduring through time. A Minkowski diagram will prove to be a helpful tool, but it neither depicts reality nor implies an ontology. A good representative of this original Einsteinian perspective is the French physicist Henri Arzeliès. In his *Relativistic Kinematics*, Arzeliès asserts, "The Minkowski continuum is an abstract space of four dimensions, the sole role of which is to interpret in geometrical language statements made in algebraic or tensor form. ... The four-dimensional continuum should therefore be regarded as a useful tool, and not as a physical 'reality'" (Arzeliès 1966: 258). While it is true that relativity theory banishes the notions of absolute spatial and temporal intervals from physics, nonetheless "It is perfectly clear that in relativity, the ordinary three-dimensional space (which is Euclidian in special relativity) and the time of pre-relativistic physics is employed" (Ibid).

The Minkowskian interpretation

There is no gainsaying Arzeliès insofar as Einstein's original formulation of SR is concerned. But it is also indisputable that once having encountered Minkowski's geometrical formulation of the theory, Einstein became an outspoken realist concerning space-time (Einstein and Infeld 1938: 219;

Einstein 1961: 150; Einstein and Besso 1979: 276–77).[3] Minkowski took his space-time ontologically: it was not merely a geometrical representation of the world of space and time as described by Einstein's SR; rather it *was* the world. When he said, "A point of space at a point of time, that is, a system of values *x*, *y*, *z*, *t*, I will call a *world-point*. The Multiplicity of all thinkable *x*, *y*, *z*, *t* systems of values we will christen the *world*," (Minkowski 1952: 76) he was making self-consciously a metaphysical statement, proposing a new ontology. Heralding "a metamorphosis of our concept of nature," Minkowski declared, "Henceforth, space by itself, and time by itself, are doomed to fade away into mere shadows, and only a kind of union of the two will preserve an independent reality" (Ibid., 75, 76). On this second interpretation of SR, the notions of reference frames, invariant velocity of light, distant simultaneity, relative motion or rest, and so forth, so central to the Einsteinian interpretation, play no role.[4] Rather the central feature of this interpretation is the light cone structure at any space-time point, which determines the geometrical properties of space-time. In 1911 A. A. Robb was able to recover all the geometric structure of Minkowski space-time on the basis of the single relation *after* among its points, conjoined with several conditions of that relation (Robb 1913). Taking Robb's relation to be extensionally equivalent to some sort of causal relation, recent theorists have defined causally the Lorentz group of transformation equations (Zeeman 1964: 490–93), orthogonality to a time-line in Minkowski space-time (Malament 1977: 293–300), and the metrical congruence of intervals in that space-time (Winnie 1977: 134–205). Space-time realists debate intramurally whether causality is truly constitutive of, rather than merely (at best) co-extensive with, Robb's fundamental relation,[5] but the point remains that the familiar physical entities of the Einsteinian interpretation make no appearance in the space-time interpretation. These two interpretations of relativity theory thus present strikingly different metaphysical visions of reality; they are as radically divergent in their ontologies as is relativity theory itself in comparison with the Newtonian physics of absolute time and space. Minkowski's space-time approach to relativity theory, especially with the development of the General Theory of Relativity (GR), has come to be the dominant mode of presentation and discussion of relativity.

The Lorentzian interpretation

It is an interesting historical fact that neither of the giants of late nineteenth century physics to whom Einstein looked for inspiration in his work on SR, H. A. Lorentz and Henri Poincaré, was ever convinced, despite being fully apprised of the empirical facts, of the truth of the Einsteinian or Minkowskian interpretations of the Lorentz transformations. Well after Einstein had formulated his SR and as he struggled to craft a GR, Lorentz in particular continued to study and lecture on problems of relativity, often in connection with Einstein. By 1908 Lorentz had already realized the incompatibility of his electron theory with Planck's quantum hypothesis, and by the 1911 Solvay Congress there was a general sense that the electron theory would have to be radically reformed in light of the advent of quantum physics (McCormach 1970: 486–88). Nonetheless, since Lorentz's attempted explanation of the phenomenon of length contraction in terms of the deformable electron was not essential to his basic physical interpretation of SR, Lorentz continued to adhere to an approach to relativity theory which preserved the classical notions of space and time. A theory may be classified as Lorentzian just in case it affirms that (i) physical objects are *n*-dimensional spatial entities which endure through time; (ii) the round trip vacuum propagation of light is isotropic in a preferred (absolute) reference frame R_o (with speed $c = 1$) and independent of the velocity of the source; and (iii) lengths contract and time rates dilate in the customary special relativistic way only for systems in motion with respect to R_o (Maciel and Tiomno 1989: 507–8).

Lorentz always spoke appreciatively of Einstein's alternate approach and lectured sympathetically on both SR and GR, while remaining finally unconvinced that Einstein had abolished the classical conceptions of time and space. Writing in 1910, he contrasted his view with Einstein's:

> Assume there were an aether; then there would be among all systems x, y, z, t one singled out in that the coordinate axes as well as the clock is at rest in the aether. If one conjoins with this the idea ... that space and time are something wholly different and that there is a 'true time' (simultaneity would then exist independently of location, in accord with the circumstance that it is possible for us to conceive of infinitely great velocities), then one easily sees that this true time would have to be indicated just by clocks which are at rest in the aether. If, then, the principle of relativity were generally valid in nature, then one would not be in a position to determine whether the coordinate system employed is that distinguished one. One thus comes to the same results as when one in agreement with Einstein and Minkowski denies the existence of the aether and the true time and treats all coordinate systems as equivalent. Which of the two modes of thought one may agree with is best left to the individual.
>
> (Lorentz 1934: 211)[6]

Lorentz, realizing that his aether compensatory interpretation is empirically equivalent to the Einstein-Minkowski interpretations, leaves it up to the individual to choose which he shall adopt. But Lorentz preferred the classical conceptions of time and space on metaphysically intuitive grounds, as he made clear in his 1922 lectures at Cal Tech:

> All our theories help us form pictures, or images, of the world around us, and we try to do this in such a way that the phenomena may be coordinated as well as possible, and that we may see clearly the way in

> which they are connected. Now in forming these images we can use the notions of space and time that have always been familiar to us, and which I, for my part, consider as perfectly clear and, moreover, as distinct from one another. My notion of time is so definite that I clearly distinguish in my picture what is simultaneous and what is not.
>
> (Lorentz 1927: 221)

Here Lorentz refuses to jettison what he took to be the intuitively obvious reality of absolute simultaneity among events in the world just because one cannot determine which spatially separated events are simultaneous or because Einstein's operationally re-defined notion of simultaneity is relative to reference frames. Moreover, he sees no good reason to scrap the intuitive distinctness of space and time in favor of Minkowski's unified reality, space–time.

A major reason that Lorentz remained unconvinced was that he was not a positivist. In 1913 he wrote,

> According to Einstein it has no meaning to speak of motion relative to the aether. He likewise denies the existence of absolute simultaneity.
>
> It is certainly remarkable that these relativity concepts, also those concerning time, have found such a rapid acceptance.
>
> The acceptance of these concepts belongs mainly to epistemology ... It is certain, however, that it depends to a large extent on the way one is accustomed to think whether one is attracted to one or another interpretation. As far as this lecturer is concerned, he finds a certain satisfaction in the older interpretations, according to which the aether possesses at least some substantiality, space and time can be sharply separated, and simultaneity without further specification can be spoken of. In regard to this last point, one may perhaps appeal to our ability of imagining arbitrarily large velocities. In that way, one comes very close to the concept of absolute simultaneity.
>
> Finally, it should be noted that the daring assertion that one can never observe velocities larger than the velocity of light contains a hypothetical restriction of what is accessible to us, [a restriction] which cannot be accepted without some reservation.
>
> (Lorentz 1920a: 23)

Here Lorentz clearly discerns the foundational role played by Einstein's verificationist theory of meaning in his formulation of SR and rejects it. In defense of absolute simultaneity, Lorentz appeals to the use of arbitrarily fast signals, even though they are not presently observable. He disregards the assumption that it is meaningless to speak of such unobservables. Elsewhere Lorentz affirms that it makes sense, if there is an aether, to speak of motion relative to it even if observers could not detect such motion (Lorentz 1914: 26).[7] He writes,

> But it needs to be clearly recognized that **A** could never assure himself of the immobility in the ether which we have attributed to him by supposition and that physicist **B** could with the same right, or rather with the same absence of right, claim that it is he who finds himself in these privileged circumstances. This incertitude, this impossibility of even disclosing a movement in relation to the aether, led Einstein and numerous other modern physicists to abandon completely the notion of an aether.
>
> There, it seems to me, is a question toward which each physicist must take a position which best accords with the manner of thinking to which he is accustomed.
>
> (Lorentz 1934: 7: 165)

Lorentz's conception of the aether was virtually equivalent to space itself. His aether was so dematerialized that Einstein, lecturing at the University of Leiden in 1920, could tease the Dutch physicist by declaring, "As regards the mechanical nature of Lorentz's aether one might say of it, with a touch of humor, that immobility was the only mechanical property which H. A. Lorentz left it" (Einstein 1920: 7). For Lorentz the aether is just the privileged spatial frame.

Lorentz thus accepts a 3+1 ontology of spatial objects enduring through a privileged time, a metaphysic which he felt no obligation to abandon out of deference to a verificationist epistemology.

Assessment of the interpretations

Metaphysical underpinnings of the classical concept of time

In another place I have argued that the collapse of positivism during the second half of the twentieth century has re-opened the discussion of the metaphysical foundations of the classical concept of time and that only Lorentz's approach to problems of space and time, in contrast to Einstein and Minkowski's, has remained unshaken by this epistemological revolution (see Craig 2002: 129–52). Specifically, I sought to expose the theological foundations of the classical concept of time as enunciated by Isaac Newton in his epochal *Philosophiae naturalis principia mathematica.*

By way of review, the *locus classicus* of Newton's exposition of his concepts of time and space is the *Scholium* to his Definitions in the *Principia.* In order to overcome "common prejudices" concerning such quantities as time, space, place, and motion, Newton draws a dichotomy with respect to these quantities between "absolute and relative, true and apparent, mathematical and common." With regard to time he asserts:

> Absolute, true, and mathematical time, of itself, and from its own nature, flows equably without relation to anything external, and by

> another name is called duration: relative, apparent, and common time, is some sensible and external (whether accurate or unequable) measure of duration by the means of motion, which is commonly used instead of true time; such as an hour, a day, a month, a year.
>
> (Newton 1966 1:6)

Newton's much misunderstood and greatly maligned distinction between absolute and relative time deserves our thoughtful consideration.

The most evident feature of this distinction is the independence of absolute time from the relative measures thereof. Absolute time, or simple duration, exists regardless of the sensible and external measurements which we try, more or less successfully, to make of it. Newtonian time is thus first of all absolute in the sense that time itself is distinct from our measures of time.

But, of course, Newton also conceived time as absolute in a more profound sense, namely, he held that time is absolute in the sense that it exists independently of any physical objects whatsoever. Usually, this is interpreted to mean that time would exist even if nothing else existed, that there exists a possible world which is completely empty except for the container of absolute space and the flow of absolute time. But here we must be very careful. Modern scholars tend frequently to forget how ardent a theist Newton was and how central a role this theism played in his metaphysical outlook. Noting that Newton considered God to be temporal and therefore time to be everlasting, David Griffin observes that "Most commentators have ignored Newton's heterodox theology, and his talk of 'absolute time' has been generally misunderstood to mean that time is not in any sense a relation and hence can exist apart from actual events" (Griffin 1986: 6–7).

In fact, Newton makes quite clear in the *General Scholium* to the *Principia*, which he added in 1713, that absolute time and space are constituted by the divine attributes of eternity and omnipresence:

> He is eternal and infinite . . .; that is, his duration reaches from eternity to eternity; his presence from infinity to infinity . . . . He is not eternity and infinity, but eternal and infinite; he is not duration or space, but he endures and is present. He endures forever, and is everywhere present; and, by existing always and everywhere, he constitutes duration and space. Since every particle of space is *always*, and every indivisible moment of duration is *everywhere*, certainly the Maker and Lord of all things cannot be *never* and *nowhere*.
>
> (Newton 1966: 2: 545)

Because God is eternal, there exists an everlasting duration, and because He is omnipresent, there exists an infinite space. Absolute time and space are therefore relational in that they are contingent upon the existence of God.

In his treatise *De gravitatione* (Newton 1962: 89–156), Newton declares explicitly that space is *not* in itself absolute (*non absoluta per se*) and therefore not a substance. Rather it is an eminent – or emanative – effect of God (*Dei effectus emanativus*). It is uncreated and co-existent with God and yet ontologically dependent upon him for its being. God's infinite being has as its consequence infinite time and space, which represent the quantity of his duration and presence. In the Neo-Platonic tradition the doctrine of emanation is associated with pantheism or panentheism. But, as Newton makes clear, he does not conceive of space or time as in any way aspects of God himself, but rather, as he says, concomitant effects of God.

It is evident that when Newton speaks of divine eternity, he does not, like scholastic theologians in the Augustinian tradition, mean a state of timelessness, but rather infinite and everlasting temporal duration. In a preliminary draft of the *General Scholium*, Newton had explicitly rejected the conception of God's eternity as an eternal now: "His duration is not a *nunc stans* without duration, nor is his presence nowhere" (cited in McGuire 1990: 93). Far from being atemporal, God's now or present is the present of absolute time. Since God is not "a dwarf-god" located at a place in space (Newton 1978: 114–29) but is omnipresent, every indivisible moment of duration is everywhere, as Newton states in the *General Scholium*. There is thus a worldwide moment which is absolutely present. Newton's temporal theism thus provides the foundation for both absolute simultaneity and absolute becoming. These are features first and foremost of God's time or metaphysical time and derivatively of measured or physical time.

Now Newton provides virtually no argument to think that God is temporal; he simply asserts it. But I have argued elsewhere that on a tensed or (to borrow McTaggart's convenient terminology) an A-Theory of time, according to which tense and temporal becoming are objective features of reality, God must be temporal in virtue of His knowledge of tensed facts and His causal relation to the world (Craig 1998b: 221–50). Hence, if God exists and an A-theory of time is correct, Newton is justified so far forth in holding that there is absolute time in the sense of a metaphysical time which is independent of physical time.

Newton freely grants that although absolute time exists it may well be the case that due to the inaccuracies of our measures the true time is not disclosed to us (Newton 1966: 1:7–8). What Newton did not realize, nor could he have suspected, is that physical time is not only *relative*, but also *relativistic*, that the approximation of physical time to absolute time depends not merely upon the regularity of one's clock, but also upon its motion. Unless a clock were at absolute rest, it would not accurately register the passage of absolute time. A clock moving relatively to oneself runs slowly. This truth, unknown to Newton, only intimated by Larmor and Lorentz in the concept of "local time," was finally grasped by Einstein.

Where Newton fell short, then, was not in his analysis of absolute or metaphysical time – he had theological grounds for positing such a time –

but in his incomplete understanding of physical time. He assumed too readily that an ideal clock would give an accurate measure of time independently of its motion. If confronted with relativistic evidence, Newton would no doubt have welcomed this correction and seen therein no threat at all to his doctrine of absolute time. In short, relativity corrects Newton's concept of physical time, not his concept of absolute time.

Of course, it hardly needs to be said that there is a great deal of antipathy in modern physics and philosophy of science toward such metaphysical realities as Newtonian space and time, primarily because they are not physically detectable. But Newton would have been singularly unimpressed with this positivistic equation between physical undetectability and non-existence. The grounds for metaphysical space and time were not physical, but philosophical, or more precisely, theological. Epistemological objections fail to worry Newton because, as Lucas nicely puts it, "He is thinking of an omniscient, omnipresent Deity whose characteristic relation with things and with space is expressed in the imperative mood" (Lucas 1973: 143). Modern physical theories say nothing against the existence of such a God or the metaphysical time constituted, in Newton's thinking, by His eternity. What Einstein's relativity theory did, in effect, was simply to remove God from the picture and to substitute in His place a finite observer. "Thus," according to Holton, "the *RT* [Relativity Theory] merely shifted the focus of space-time from the sensorium of Newton's God to the sensorium of Einstein's abstract *Gedanken* experimenter – as it were, the final secularization of physics" (Holton 1973: 171). But for a man like Newton such a secular outlook impedes rather than advances our understanding of the nature of reality.

Unfortunately in our secular age physicists and philosophers of space and time rarely, if ever, give careful consideration to the difference God's existence would make for our conceptions of time and space. But in a fascinating passage in his essay "*La mesure de temps*," Poincaré does briefly entertain the hypothesis of "*une intelligence infinie*" and considers the implications of such a hypothesis. Poincaré is reflecting on the problem of temporal succession. In consciousness, the temporal order of mental events is clear. But going outside consciousness, we confront various difficulties. One of these concerns is how we can apply one and the same measure of time to events which transpire in "different worlds," that is, to spatially distant events. What does it mean to say that two psychological phenomena in two consciousnesses happen simultaneously? Or what does it mean to say a supernova occurred before Columbus saw the isle of Espanola? "All these affirmations," says Poincaré, "have by themselves no meaning" (Poincaré 1982: 228). Then he remarks,

> We should first ask ourselves how one could have had the idea of putting into the same frame so many worlds impenetrable to one another. We should like to represent to ourselves the external universe, and only by so doing could we feel that we understood it. We know we can never

> attain this representation: our weakness is too great. But at least we desire the ability to conceive an infinite intelligence for which this representation could be possible, a sort of great consciousness which should see all, and which should classify all *in its time*, as we classify, *in our time*, the little we see.
>
> This hypothesis is indeed crude and incomplete, because this supreme intelligence would be only a demigod; infinite in one sense, it would be limited in another, since it would have only an imperfect recollection of the past; it could have no other, since otherwise all recollections would be equally present to it and for it there would be no time. And yet when we speak of time, for all which happens outside of us, do we not unconsciously adopt this hypothesis; do we not put ourselves in the place of this imperfect god; and do not even the atheists put themselves in the place where God would be if he existed?
>
> What I have just said shows us, perhaps, why we have tried to put all physical phenomena into the same frame. But that cannot pass for a definition of simultaneity, since this hypothetical intelligence, even if it existed, would be for us impenetrable. It is therefore necessary to seek something else.
>
> (Ibid. 228–29)

Poincaré here suggests that, in considering the notion of simultaneity, we instinctively put ourselves in the place of God and classify events as past, present, or future according to His time. Poincaré does not deny that such a perspective would disclose to us true relations of simultaneity. But he rejects the hypothesis as yielding a definition of simultaneity because *we* could not know such relations; such knowledge would remain the exclusive possession of God Himself.

But clearly, Poincaré's misgivings are relevant to a definition of simultaneity only if one is presupposing some sort of verificationist theory of meaning, as he undoubtedly was. The fact remains that God would know the absolute simultaneity of events even if we grope in total darkness. Nor need we be concerned with Poincaré's argument that such an infinite intelligence would be a mere demigod, since it is a *non sequitur* that a being with perfect recollection of the past cannot be temporal. There is no conceptual difficulty in the idea of a being which knows all true past-tense propositions. That such a being would be temporal is evident from the fact that as events transpire, more and more past tense propositions become true, so that the content of his knowledge is constantly changing. Hence, it does not follow that if God is temporal, He cannot have perfect recollection of the past.

Poincaré's hypothesis suggests, therefore, that God's present is constitutive of relations of absolute simultaneity. Lorentz agreed. In a passage redolent of the *General Scholium* and *Opticks* of Newton, Lorentz in a letter to Einstein in January of 1915 broached considerations whereby "I cross the borderland of physics":

> A 'World Spirit' who, not being bound to a specific place, permeated the entire system under consideration or 'in whom' this system existed and who could 'feel' immediately all events would naturally distinguish at once one of the systems *U, U′, etc.* above the others.
>
> (Lorentz 1989: 274)

Such a being, says Lorentz, could "directly verify simultaneity". On this view, J. M. Findlay was wrong when he said, "... the influence which harmonizes and connects all the world-lines is not God, not any featureless, inert, medium, but that living, active interchange called ... Light, offspring of Heaven firstborn" (Findlay 1978–79: 6–7). On the contrary, the use of light signals to establish clock synchrony is a convention which finite and ignorant creatures have been obliged to adopt, but the living and active God, who knows all, would not be so dependent.

In God's temporal experience, there would be a moment which is present in metaphysical time, wholly independently of physical clock times. He would know, without any dependence on clock synchronization procedures or any physical operations at all, which events were simultaneously present in metaphysical time. He would know this simply in virtue of His knowing at every such moment the unique set of present-tense propositions true at that moment, without any need of a *sensorium* or physical observation of the universe.

How, then, would God's metaphysical time relate to our physical time? From what has been said thus far, it seems that God's existence in metaphysical time and His real relation to the world would imply that a Lorentzian interpretation of relativity is correct. Such a theory is required in view of divine temporality, for God in the "now" of metaphysical time would know which events in the universe are now being created by Him and are therefore absolutely simultaneous with each other and with His "now." This startling conclusion shows clearly that Newton's theistic hypothesis is not some idle speculation, but has important implications for our understanding of how the world is and for assessment of rival scientific theories.

Accordingly, we may argue

1. God exists.
2. An A-Theory of time is correct.
3. If an A-Theory of time is correct, there are tensed facts and temporal becoming.
4. If God exists and there are tensed facts and temporal becoming, then God knows tensed facts and is the cause of things' coming to be.
5. If God knows tensed facts and is the cause of things' coming to be, then God is temporal.
6. There are tensed facts and temporal becoming. (2, 3)
7. God exists and there are tensed facts and temporal becoming. (1, 6)
8. God knows tensed facts and is the cause of things' coming to be. (4, 7)

9. God is temporal. (5, 8)
10. If God is temporal, then a privileged reference frame exists.
11. If a privileged reference frame exists, then a Lorentzian interpretation of SR is correct.
12. A privileged reference frame exists. (9, 10)
13. A Lorentzian interpretation of SR is correct. (11, 12)

It seems to me that this argument is cogent. As a theist, therefore, I accept (13). However, I recognize that non-theistic thinkers will be little moved by the argument. Therefore, it would be interesting to ask whether for the sake of dialectical effectiveness the argument might not be freed of its theistic underpinnings. Reflection on the argument convinces me that this is, indeed, possible. For the yeoman's work in this argument is done, not the assumption of theism, but rather by the assumption of an A-theory of time. One may plausibly argue directly, I think, that if an A-Theory of time is correct, then a Lorentzian interpretation of SR is correct. In the sequel, therefore, I shall abandon the assumption that God exists and seek to assess the three competing, physical interpretations of SR given merely the fact that we have good grounds for affirming the reality of tense and temporal becoming.

The Einsteinian interpretation

The characteristic feature of the Einsteinian interpretation of the SR formalism is its attempt to wed relativity with a classical 3+1 ontology of space and time. While this original interpretation is today largely disfavored in the foundations of physics community, it nevertheless continues to enjoy widespread acceptance among philosophers, particularly those embracing a tensed, as opposed to tenseless, theory of time and those advocating an endurantist, as opposed to a perdurantist, account of temporal persistence.[8] It therefore merits our consideration. I shall argue that the Einsteinian interpretation is implausible and explanatorily deficient.

First, the pluralistic fragmentation of reality into distinct spaces and times associated with reference frames is an ontology which is, to put it bluntly, fantastic. This is a complaint voiced by many space-time realists.[9] At the root of the complaint lies the conviction that there is a single world, an objective reality independent of observers, the conviction that if we both exist then what co-exists with me co-exists with you. It is fantastic to think that you and I, occupying the same location in space and time, but in relative motion, should in virtue of that motion literally dwell in two different worlds, which intersect only at a point. Yet SR requires that even if we are merely passing each other in automobiles, our hyperplanes of simultaneity do not coincide, and at sufficient distances empirically distinguishable events and things are occurring and exist for me which are future and therefore unreal for you. Other events which are in my future and therefore unreal are already actual for you. But if I decelerate and we come to relative rest, then we share the same reality; events which were once present and real in relation to me are now non-existent and future. One can change frames and, hence, realities just by changing one's relative motion.

By contrast, on the Minkowskian interpretation, which makes no link between simultaneity and reality, events do not pop in and out of existence as I switch reference frames. All that changes is which class of events is orthogonal to my world-line in space-time at a designated point and, hence, which events I reckon to be simultaneous with my present. All the events subsist tenselessly, and different hyperplanes in space-time serve merely to mark out which events count as simultaneous relative to my inertial frame. There is a shared, objective reality which exists independently of observers or reference frames, and we all inhabit the same space-time world; we just reckon different events in that one world to stand in the relation of simultaneity with one another. On a space-time ontology, there is thus a unified, independent reality which is merely *measured* differently by observers using different coordinate systems. But on the Einsteinian interpretation, reality literally falls apart, and there is no one way the world is.

Second, the Einsteinian interpretation is explanatorily deficient. On the Einsteinian interpretation physical objects have properties of shape, mass, and duration only extrinsically, relative to inertial frames, yet why this is so is not explained (Christensen 1993: 260; Nerlich 1994: 5). Moreover, it is unclear why 3-dimensional objects enduring through time suffer relativistic effects such as length contraction and time dilation in virtue of their being in relative motion. It is important to realize that under the 3+1-dimensional ontology of the Einsteinian interpretation these relativistic effects are *just as much real, physical effects* as under aether compensatory theories such as Lorentz's. On the Einsteinian interpretation length contraction and clock retardation cannot be dismissed as merely apparent phenomena on the analogy of the mutual observation of diminishing size when two observers retreat from each other. Admittedly, since length contraction and time dilation are reciprocal and the result of merely relative motion, it does seem incredible that they could be anything more than mere appearances, just as the so-called "pure relativists" like Bergson and Dingle insisted (see Dingle 1940; Bergson 1965). But Einstein realized right from the start that these effects described in his theory were real, not apparent, and could be shown to be real by various *Gedankenexperimente.*[10] Examples could be multiplied (see Shaw 1962: 72; and further, Jánossy 1971: 128–31; Lorenz 1982: 308–12) to prove that, perhaps contrary to expectation, the Einsteinian interpretation of relativity theory involves real, physical length contraction and clock retardation, just as much as does the Lorentzian theory. Podlaha concludes,

> In the relativity theory, the length contraction and time dilatation in all frames is often viewed as a consequence of a 'perspective of observation,' similarly as a rod seems to change its length as observed under different

> angles. . . . however, it is seen that the results of relativistic experiments have their origin in the length contraction and time dilatation effects which are so real as a change of the length of a rod caused by the change of temperature.
>
> (Podlaha 1975: 74–75)

Podlaha's comments are quite correct, at least as far as the Einsteinian interpretation of SR is concerned.

The Einsteinian interpretation is explanatorily deficient in that it provides, indeed, permits, no causal account of these real, physical distortions of 3-dimensional continuants. In contrast to the Lorentzian theory, which sought for a causal explanation for these relativistic effects, in the Einsteinian interpretation they are just deduced from the theory's postulates. Lorentz himself remarked on this difference between the theories:

> His [Einstein's] results concerning electromagnetic and optical phenomena . . . agree in the main with those which we have obtained in the preceding pages. The chief difference being that Einstein simply postulates what we have deduced, with some difficulty and not altogether satisfactorily, from the fundamental equations of the electrodynamic field. By doing so, he may certainly take credit for making us see in the negative results of experiments like those of Michelson, . . . not a fortuitous compensation of opposing effects, but the manifestation of a general and fundamental principle. Yet, I think something may also be claimed in favor of the form in which I have presented the theory. I cannot but regard the ether which can be the seat of an electromagnetic field with energy and its vibrations, as endowed with a certain degree of substantiality, however different it may be from ordinary matter.
>
> (Lorentz 1952: 229–30)

According to Arthur Miller,

> The principle of relativity of Bucherer, Lorentz, and Poincaré resulted from careful study of a large number of experiments, and it was on the basis of a theory in which empirical data could be explained to have been *caused* by electrons interacting with an ether. Einstein's principle of relativity excluded the ether of electromagnetic theory and did not explain anything.
>
> (Miller 1981)

On the Einsteinian interpretation, then, such relativistic phenomena involve real deformations of 3-dimensional objects enduring through time which have no causal explanation, but are merely correlated with different reference frames. They come simply as deductions from the two postulates of the theory. By contrast, on the Minkowskian interpretation, 3-dimensional objects do not suffer length contraction or time dilation for the simple reason that 3-dimensional objects do not exist. Reality is 4-dimensional,[11] and the account of relativistic phenomena is, on this interpretation, much more akin to the change of perspective imagined by the pure relativists. Given a space-time ontology, the different length and time measurements given for specific objects and events by various observers is, as Taylor and Wheeler put it, just a matter of looking at them "from several angles" (Taylor and Wheeler 1966: 4). A 4-dimensional object viewed from a certain angle will appear fore-shortened in comparison with the object as viewed from a different angle. Less metaphorically, because coordinate time and length intervals, as opposed to proper time and length intervals, are not invariant under the Lorentz transformation equations, they vary from coordinate system to coordinate system. A Lorentz transformation may be regarded merely as a rotation of the coordinate axes in 4-dimensional space-time, with the result that different coordinate lengths and temporal intervals will be assigned to the same 4-dimensional object. Although this rotation of the axes is obscured in a 2-dimensional Minkowski diagram, it may be more perspicuously exhibited by means of a Loedel diagram.[12] Thus length contraction is not the mysterious (reciprocal!) shrinking of enduring, 3-dimensional objects in relative, uniform motion, but merely the result of our applying different coordinate systems to the unchanging, 4-dimensional, spatio-temporal object and subtracting the values of the spatial coordinates of its endpoints. Similarly, time dilation does not involve a literal slowing down of relatively moving clocks as they endure through time, but rather results from the application of different coordinate systems to the changeless 4-dimensional object and calculating the difference between the temporal coordinates of two events. Moreover, in the pseudo-Euclidian Minkowski space-time a curved world-line between two points is the shortest distance, and hence, the clock of an observer tracing out such a path through space-time will record less time than a clock following a straight path. Thus in the Twin Paradox, although the path of the traveling twin appears in a Minkowski diagram to be longer than that of his stationary counterpart, that impression is due to the 2-dimensionality of the diagram. In 4-dimensional space-time, the path of the traveling sibling is actually shorter than that of the earth-bound twin (see Bondi 1964: 67; Taylor and Wheeler 1966: 34).[13] Therefore, it is not surprising that his clock records less time and that he is younger than his brother at their reunion. Since length measurements are inherently dependent upon time measurements (lengths being a matter not simply of computing $x_2 - x_1$, but $x_2(t_2) - x_1(t_1)$ on the assumption that $t_2 = t_1$), it is not surprising that a relatively moving rod, as viewed by a particular observer, should be calculated to suffer contraction. But, in a sense, the contraction of rods and the retardation of clocks is on the space-time interpretation ultimately a chimera, since 3-dimensional, enduring objects like rods and clocks do not exist.

Einsteinians and Lorentzians are apt to regard the space-time account as a pseudo-explanation masquerading as a genuinely explanatory account. But that is to fail to take seriously the 4-dimensional ontology of the Minkowskian. Given that such an ontology of physical objects is correct, as opposed to the ontology of continuants presupposed by proponents of the Einsteinian and Lorentzian interpretations, one can appreciate why Nerlich, an enthusiastic Minkowskian, finds "the elegant spacetime interpretation of the Fitzgerald contraction and the 'slowing of moving clocks' extraordinarily insightful and intuitive".[14] By contrast the Einsteinian interpretation is markedly explanatorily deficient.

In summary, the Einsteinian interpretation of the SR mathematical formalism, with its pluralistic ontology and contracting and retarded 3-dimensional continuants, is fantastic and explanatorily impoverished.

The Minkowskian interpretation

The great advantage of the Einsteinian interpretation is that by retaining a classical 3+1-dimensional ontology of space and time, it makes room for the objective reality of tense and temporal becoming. By contrast, the Minkowskian interpretation through its commitment to space-time realism precludes temporal becoming and seems to permit no room for tensed properties or modes of existence like presentness. The Minkowskian interpretation thus flies in the face of our experience of time. Psychologist William Friedman, who has made a career of the study of time consciousness, reports, "Like [temporal] order and the causal priority principle, the division between past, present, and future so deeply permeates our experience that it is hard to imagine its absence" (Friedman 1990: 92). We have, he writes,

> an irresistible tendency to believe in a present. Most of us find quite startling the claim of some physicists and philosophers that the present has no special status in the physical world, that there is only a sequence of times, that the past, present, and future are only distinguishable in human consciousness.
>
> (Ibid. 2)

As a result, virtually all philosophers of space and time, including proponents of theories of tenseless time, admit that the view of the common man is that time involves a past, present, and future which are objectively real and that things or events really do come to be and pass away in time. For example, Oaklander grants that "non-philosophers have never doubted that there is such a phenomenon as temporal flow or passage. ... " (Oaklander 1984: 1). Horwich muses that "The quintessential property of time, it may seem, is the difference between past and future" (Horwich 1987: 15). Coburn confesses, "If the existence of A-facts [i.e., tensed facts] is an illusion, it is one of our most stubborn ones" (Coburn 1990: 118). Nerlich is forced to conclude, "Something in experience deludes us about time, something obvious and pervasive – there obtrusively in every thought and perception, yet elusive to analysis" (Nerlich 1998: 130). Mellor acknowledges, "the experienced presence of experience, is the crux of the tensed view of time. . . . and the tenseless camp must somehow explain it away" (Mellor 1981: 6).

Powerful considerations in favor of the objectivity of tensed facts and temporal becoming include (i) the indispensability and irreducibility of linguistic tense, which mirrors the tensed facts which are characteristic of reality;[15] and (ii) our experience of presentness and temporal becoming, which, like our experience of the external world, may be plausibly taken to ground properly basic beliefs in the reality of tensed time.[16] Powerful objections can, in turn, be lodged against theories of tenseless time, including (i) in the absence of objective distinctions between past, present, and future, the relations ordering events are only gratuitously regarded as genuinely temporal relations of earlier/later than;[17] (ii) even the subjective illusion of temporal becoming involves itself an objective temporal becoming of contents of consciousness;[18] and (iii) space-time realism implies perdurantism, the view that objects have spatio-temporal parts, a doctrine which is metaphysically counter-intuitive, incompatible with moral accountability, and entails the bizarre counterpart theory of transworld identity.[19] Discussion of these several points lies far beyond the scope of this paper, but I think it is evident that any interpretation of reality which finds itself obliged to deny the reality of tensed facts and temporal becoming faces *prima facie* a disadvantage.

Certain Minkowskians have therefore argued that their interpretation of SR can be combined with a theory of tensed time. Thus, Howard Stein (Stein 1968: 14; 1991: 147–67; see also Capek 1976: 501–24; Denbigh 1978: 309) in response to Putnam's claim that SR has eliminated the notion of becoming, proposes to relativise becoming to space-time points:

> For an event at spacetime point *a*, those events, and only those, *have already become* (real or determinate), which occur at points in the topological closure of the past of *a*.

According to this explication, reality is relative to space-time points and consists in all those events lying inside or on the past-directed light cone of *a*, including *a* itself. According to Stein, in the context of SR we can only think of the temporal evolution of the world from the chronological perspective of each space-time point. There exists no knife-edge of becoming on this view, even for individual reference frames, but mere pinpoints of becoming constituted by the vertex of any hypothetical observer's past light cone, none of which is privileged (see Clifton and Hogarth 1995: 355–87). Putnam's error lay in thinking that the relation *R* "is real to" is transitive, such that an event at space-time point *c*, which is real for an observer at space-time point *b*, is also real to an observer at space-time point *a*, if events at *b* are real for *a*. Just as "is simultaneous with" is not a transitive relation, neither is *R*.

One of the major difficulties with such an attempt to combine space-time realism with temporal becoming is that it runs smack into McTaggart's famous paradox (see Craig 1998a: 122–27). Space-time events would have somehow to remain identical over time even though they possess at different times different tense determinations like pastness and presentness, an apparently impossible feat. Moreover, this proposal relapses into the same fragmentation of reality characteristic of the Einsteinian interpretation. Worse, on this view present reality collapses to what is here at this point-instant. If this is not self-evidently absurd, Mellor also points out that if I-now am never located in London, then the bizarre conclusion follows that the tensed sentence "It is raining in London" is never true (Mellor 1973–74: 75–76; cf. Le Poidevin 1997: 541–46). Sklar notes the further strange consequence of this alternative that we must say that there will be events which are now such that they will be in my real past at some future time, but which will never have a present reality to me at all! (Sklar 1981: 140; cf. Dorato 1995: Chap. 12). That is because they are now in my future light cone and will be in my past light cone, but have no reality until they are past. Since neither future nor distant present events exist, how is it that such events become past without ever having been actualized in the present? Indeed, is it not self-contradictory to assert that the sentence "The Battle of Waterloo occurred" is true and yet that it was never true that "The Battle of Waterloo is occurring"? Thus, the attempt to combine a theory of tensed time with a Minkowskian interpretation of relativity seems abortive.

The Minkowskian interpretation of SR thus faces the terrible dilemma of either denying the objective reality of tensed facts and temporal becoming or else succumbing to the same fragmentation of reality that plagues the Einsteinian interpretation and, if McTaggart is right, to incoherence. What is needed is a physical interpretation of the SR formalism predicated upon a coherent metaphysic which permits us to affirm both a theory of tensed time and a unified view of reality.

The Lorentzian interpretation

The Lorentzian interpretation seems to meet these *desiderata*. Lorentz's 3+1-dimensional ontology preserves the classical notions of space and time and so affords room for tensed facts and temporal becoming, and his aether compensatory approach to SR ensures the existence of absolute simultaneity and length without compromising empirical adequacy. Although in semi-popular expositions of SR scorn is usually poured upon the Lorentzian interpretation, the fact is that a small minority of physicists, including such notable figures as H. E. Ives, Geoffrey Builder, and S. J. Prokhovnik, have continued the Lorentzian research program down to the present day, so that the correct interpretation of relativity theory remains a matter of debate.[20] In fact, that debate has intensified in recent years, so that an aether compensatory approach to SR "no longer has," in Balashov's words, the "distinctively pseudo-scientific flavor it has had until very recently" (Balashov 2000). There are a number of modern equivalents of an aether which serve to mark off a privileged frame with respect to which relations of absolute simultaneity can be defined:

(i) The cosmological fluid. The Robertson-Walker metric for our expanding universe is expressed in such a way that every fundamental particle has a fixed set of coordinates which do not vary with time. The "gas" of fundamental particles is itself a sort of ether in that it is co-extensive with and at rest with respect to space. Heller, Klimek, and Rudnicki remark, "We may talk of symmetries in the energy-mass distribution. ... only after distinguishing a certain universal frame of reference in which these symmetries appear in a natural way. The existence of such a particular frame of reference resembles the concept of the aether in classical electrodynamics" (Heller *et al.* 1974: 3; see also Prokhovnik 1985, 1987, 1988). Although the cosmological fluid is radically different from the classical aether in that it exists in a systematically expanding frame, still this modern ether allows in a natural way for the existence of (1) universal cosmic time; (2) 3-spaces of constant curvature orthogonal to the time lines; and (3) a frame of reference co-moving with the substratum (Heller et al. *1974:* 4–5). Therefore, just as the classical aether was regarded as the physical realization of the fundamental reference frame, so the gas of fundamental particles serves to distinguish physically an equally fundamental frame.

(ii) The microwave background radiation. The cosmic microwave background radiation fills all of space and is remarkably isotropic for any observer at rest with respect to the expansion of space. The radiation background will be anisotropic for any observer in motion with respect to an observer whose spatial coordinates remain fixed. It is therefore a sort of ether, serving to distinguish physically a fundamental universal reference frame (Ibid.: 4). What is especially astounding is that tests have actually been able to detect the earth's motion relative to the radiation background, thus fulfilling nineteenth century physics' dream of measuring the aether wind! Smoot, Gorenstein, and Muller discovered that the earth is moving relative to the radiation background with a velocity of 390+60 km/sec in the direction of the constellation Leo. They comment, "The cosine anisotropy is most readily interpreted as being due to the motion of the earth relative to the rest frame of the cosmic blackbody radiation – what Peebles calls the 'new aether drift'" (Smoot *et al.* 1977: 899).[21] What Michelson and Morley failed to detect using visible light radiation, twentieth century physicists discovered using microwave radiation.[22]

(iii) The quantum mechanical vacuum: Underlying all of physical reality is the quantum mechanical vacuum, which is a sea of evanescent particles forming by fluctuations of the energy field and returning almost immediately to it. The quantum realm supplies a modern equivalent of the aether in various ways, among which we may mention two:

(a) Quantum electrodynamics. In SR Einstein dispensed with the aether medium, so that electromagnetic energy was regarded as traveling through an empty vacuum. As quantum theory penetrated electrodynamics, however, the vacuum was discovered to be anything but empty and to be the seat of fluctuating electro-magnetic fields. Dirac, who pioneered the development of relativistic, quantum electrodynamics, regarded the idealized state of the quantum vacuum as a suitable candidate for an ether (Dirac 1951: 906–7). Dirac's ether differs from the classical aether in that it is not amenable to a mechanical description; but it nonetheless plays a similar role to its classical predecessor in that it defines an inertial system within special relativity (Rompe and Treder 1984: 603–4). As a physically real, universal, inertial system, Dirac's quantum vacuum serves to delineate a fundamental reference frame with respect to which, like the classical aether, privileged velocities occur and, hence, privileged spatial and temporal relations may be established.
(b) The EPR experiment and Bell's Inequalities. A second way in which quantum mechanics serves to disclose a privileged reference frame and absolute simultaneity concerns the startling experimental results obtained on what was originally a thought experiment proposed by Einstein, Podolsky, and Rosen aimed at exposing the deficiencies of Bohr's Copenhagen Interpretation of quantum measurement situations. According to one recent commentator, the experimental verification of violations of Bell's Inequalities "constitutes the most significant event of the last half-century" for those interested in the fundamental structure of the physical world (Maudlin 1994: 4). In 1964, J. S. Bell showed that any hidden variables theory which preserved locality – that is to say, prohibited action at a distance – must make statistical predictions which disagree with those made by quantum mechanics (Bell 1983: 403–8). Suddenly, the EPR experiment was seen to lead to testable results. Since Bell's Theorem was explicated, a number of EPR-type experiments have been run, and the most precise of these, notably the experiments of Alain Aspect at the Institut d'Optique d'Orsay and the long-distance tests of Tittel, Brendel, Zbinden, and Gisin of the University of Geneva, have fully vindicated the predictions of quantum mechanics (see Aspect and Grangier 1986: 1–15; Tittel *et al.* 1999: 4150–63). The breaching of the Bell Inequalities therefore necessitates abandonment of the locality assumption which underlay the EPR thought experiment.

The demonstration that reality is non-local seems to present us with a dilemma, either horn of which has, in turn, important implications for the existence of a privileged frame. Our first alternative is to hold that adjustments in the measuring device at point *A* are somehow causally linked to the state of the photon at point *B*. Since the collapse of the wave function occurs instantaneously over arbitrarily large distances, quantum theory on this interpretation requires that a measurement at *A* causes an instantaneous change at *B*. But such an instantaneous influence establishes absolute simultaneity and thus requires a re-interpretation of quantum theory along neo-Lorentzian lines. In his 1964 paper, Bell concluded, " ... there must be a mechanism whereby the setting of one measuring device can influence the reading of another instrument, however remote. Moreover, the signal involved must propagate instantaneously, so that such a theory could not be Lorentz invariant" (Bell 1983: 107). Such instantaneous causal connections serve to establish an absolute reference frame in which the events at *A* and *B* are simultaneous. Hence, Bell, later pondering the implications of Aspect's experiments, comments,

> I think it's a deep dilemma, and the resolution of it will not be trivial; it will require a substantial change in the way we look at things. But I would say that the cheapest resolution is something like going back to relativity as it was before Einstein, when people like Lorentz and Poincaré thought that there was an aether – a preferred frame of reference – but that our measuring instruments were distorted by motion in such a way that we could not detect motion through the aether. ... that is certainly the cheapest solution. Behind the apparent Lorentz invariance of the phenomena, there is a deeper level which is not Lorentz invariant. ... what is not sufficiently emphasized in textbooks, in my opinion, is that the pre-Einstein position of Lorentz and Poincaré, Larmor and Fitzgerald was perfectly coherent, and is not inconsistent with relativity theory. The idea that there is an aether, and these Fitzgerald contractions and Larmor dilations occur, and that as a result the instruments do not detect motion through the aether – that is a perfectly coherent point of view. ... The reason I want to go back to the idea of an aether here is because in these EPR experiments there is the suggestion that behind the scenes something is going faster than light. Now if all Lorentz frames are equivalent, that also means that things can go backward in time. ... [this] introduces great problems, paradoxes of causality, and so on. And so it is precisely to avoid these that I want to say there is a real causal sequence which is defined in the aether.
>
> ("John Bell," (1986) interview in Davies and Brown; cf. Bell 1987: 279; see also Bell 1984: 66–76)

If, then, we allow for causal connections between the events situated at points *A* and *B*, the EPR experiment implies the existence of absolute simultaneity and an ether frame.

And, in fact, such a theory exists in the form of the de Broglie-Bohm pilot wave model (Cushing *et al.* 1996). According to one recent discussant there has been "a sea change" in the foundations of physics community with respect to the possibility of relations of absolute simultaneity due to a new appreciation of the de Broglie-Bohm approach (Kennedy 1997). Bohmian quantum mechanics is mathematically consistent and consonant with all experimental results but is a deterministic theory featuring super-luminal causal influences. Callender and Weingard have applied Bohmian quantum mechanics cosmologically and report that "when cosmological factors are considered, the de Broglie-Bohm interpretation remains the only satisfactory interpretation of quantum theory" (Callender and Weingard 1994: 218). This model enables one to resolve the problem of time without having to split worlds, multiply minds, or even worry about observers' collapsing wave functions. Time in the Bohmian model remains essentially the immutable, external, unobservable, unique time of Newtonian mechanics, in which the dynamic variables of cosmology, like the radius of the universe and the scalar field, evolve. Adoption of such an interpretation would require a Lorentzian approach to relativity theory, but an increasing number of theorists today are willing to give a serious look at Bohm's interpretation. Even if the details of his theory undergo revision and development, the commitment to a preferred time will remain one of the defining features of such an approach.

Suppose, on the other hand, that we reject the existence of causal connections between events occurring at *A* and *B* in favor of some interpretation according to which the photons at *A* and *B* are somehow correlated, but not causally connected. On this interpretation, the composite state consisting of the two photons with their respective measuring devices constitutes a single system, which is in a definite state (see Bohr 1964: 145–51). To affect the behavior of one photon via measurement is to disturb the *whole* system. When the wave function of a photon at *A* collapses, there is an immediate and correlated collapse of the wave function of its counterpart at *B*. But clearly, even though this interpretation denies the super-luminal causal influence from *A* to *B* posited by the de Broglie-Bohm theory, it still just as effectively abrogates the relativity of simultaneity, since the collapse of the paired wave functions is simultaneous. Therefore, asseverations that such an interpretation would not run contrary to the received interpretation of SR because that theory prohibits only *signals* of super-luminal velocities, not instantaneous correlations, are quite beside the point.[23] The point is that such correlations furnish the means of establishing relations of absolute simultaneity and a fundamental frame.[24] As Maudlin points out

> In Minkowski spacetime this theory of wave collapse no longer makes sense. The collapse can be instantaneous in at most one reference frame, leading to two possibilities: either some feature of the situation picks out a preferred reference frame, with respect to which the collapse is instantaneous, or the collapse is not instantaneous at all.
>
> (Maudlin 1994: 196; cf. pp. 137–38, 144).[25]

The problem posed by instantaneous collapse of the wave function for relativity theory can be clarified by realizing that since, according to SR, simultaneity is relative to reference frames, the collapse of the wave function for spatially separated photons will itself become relative to a reference frame. The problem is that properties like polarization and spin are not relational, but intrinsic properties of photons (Ryff 1992: 249). If the universe contained a single photon, it would have a definite polarization. Therefore the possession of such properties should not be made hyper-plane dependent. After surveying various attempts to integrate Bell's Inequalities with relativity theory, Maudlin concludes that these attempts "entail such severe dislocations of our physical view that one must seriously consider whether our grounds for adhering to Relativity are really strong enough to justify such extreme measures" (Maudlin 1994: 239; cf. Eberhard 1978: 392–419). "Indeed, the cost exacted by those theories which retain Lorentz invariance is so high that one might rationally prefer to reject Relativity as the ultimate account of space-time structure" (Maudlin 1994: 220). The postulation of preferred hyper-planes of simultaneity in the structure of space-time is, in fact, the only position which does not face severe difficulties. The only objection to such an approach to quantum theory and EPR, which is essentially Lorentzian, is that it requires one to reject SR – or more strictly speaking, to reject those theories which are composed of the customary SR mathematical formalism conjoined with the physical interpretation of Einstein or Minkowski, respectively.

The above three physical realities – the cosmological fluid, the microwave background radiation, and the quantum mechanical vacuum – all serve to revitalize in a new guise the concept of the aether.[26] Bohm himself, maintaining that there is "a unique spacetime frame, in terms of which 'simultaneous contact' would be specified," comments, "Empirically, this should be close to the frame in which the mean velocity of the 3°K radiation background in space is zero."[27] Since the microwave background radiation is isotropic only with respect to the frame in which the fundamental particles of the expanding universe are at rest, the fundamental frame of the cosmic expansion fills the role of the preferred frame required by quantum theory. According to Prokhovnik

> The notion of non-local causality, discussed by Bell, requires a criterion of simultaneity which has some absolute significance; it is seen that a cosmological basis for a universal measure of (cosmic) time resolves this problem. ... the existence of an observationally-based fundamental frame is invaluable not only for the understanding of our universe and of Special Relativity, but also to make sense of quantum theory along the lines proposed, for example, by Dirac. ... and John Bell.
>
> (Prokhovnik preprint)

While differing markedly from the classical aether, the new ethers play the same essential role in marking off a unique, fundamental reference frame, the frame of expanding physical space, in which privileged relations of simultaneity may be established.

Empirical considerations may thus actually favor a Lorentzian interpretation of the mathematical formalism of SR. Lorentz's prediction seems to have been vindicated when he mused, "In my opinion it is not impossible that in the future this road, indeed abandoned at present, will once more be followed with good results, if only because it can lead to the thinking out of new experimental tests" (Lorentz 1920b: 61–62). Certainly empirical considerations do not disqualify a Lorentzian approach. Are there, then, overriding non-empirical considerations for preferring the received interpretation of SR?

It is often asserted that the received interpretation is simpler and therefore to be preferred to a Lorentzian interpretation. However, as is well known, one must be very cautious about the connection between the simplicity of a theory and its *truth*, as opposed to its *utility*. In any case, the claim that the received interpretation is simpler is mistaken. Although Lorentz's own theory was more complicated than Einstein's, H. E. Ives was able to derive the Lorentz transformation equations from (i) the laws of conservation of energy and momentum; and (ii) the laws of transmission of radiant energy. "The space and time concepts of Newton and Maxwell are retained without alteration," he wrote, "It is the dimensions of the material instruments for measuring space and time that change, not space and time that are distorted" (Ives 1979: 247, 255). On Ives's achievement, Martin Ruderfer comments that he succeeded in elevating Lorentz's *ad hoc* theory to an equal status with SR and did so with the same number of basic assumptions as Einstein, so that his theory has the same "beauty". "The Ives and Einstein interpretations represent two different, but equally valid, views of the same set of observations" (Ruderfer 1979: 243). Hence, assertions that the received interpretation is simpler than a Lorentzian interpretation are simply incorrect.

Probably at the root of many physicists' rejection of a Lorentzian interpretation of relativity theory is the deep-seated conviction that comes to expression in Einstein's aphorism: "Subtle is the Lord, but malicious He is not."[28] That is to say, if there exists a fundamental asymmetry in nature, then nature will not conspire to conceal it from us by precisely countervailing effects. D'Abro expresses this sentiment clearly:

> If Nature was blind, by what marvelous coincidence had all things been so adjusted as to conceal a velocity through the ether? And if Nature was wise, she had surely other subjects to attend to, more worthy of her consideration, and would scarcely be interested in hampering our feeble attempts to philosophize. In Lorentz's theory, Nature, when we read into her system all these extraordinary adjustments *ad hoc*, is made to appear mischievous; it was exceedingly difficult to reconcile one's self to finding such human traits in the universal plan.
>
> (d'Abro 1950: 138)

One difficulty with this objection is that it seems to be guilty of greatly over-exaggerating the extent of the alleged conspiracy. After all, SR is a restricted theory of relativity: it is only uniform motion relative to a privileged frame that fails to manifest itself. But in all other cases of motion, the absolute character of that motion is disclosed. This is not to say that acceleration or rotation proves the reality of privileged space, but it is to say that, given the classical concepts of time and space, nature does not at all conspire to conceal either absolute motion or the privileged space from us. Moreover, as we have seen, there are modern equivalents of the classical aether which serve plausibly to disclose a privileged frame. Indeed, when non-Lorentzians complain that no evidence of a privileged space and time exist, one wonders what it would take to convince them of the contrary. If no empirical evidence of a fundamental frame is incapable of being explained away, then the supposed failure of nature to disclose such a frame to us becomes trivial. The more difficult it is for nature to provide evidence of the existence of a privileged frame, the less compelling the charge that she is conspiring against us to conceal it.

But even considered in abstraction from these considerations, why accept the assumption that fundamental asymmetries in nature must disclose themselves to us? This assumption is by no means obvious, as Martin Carrier explains:

> Science would be an easy matter if the fundamental states of nature expressed themselves candidly and frankly in experience. In that case we could simply collect the truths lying ready before our eyes. In fact, however, nature is more reserved and shy, and its fundamental states often appear in masquerade. Put less metaphorically, there is no straightforward one-to-one correspondence between a theoretical and an empirical state. One of the reasons for the lack of such a tight connection is that distortions may enter into the relation between theory and evidence, and these distortions may alter the empirical manifestation of a theoretical state. As a result, it is in general a nontrivial task to excavate the underlying state from distorted evidence. (Carrier 1993: 3)[29]

On a Lorentzian theory, Carrier's general remarks on distortions of a theoretical state in its empirical manifestation are quite literally true, for the result of uniform motion relative to privileged space is length contraction and clock retardation, phenomena which, it will be recalled, are every bit as real under the Einsteinian interpretation.[30] If it is in general difficult to excavate the underlying state of nature from distorted evidence, if nature's fundamental states often appear in masquerade, then why are the relativistic phenomena which mask the privileged frame improbable on a classical ontology?

As for d'Abro's concern with finding "human traits in the universal plan," the Lorentzian might plausibly appeal to the Anthropic Principle in response.[31] According to that principle, features of the universe can only be judged in their correct perspective when due allowance has been made for the fact that certain features of the universe are necessary if it is to contain observers like ourselves (Barrow and Tipler 1986: 15). Since our existence depends on the maintenance of equilibrium states within us, the Lorentz-Fitzgerald contraction and clock retardation are necessary pre-requisites of our existence as observers. Thus, nature's alleged conspiracy, when seen in anthropic perspective, seems much less mischievous. Therefore, this objection also falls to the ground.

Some Minkowskians might argue that a Lorentzian interpretation of SR is disfavored because of the excessive space-time structure which it posits with respect to the laws of motion. Recognizing the failure of verificationist critiques of absolute space on the basis of its unobservability, Earman, for example, suggests that "this objection of unobservability is more accurately stated as an objection based on Occam's razor" (Earman 1989: 48). What Earman has in mind are criteria which would serve to establish what sort of structure space-time is endowed with. He proposes two "symmetry principles" which he presents as conditions of adequacy on theories of motion:

> SP1: Any dynamical symmetry of a theory *T* is a space-time symmetry of *T*.
> SP2: Any space-time symmetry of a theory *T* is a dynamical symmetry of *T*.

What justification is there for these criteria? Earman explains, "Behind both principles lies the realization that laws of motion cannot be written on thin air alone, but require the support of various space-time structures. The symmetry principles then provide standards for judging when the laws and the space-time structure are appropriate to one another" (Earman 1989: 486). With respect to (SP1), which will be crucial in the case at hand, Earman writes,

> The motivation for (SP1) derives from combining a particular conception of the main function of laws of motion with an argument that makes use of Occam's razor. Laws of motion, at least in so far as they relate to particles, serve to pick out a class of allowable or dynamically possible trajectories. If (SP1) fails, the same set of trajectories can be picked out by the laws working in the setting of a weaker space-time structure. The theory that fails (SP1) is thus using more space-time structure than is needed to support the laws, and slicing away this superfluous structure serves to restore (SP1).
>
> (Earman 1989: 46–47)

A prime example of (SP1) at work is the demonstration that Newtonian mechanics does not, in fact, require Newtonian space-time, but only Galilean space-time. While the space-time symmetries of the theory are Newtonian, the dynamic symmetries are Galilean, a "clear violation" of (SP1) (Earman 1989: 48). By excising absolute space, one produces a theory whose space-time symmetries and dynamic symmetries are both Galilean, in compliance with (SP1). Now it might similarly be argued – though Earman does not do so – that a Lorentzian theory also violates (SP1) in that it posits more structure to space-time, such as hyperplanes of absolute simultaneity, than is necessary to explain the symmetries required by relativistic laws of motion, that is, the symmetries of the Lorentz-Poincaré group. Therefore, such structures should be excised by Ockham's razor.

Now Earman is undoubtedly on to something important here. One should like to have some sort of constraint upon the postulation of gratuitous space-time structures. But the difficulty with (SP1) and (SP2) is that they are too restrictive, or to put the point another way, they can be overridden by considerations broader than the laws of motion. Indeed, one of the central lessons of our present investigation is that in questions of time and space metaphysical considerations cannot be ignored.[32] We have seen, for example, that the 4-dimensional Minkowskian world is incompatible with the reality of tense and temporal becoming and also faces a variety of significant objections. Such metaphysical considerations qualify any force which Earman's symmetry principles might possess.

More fundamentally, Earman's principles presuppose a space-time ontology and therefore cannot be employed to justify a space-time interpretation of the formalism of SR over a Lorentzian space and time interpretation. Earman employs a space-time approach to all the theories he considers, speaking not only of Newtonian space-time but even of Machian space-time. This involves him in frequent anachronisms: he speaks of Newton's "insistence that the structure of space-time is immutable;" he says that "If we take Newton at his word in the Scholium, the space-time setting for the theory of motion and gravitation of the *Principia* is supposed to be a full Newtonian space-time ... ;" he explains that for Newton "the space-time is given once and for all as an emanative effect of God. ... " (Earman 1989: 35–36, 45, 48). Of course, Newton himself insisted on and believed none of this, since his was a 3+1 ontology involving physical objects enduring through time. His laws of motion are not supported by various space-time structures, as Earman assumes, for there are on Newton's view no such structures. Time and space have intrinsic structure, and motion is to be explained, not in terms of the geometrical structure of space-time, but in terms of forces acting on bodies in space enduring through time. Earman's principles, then, already take for granted a space-time realism. At best, therefore, with respect to SR, Earman's symmetry principles could only serve to justify a Minkowskian space-time realism over a Lorentzian space-time realism. But if, so to speak, only a single slice of space-time actually exists, as Einsteinians and Lorentzians hold, then the question becomes whether the Einsteinian interpretation is preferable to the

Lorentzian interpretation. Lorentzians and Minkowskians concur that the Einsteinian interpretation is untenable. In short, one's ontology may warrant postulating more structure to space and time than symmetry principles alone would allow.

Occasionally arguments are offered on behalf of a Minkowskian interpretation of relativity theory which are relevant to a Lorentzian approach, and these merit our attention. For example, oft-times space and time theorists have accused space-time realists of a gratuitous hypostatization of what is only a geometrical representation of space and time. There is no reason to think that because a thing's spatio-temporal location can be plotted on a Minkowski diagram, therefore an entity *space-time* exists anymore than because temperature and pressure can be similarly related in a diagrammatic way, therefore an entity *temperature-pressure* exists. In response to this charge D'Abro points to the 4-dimensional metric of Minkowski space-time as its distinguishing feature, in virtue of which it is not like the artificial temperature-pressure continuum (D'Abro 1950: 347). The temperature-pressure continuum can be separated into two distinct elements, but Minkowski space-time cannot. D'Abro concludes,

> In short, the reality of space-time arises from Minkowski's discovery that it was possible to discover an invariant distance between two points in space-time, holding for all observers; and that it was impossible to define any such invariant distance in space alone or in time alone, showing that space and time by themselves were phantoms. These last two concepts must henceforth be considered jointly, and no longer as separate entities.
>
> (D'Abro 1950: 447)

One can, of course, adopt a space-time approach even to Newtonian theory. But D'Abro's point is that there is no special 4-dimensional metric for such a continuum; it reduces to a mere geometric representation of the Galilean transformations (D'Abro 1950: 472–73). Like the temperature-pressure continuum, no real unification takes place.

This is an odd argument on behalf of space-time realism, since the would-be Newtonian space-time realist must be dismissed along with the temperature-pressure realist, even though he holds to a space-time realism, too. His space-time is a mathematical fiction, we are told, whereas Minkowski space-time is not, since the latter alone has a 4-dimensional metric. But the reason it has such a 4-dimensional metric is precisely because of its preclusion of relations of absolute simultaneity. In the limit case where $c \rightarrow \infty$, the "absolute elsewhere" region in space-time is squeezed out, and the Newtonian dichotomy between a universal, absolute past and future is recovered (Costa de Beauregard 1987: 59–60; cf. Lucas and Hodgson 1990: 228–38). Thus the very existence of a space-time metric and a light cone structure in space-time depends upon there not existing such a global, absolute separation of past and future. The Minkowskian space-time realist can hardly justify relativistic space-time over the Newtonian realist's classical space-time on the grounds that only the former's has a space-time metric, since it is a necessary condition of relations of absolute simultaneity, which are the distinguishing mark of classical space-time, that no such metric exists in Newtonian space-time. But if the presence of such a metric serves merely to differentiate Minkowskian from Newtonian space-time, rather than to justify realism about the former over realism about the latter, then neither can it justify Minkowskian space-time realism over Newtonian space and time realism, since the latter does not differ metrically from Newtonian space-time realism.

What other reasons might be offered for preferring space-time realism? Earman and Friedman, in defense of space-time realism over the classical ontology, have asserted that "neglect of space-time structure in the classical context has led to a number of philosophical errors and oversights" (Earman and Friedman 1973: 329; cf. Earman 1970: 259–77). Presumably, they have reference, for example, to classical theorists' overlooking Galilean space-time, which preserves absolute time without absolute space. But such examples serve to show only the admitted heuristic utility of a space-time methodological approach to questions of space and time. In the same way that possible worlds semantics is an illuminating and perhaps indispensable tool for exploring questions of modality but does not commit one *ipso facto* to modal realism, so also space-time methodology can enlighten without implying a space-time ontology. The same point may be made concerning the avoidance of philosophical errors – though here one must add that a space-time approach might also lead one to commit philosophical errors which a space and time approach would have avoided. For example, attempts to relativise acceleration and rotation by appeal to 4-dimensional structures come to mind, representations which may be of heuristic value but have also led some thinkers to conclude erroneously that absolute motions such as Newton pointed to have been eliminated. Or again, due to the isomorphism of the space and time dimensions in space-time, which differ only in sign, erroneous philosophical conclusions about the interchangeability of time and space, such as obtains in the Schwarzschild metric or the Hartle-Hawking cosmology, might be drawn, which a space and time approach would perhaps prevent. To return to oversights, the Lorentzian will charge that a perspectival, 4-dimensional approach to relativistic phenomena has resulted in the neglect of the search for the dynamic causes of time dilation and length contraction, a major oversight. The obvious difficulty here is that one man's insights are another man's oversights, and it is difficult to tell which one is correct without independent justification of one point of view.

Earman elsewhere argues that even if space-time can be ignored in the context of classical physics, one cannot neglect it in the context of relativistic physics, since space-time is not uniquely separable into space and time

(Earman 1970: 259–60; cf. Earman and Friedman 1973: 352). In fact, he points out, there are relativistic space-times which resist every effort to separate them into a 3-dimensional space enduring through a 1-dimensional time. But Earman is speaking here of GR space-times, not SR space-time. So far as SR is concerned, Einstein's original formulation of the theory shows that a space and time approach is necessarily equivalent to a space-time approach. Moreover, one cannot automatically assume that just because a particular space-time can be modeled that such a space-time represents a realistic possibility (think, for example, of imaginary space-times). One might well question whether models such as Gödel's, alluded to by Earman, which permit pathological results like closed time-like loops, are anything more than mathematical curiosities.

In sum, even if, as we have seen, a Minkowskian interpretation of the SR formalism is in some respects, at least, superior to an Einsteinian interpretation, there do not seem to be comparably good reasons, empirical or philosophical, to prefer a Minkowskian interpretation of SR over a Lorentzian conception of space and time. If anything, the Lorentzian approach seems to have the empirical edge.

On the other hand, in addition to the considerations pertinent to tensed time mentioned earlier, I think we do have good reasons for rejecting space-time realism and, therefore, a Minkowskian interpretation of the formalism of SR. Inherent to the concept of space-time is the indissoluble unification of space and time into a 4-dimensional continuum. Hence, Minkowski's pronouncement that "Henceforth, space by itself, and time by itself, are doomed to fade away into mere shadows, and only a kind of union of the two will preserve an independent reality" (Minkowski 1952: 75). It seems to me, however, that we have powerful metaphysical grounds for believing that time can exist independently of space. In making his pronouncement, Minkowski forgot Kant's insight that time is a form applicable not only to the external world, but to consciousness as well (Kant 1970, A34/B51, p. 77). Concerning Minkowski space-time, Wenzl cautions:

> From the standpoint of the physicist, this is a thoroughly consistent solution. But the physicist will (doubtless) understand the objection, raised by philosophy, that time is by no means a merely physical matter. Time is, as Kant put it, the form not merely of our outer sense but also of our inner sense. … Should our experiences of successiveness and of memory be mere illusion … ?
>
> (Wenzl 1949: 587–88)

A series of mental events alone is sufficient to set up a temporal sequence.[33] Thus, for example, we can imagine God's counting down to the moment of creation: " …, 3, 2, 1, *Fiat lux*!" The beginning of space-time would then be preceded by a metaphysical time associated with the mental events of counting which would be wholly independent of space.

In addition, space-time realism confronts a host of objections due to its implication of a tenseless view of time, as previously mentioned. Thus, with little to commend space-time realism over a Lorentzian conception of space and time, good philosophical and empirical grounds for preferring a Lorentzian approach over a Minkowskian approach, and powerful objections lodged against space-time realism, we may conclude that there is good reason to prefer a Lorentzian approach to SR over a Minkowskian interpretation of the SR formalism.

Conclusion

It seems to me, therefore, that despite the widespread aversion to a Lorentzian physical interpretation of the mathematical formalism of SR, such antipathy is really quite unjustified. Admitted on all sides to be at least empirically equivalent to the Einsteinian and Minkowskian interpretations, the Lorentzian interpretation is the only one of the trio which can give us both the objective reality of tense and temporal becoming and a unified picture of the world. It is neither more complicated than its rivals nor unacceptably contrived. It is the one approach to relativity theory which promises to mesh with recent advances in cosmology and quantum mechanics and electrodynamics. At the same time it averts all the problems of space-time realism inherent in the Minkowskian interpretation. Therefore, we have good grounds – wholly apart from the theological considerations which motivated the classical concept of time – for accepting a Lorentzian physical interpretation of the mathematical formalism of the Special Theory of Relativity.

Notes

1 Nerlich calls such an interpretation of SR the relativity interpretation (Nerlich 1994: 63).
2 Tenseless time theorist D. H. Mellor does attempt to combine a 3+1 ontology with space-time realism (Mellor 1981: 105), but the combination of endurantism and space-time realism is as incoherent as the wedding of space-time realism with the objectivity of tense and temporal becoming, and for the same reason, since endurantism and a tensed theory of time are logically equivalent; see note 54 and relevant discussion in the text. See further Carter and Hestevold 1994: 269–83.
3 The seriousness with which Einstein took this conception may be seen in the fact that when his life-long friend Michael Besso died, Einstein sought to comfort his bereaved family by reminding them that for physicists Besso had not ceased to exist, but exists tenselessly as a permanent feature of space-time reality (Hoffman and Dukas 1972: 258).
4 Nerlich calls this the space-time interpretation (Nerlich 1998: 128–29).
5 On this score Nerlich finds himself at odds with theorists such as Reichenbach, Grünbaum, van Fraassen, Salmon, and Winnie. On Nerlich's view the heart of SR is the second-order constraint of Lorentz invariance on all physical laws, which springs from the inherent symmetries of Minkowski space-time. Temporal relations in that space-time cannot be reduced to causal relations because the latter presuppose the former.

6 Notice that Lorentz eschews operational definitions of time and does not equate time with clock readings. This is particularly evident in his failure to ascribe at first physical reality to his local times (Lorentz 1934: 262). Eventually Lorentz came to see the reality of clock retardation. But even the readings of a clock at rest in the aether are but a sensible measure of true time, not definitive of true time.

7 See also the outstanding historical study by Illy (1989: 272), who comments, "Lorentz distinguishes the existence of a state of motion from its observability – a distinction void of meaning according to Einstein."

8 Indeed, I should venture to surmise that it is the majority view among A-Theorists and endurantists. Process philosophers also often appear to favor this view.

9 See the seminal article by Putnam (1967: 240–47), which is, however, a very incoherent presentation. For more cogent, recent versions of the complaint, see Savitt (1998); Callender (1998); and especially Balashov (2000: 129–66).

10 See the *Auseinandersetzung* between V. Varicak (1911: 169, 509–10). For helpful discussion see Winnie (1972: 1091–94).

11 It is important to note that accepting the superiority of the space-time interpretation does not without further ado commit one to space-time substantivalism. Space-time realists disagree among themselves whether space-time should be construed substantivally or relationally, substantivalists maintaining that space-time is a substance existing independently of objects and events located in it, relationalists holding that space-time is not a substance, but depends for its existence on the existence of objects and events. Both are united in opposition to the Einsteinian interpretation in contending that all space-time points are equally real. For discussion, see Horwich (1978: 397–419).

12 For an illustration see Angel (1980: 89).

13 Marder (1971: 78) paradoxically endorses Rindler's interpretation that the traveler's path is Lorentz-contracted.

14 Nerlich, personal communication, June 20, 1991.

15 For an outstanding defense of this point, see Smith (1993); see also Craig (1996a: 5–26; 1996b: 249–69; 2000a: 165–85).

16 One of the most eloquent spokesmen for this point of view has been Schlesinger (1980: 34–39, 138–39). See also Craig (1999a: 515–37; 1999b: 107–20; 2001: 159–66).

17 There is no good treatment of this yet, but see Gale (1968: 90–97) and Mellor (1981: 140). See also Craig (2000b: Chap. 7).

18 Again, this point needs to be better developed, but see Geach (1972: 306) and McGilvray (1979: 275–99). See also Craig (2000b: Chap. 8).

19 See the excellent study by Merricks (1994: 165–84); see also Lewis (1986: 305–9) and the incisive piece by Van Inwagen (1990: Chap. 9).

20 It is noteworthy that in Zahar's analysis, SR did not by itself supersede Lorentz's program. Zahar can claim that Einstein's research program superseded Lorentz's only by taking GR to be continuous with SR and by then appealing to the empirical confirmation of GR (Zahar 1973: 95–123; 223–62). But as is well known, GR is not an extension of SR; it is a theory of gravitation, not a theory of relativity. It therefore follows that Einstein's program never in fact did supersede Lorentz's, in the sense of demonstrating its superiority over Lorentz's.

21 Cf. Kanitscheider's remark: "The cosmic background radiation ... furnishes a reference frame, relative to which it is meaningful to speak of absolute motion" (Kanitscheider 1984: 256).

22 One can only speculate whether, had this microwave background radiation and the measure of our motion relative to it been known to Einstein prior to 1905, he would have claimed that no fundamental frame exists relative to which all local inertial frames are in motion. See Miller's paper (1980: 66–91) in connection with the remarks of Dirac (1980: 110–11). Cf. the remark of Munsouri and Sexl: "The

discovery of the cosmic background radiation has shown that cosmologically a preferred system of reference does exist. This system is defined and singled out much more unambiguously to be a candidate for a possible 'ether frame' than was the solar rest frame in Einstein's day" (Munsouri and Sexl 1997: 497–98).

23 Such asseverations may not even be true. It is very difficult to see why the entanglement-based quantum cryptography suggested by the experiments of the Geneva Group (Clifton and Hogarth 1995: 355–97) would not involve the instantaneous transmission of information even in the absence of super-luminal propagation of causal influences.

24 See Stapp (1988: 71–72). See further Popper (1982: xviii, 20). Even Popper's use of the expression "infinite velocity" is misleading, since the salient point is the simultaneous collapse of the correlated two wave functions, as if they were joined by an influence of infinite velocity. See Shimony (1978: 13).

25 Although Maudlin interprets the counterfactual dependence relation which exists between the separated polarization events in EPR as a causal connection, his emphasis lies on the fact that relativity cannot make physical sense of instantaneous collapse of the wave function.

26 On their coincidence see Cushing (1996: 175); Popper (1982: 30).

27 David Bohm to D. R. Griffin, May 17, 1992, cited in Griffin, "God and Relativity Physics," p. 110. When Callender and Weingard contrast the cosmic time of Bohmian cosmology with "the arbitrary parameter found in general relativity," the contrast concerns the arbitrariness permitted by GR taken *in abstracto* (Callender and Weingard 1994: 227), not taken with the existing boundary conditions of isotropy and homogeneity. GR cosmic time and Bohmian cosmic time may well be extensionally equivalent, even though intentionally diverse.

28 "Raffiniert ist der Herr Gott, aber boshaft ist er nicht," remark of Albert Einstein during a visit to Princeton, upon being informed that D. C. Miller, a former colleague of Michelson, had claimed to have detected the ether wind (cited in Pais 1982: 113–14).

29 Tim Maudlin has emphasized and illustrated Carrier's point. He surveys what he characterizes as the "teratological collection" of theories attempting to explain the Bell Inequalities and integrate the EPR results with relativity theory and concludes, "One way or another God has played us a nasty trick" (Maudlin 1994: 241). One cannot dismiss neo-Lorentzian relativity on the grounds that it would be deceptive, since partisans of each theory could say the same of rival positions. " … the real challenge falls to the theologians of physics, who must justify the ways of a Deity who is, if not evil, at least extremely mischievous" (Ibid. p. 242).

30 It is perhaps noteworthy that D'Abro mistakenly holds that "The Fitzgerald contraction is no longer a real physical contraction, as it was assumed to be in Lorentz's theory" (D'Abro 1950: 151).

31 I owe this point to Robin Collins.

32 Notice that the foundations of Newtonian time and space, for example, were laid in Newton's theism (as Earman acknowledges Earman 1989: 10), but Earman simply ignores such metaphysical considerations, choosing instead to put a "modern gloss" on Newton's views.

33 A point emphasized by Reichenbach:

> An act of thought is an event and therefore defines a position in time. If my experiences are always produced within the framework of a 'now,' that means that each act of thought defines a point of reference. We cannot escape the 'now' because the attempt to escape signifies an act of thought and therefore defines a 'now'.
>
> (Reichenbach 1952: 157)

References

Angel, Roger B. (1980) Relativity: The Theory and Its Philosophy, Foundations and Philosophy of Science and Technology, Oxford: Pergamon Press.

Arzeliès, Henri (1966) Relativistic Kinematics, rev. ed., Oxford: Pergamon Press.

Aspect, Alain and Philippe Grangier (1986) 'Experiments on Einstein-Podolsky-Rosen-type correlations with pairs of visible photons', in R. Penrose and C. J. Isham *(eds.)*, Quantum Concepts in Space and Time, Oxford: Clarendon Press, pp. 1–15.

Balashov, Yuri (2000) 'Enduring and Perduring Objects in Minkowski Space-Time', Philosophical Studies 99, 129–66.

Barrow, John D., and Frank J. Tipler (1986) The Anthropic Cosmological Principle, Oxford: Clarendon Press.

Bell, J. S. (1983) 'On the Einstein Podolsky Rosen Paradox', Physics (1964) 1, 195–200, reprinted in Quantum Theory and Measurement, John Archibald Wheeler and Wojciech Hubert Zurek (eds.), Princeton Series in Physics, Princeton: Princeton University Press, pp. 403–8.

—— (1984) 'Quantum, Space and Time – The Quest Continues', in Asim O. Barut, Alwyn van der Merwe, and Jean-Pierre Vigier (eds.), *Cambridge Monographs on Physics*, Cambridge: Cambridge University Press.

—— (1986) 'John Bell', interview in P. C. W. Davies and J. R. Brown, The Ghost in the Atom, Cambridge: Cambridge University Press.

—— (1987) 'The paradox of Einstein, Podolsky, and Rosen: action at a distance in quantum mechanics?' Speculations in Science and Technology 10, 279.

Bergson, Henri (1965) Duration and Simultaneity, trans. Leon Jacobson, Library of Liberal Arts, 1922; rep. ed., Indianapolis: Bobbs-Merrill.

Bohr, Niels (1964) 'Can Quantum-Mechanical Description of Physical Reality Be Considered Complete?' Physical Review 48, 696–702, reprinted in Quantum Theory and Measurement, John Archibald Wheeler and Wojciech Hubert Zurek (eds.), Princeton Series in Physics, Princeton: Princeton University Press, pp. 145–51.

Bondi, Herman (1964) Relativity and Common Sense, New York: Dover Publications.

Builder, G. (1971) 'The Constancy of the Velocity of Light', Australian Journal of Physics (1958) 11, 457–80; reprinted in Speculations in Science and Technology 2, 422.

Burge, Tyler (1992) 'Philosophy of Language and Mind: 1950–90', Philosophical Review 101, 3–51.

Callender, Craig (1998) 'Is Presentism Worth its Price?' Paper presented at the symposium 'The Prospects for Presentism in Spacetime Theories', Philosophy of Science Association biennial meeting, October, 1998, Kansas City.

Callender, Craig and Robert Weingard (1994) 'The Bohmian Model of Quantum Cosmology', in David Hull, Micky Forbes, and Richard M. Burian *(eds.)*, PSA 1994, East Lansing, Mich.: Philosophy of Science Association, p. 218.

Capek, Milic (1976) 'The Inclusion of Becoming in the Physical World', in M. Capek *(ed.)*, Concepts of Space and Time, Boston Studies in the Philosophy of Science 22, Dordrecht: D. Reidel, pp. 501–24.

Carrier, Martin (1993) 'Physical Force or Geometrical Curvature?' in John Earman, Allen I. Janis, Gerald J. Massey, and Nicholas Rescher (eds.), Philosophical Problems of the Internal and External Worlds, Pittsburgh: University of Pittsburgh Press, p. 3.

Carter, William R. and H. Scott Hestevold (1994) 'On Passage and Persistence', American Philosophical Quarterly 31, 269–83.

Christensen, F. M. (1993) Space-like Time, Toronto: University of Toronto Press.

Clifton, Rob and Mark Hogarth (1995) 'The Definability of Objective Becoming in Minkowski Spacetime', Synthese 103, 355–87.

Coburn, Robert C. (1990) The Strangeness of the Ordinary: Problems and Issues in Contemporary Metaphysics, Savage, Maryland: Rowman & Littlefield.

Costa de Beauregard, Oliver (1987) Time, the Physical Magnitude, Boston Studies in the Philosophy of Science 99, Dordrecht: D. Reidel.

Craig, William Lane (1996a) 'Tense and the New B-Theory of Language', *Philosophy* 71, 5–26.

—— (1996b) 'The New B-Theory's Tu Quoque Argument', Synthese 107, 249–69.

—— (1998a) 'McTaggart's Paradox and the Problem of Temporary Intrinsics', Analysis 58, 122–27.

—— (1998b) 'The Tensed vs. Tenseless Theory of Time: A Watershed for the Conception of Divine Eternity', in R. LePoidevin (ed.), Questions of Time and Tense, Oxford: Oxford University Press, pp. 221–50.

—— (1999a) 'On Our Differential Experience of the Past and Future', Southern Journal of Philosophy 37, 515–37.

—— (1999b) 'The Presentness of Experience', in Mogens Wegener (ed.), Time, Creation, and World Order, Aarhus, Denmark: University of Aarhus, pp. 107–20.

—— (2000a) 'On the Truth Conditions of Tensed Sentence Types', Synthese 120, 165–85.

—— (2000b) *The Tenseless Theory of Time: a Critical Examination*, Synthese Library, Dordrecht: Kluwer Academic Publishers.

—— (2001) 'On Wishing It Were Some Other Time', Philosophy and Phenomenological Research 62, 159–66.

—— (2002) 'The Elimination of Absolute Time by the Special Theory of Relativity', in Gregory E. Ganssle and David M. Woodruff (eds.), God and Time, Oxford: Oxford University Press, pp. 129–52.

Cushing, James T. (1996) 'What Measurement Problem?' in Rob Clifton *(ed.)*, Perspectives on Quantum Reality, University of Western Ontario Series in Philosophy of Science 57, Dordrecht: Kluwer Academic Publishers, p. 175.

Cushing, James T, Arthur Fine, and Sheldon Goldstein (1996) Bohmian Mechanics and Quantum Theory: an Appraisal, Boston Studies in the Philosophy of Science 184, Dordrecht: Kluwer Academic Publishers.

D'Abro, A. (1950) The Evolution of Scientific Thought, 2nd rev. ed. (1927); rep. ed.: n.p.: Dover Publications, p. 138.

Denbigh, K. G. (1978) 'Past, Present, and Future', in J. T. Fraser, N. Lawrence, and D. Park (eds.), The Study of Time III, Berlin: Springer Verlag, p. 309.

Dingle, Herbert (1940) The Special Theory of Relativity, London: Methuen.

Dirac, P. A. M. (1951) 'Is There an Aether?' Nature 168, 906–7.

—— (1980) 'Discussion', in Harry Woolf *(ed.)*, Some Strangeness in the Proportion, Reading, Mass.: Addison-Wesley Publishing Co., pp. 110–11.

Dorato, Mauro (1995) Time and Reality: Spacetime Physics and the Objectivity of Temporal Becoming, Collana di Studi Epistemologici 11, Bologna: CLUEB.

Earman, John (1970) 'Space-Time, or How to Solve Philosophical Problems and Dissolve Philosophical Muddles without Really Trying', Journal of Philosophy 67, 259–77.

—— (1989) World Enough and Space-time, Cambridge, Mass.: MIT Press.
Earman, J. and M. Friedman (1973), 'The Meaning and Status of Newton's Law of Inertia and the Nature of Gravitational Forces', Philosophy of Science 40: 329.
Eberhard, P. H. (1978) 'Bell's Theorem and the Different Concepts of Locality', Il Nuovo Cimenta 46B, 392–419.
Einstein, A. (1911) 'Zum Ehrenfestschen Paradoxen', Physikalische Zeitschrift 12, 509–10.
—— (1920) Äther und Relativitätstheorie, Berlin: Julius Springer Verlag.
—— (1961) Relativity: The Special and the General Theory, New York: Crown Trade Paperback.
—— (1981) 'On the Electrodynamics of Moving Bodies', trans. Arthur I. Miller, Appendix to Arthur I. Miller, Albert Einstein's Special Theory of Relativity, Reading, Mass.: Addison-Wesley.
Einstein, Albert and Michele Besso (1979) Correspondance 1903–55, trans. with Notes and an Introduction by Pierre Speziali, Paris: Hermann, pp. 276–77.
Einstein, Albert and Leopold Infeld (1938) The Evolution of Physics, New York: Simon & Shuster.
Findlay, J. M. (1978–79) 'Time and Eternity', Review of Metaphysics 32, 6–7.
Friedman, William (1990) About Time, Cambridge, Mass.: MIT Press.
Gale, Richard (1968) The Language of Time, International Library of Philosophy and Scientific Method, London: Routledge & Kegan Paul.
Geach, Peter (1972) 'Some Problems about Time', in Logic Matters, Berkeley: University of California Press, p. 306.
Griffin, David Ray (1986) 'Introduction: Time and the Fallacy of Misplaced Concreteness', in David R. Griffin (ed.), Physics and the Ultimate Significance of Time, Albany, N.Y.: State University of New York Press, pp. 6–7.
Heller, Michael, Zbigniew Klimek, and Konrad Rudnicki (1974), 'Observational Foundations for Assumptions in Cosmology', in M. S. Longair (ed.), Confrontation of Cosmological Theories with Observational Data, Dordrecht: D. Reidel, p. 3.
Hoffman, Banesh with Helen Dukas (1972) Albert Einstein: Creator and Rebel, London: Hart-Davis, MacGibbon.
Holton, Gerald (1973) 'On the Origins of the Special Theory of Relativity', in Gerald Holton *(ed.)*, Thematic Origins of Scientific Thought: Kepler to Einstein, Cambridge, Mass.: Harvard University Press, p. 171.
Horwich, Paul (1978) 'On the Existence of Time, Space, and Space-Time', Noûs 12, 397–419.
—— (1987) Asymmetries in Time, Cambridge, Mass.: MIT Press.
Illy, Jozsef (1989) 'Einstein Teaches Lorentz, Lorentz Teaches Einstein. Their Collaboration in General Relativity, 1913–20', Archive for History of Exact Sciences 39, 272.
Ives, Herbert E. (1979) 'Derivation of the Lorentz Transformations', Philosophical Magazine (1945) 36, 392–401, reprinted in Speculations in Science and Technology 2, 247, 255.
Jánossy, L. (1971) Theory of Relativity Based on Physical Reality, Budapest: Akadémiae Kiadó.
Kanitscheider, Benulf (1984) Kosmologie, Stuttgart: Philipp Reclam.
Kant, Immanuel (1970) Immanuel Kant's 'Critique of Pure Reason,' trans. Norman Kemp Smith, London: Macmillan, A34/B51, p. 77.
Kennedy, J. B. (1997) Paper presented at the Central Division Meeting of the American Philosophical Association, April 23–26, 1997, Pittsburgh, Pennsylvania.

Le Poidevin, Robin (1997) 'Relative Realities', Studies in History and Philosophy of Modern Physics 28B, 541–46.
Lewis, Delmas (1986) 'Persons, Morality, and Tenselessness', Philosophy and Phenomenological Research 47, 305–9.
Lorentz, H. A. (1914) 'Het relativiteitsbeginsel. Voordrachten in Teyler's Stichtung', Arch. Musée Teyler 2, 26.
—— (1920a) Das Relativitätsprinzip, Pais translation, Leipzig: Teubner.
—— (1920b) The Einstein Theory of Relativity, New York: Brentano's.
—— (1927) Problems of Modern Physics, R. Bateman (ed.), Boston: Ginn, p. 221.
—— (1934) 'Alte und neue Fragen der Physik', Physikalische Zeitschrift (1910) 11, 1234ff, in P. Zeeman and A. D. Fokker (*eds.*), Collected Papers, The Hague: Martinus Nijhoff, p. 211.
—— (1952) The Theory of Electrons [1909] (rep. ed.), New York: Dover, pp. 229–30.
Lorentz, Dieter (1982) 'Über die Realität der FitzGerald-Lorentz Kontraction', Zeitschrift für allgemeine Wissenschaftstheorie 13/2, 308–12.
Lucas, J. R. Lucas (1973) A Treatise on Time and Space, London: Methuen.
Lucas, J. R. and P. E. Hodgson (1990) Spacetime and Electromagnetism, Oxford: Clarendon Press.
McCormach, Russell (1970) 'H. A. Lorentz and the Electromagnetic View of Nature', Isis 61, 486–88.
McGilvray, James (1979) 'A Defense of Physical Becoming', Erkenntnis 14, 275–99.
McGuire, J. E. (1990) 'Predicates of Pure Existence: Newton on God's Space and Time', in Phillip Bricker and R. I. G. Hughes (eds.), Philosophical Perspectives on Newtonian Sciences, Cambridge, Mass.: MIT Press, p. 93.
Maciel, A. K. A. and J. Tiomno (1989) 'Analysis of Absolute Space-Time Lorentz Theories', Foundations of Physics 19, 507–8.
Malament, David (1977) 'Causal Theories of Time and the Conventionality of Simultaneity', Nous 11, 293–300.
Marder, L. (1971) Time and the Space-Traveller, London: George Allen & Unwin.
Maudlin, Tim (1994) Quantum Non-Locality and Relativity, Oxford: Blackwell.
Mellor, D. H. (1973–74) 'Special Relativity and Present Truth', Analysis 34, 75–76.
—— (1981) Real Time, Cambridge: Cambridge University Press.
Merricks, Trenton (1994) 'Endurance and Indiscernibility', Journal of Philosophy 91, 165–84.
Miller, Arthur I. (1980) 'On Some Other Approaches to Electrodynamics in 1905', in Harry Woolf (ed.), Some Strangeness in the Proportion, Reading, Mass.: Addison-Wesley Publishing Co., pp. 66–91.
—— (1981) Albert Einstein's Special Theory of Relativity, Reading, Mass.: Addison-Wesley.
Minkowski, H. (1952) 'Space and Time', in W. Perrett and G. B. Jeffery (eds.), The Principle of Relativity, New York: Dover Publications, p. 76.
Munsouri, Reza and Roman U. Sexl (1997) 'A Test Theory of Special Relativity: I: Simultaneity and Clock Synchronization', General Relativity and Gravitation 8, 497–98.
Nerlich, Graham (1994) What Spacetime Explains, Cambridge: Cambridge University Press.
—— (1998) 'Time as Spacetime', in R. LePoidevin (ed.), Questions of Time and Tense, Oxford: Oxford University Press, pp. 128–29.
Newton, Isaac (1962) 'On the Gravity and Equilibrium of Fluids', [De gravitatione et aequipondio fluidorum] in A. Rupert Hall and Marie Boas Hall (eds.),

Unpublished Scientific Papers of Isaac Newton, Cambridge: Cambridge University Press, pp. 89–156.
Newton, Isaac (1966) Sir Isaac Newton's 'Mathematical Principles of Natural Philosophy' and his 'System of the World,' trans. Andrew Motte, rev. with an Appendix by Florian Cajori, Los Angeles: University of California Press, p. 6.
—— (1978) 'Place, Time, and God', in J. E. McGuire (ed.), 'Newton on Place, Time, and God: An Unpublished Source', British Journal for the History of Science 11, 114–29.
Oaklander, L. Nathan (1984) Temporal Relations and Temporal Becoming, Lanham, Maryland: University Press of America.
Pais, Abraham (1982) 'Subtle is the Lord . . . ': The Science and Life of Albert Einstein, Oxford: Oxford University Press.
Podlaha, M. F. (1975) 'Length contraction and time dilatation in the special theory of relativity – real or apparent phenomena?' Indian Journal of Theoretical Physics 25, 74–75.
Poincaré, Henri (1982) 'The Measure of Time', in The Foundations of Science, trans. G. B. Halstead, 1913; (rep. ed.), Washington, D.C.: University Press of America, p. 228.
Popper, Karl (1982) Quantum Theory and the Schism in Physics, ed. W. W. Barley, III, Totowa, N. J.: Rowman & Littlefield, pp. xviii, 20.
Prokhovnik, Simon J. (preprint) 'A Cosmological Basis for Bell's Views on Quantum and Relativistic Physics'.
Prokhovnik, Simon J. (1985) Light in Einstein's Universe, Dordrecht: D. Reidel.
—— (1987) 'The Twin Paradoxes of Special Relativity – Their Resolution and Implications', Foundations of Physics 19 (5), 541–52.
—— (1988), 'The Logic of the Clock Paradox', paper presented at the International Conference of the British Society for Philosophy of Science, 'Physical Interpretations of Relativity Theory', 16–19 September, 1988, Imperial College of Science and Technology, London.
Putnam, Hilary (1967) 'Time and Physical Geometry', Journal of Philosophy 64, 240–47.
Reichenbach, Hans (1952) 'Les fondements logiques de la mécanique des quanta', Annales de l'Institut Henri Poincaré 13, 157.
Robb, A. A. (1913) A Theory of Time and Space, Cambridge: Heffer & Sons.
Rompe, R. and H.-J. Treder (1984) 'Is Physics at the Threshold of a New Stage of Evolution?' in Asim O. Barut, Alwyn van der Merwe, and Jean-Pierre Vigier (*eds.*), Quantum, Space and Time – The Quest Continues, Cambridge Monographs on Physics, Cambridge: Cambridge University Press, pp. 603–4.
Ruderfer, Martin (1979) 'Introduction to Ives' "Derivation of the Lorentz Transformations"', Speculations in Science and Technology 2, 243.
Ryff, Luiz Carlos (1992) 'Gedanken Experiments on Duality', in Franco Selleri (ed.), Wave-Particle Duality, New York: Plenum Press, p. 249.
Savitt, Steven F. (1998) 'There's No Time like the Present (in Minkowski Spacetime)', paper presented at the symposium 'The Prospects for Presentism in Spacetime Theories,' Philosophy of Science Association biennial meeting, October, 1998, Kansas City.
Schlesinger, George (1980) Aspects of Time, Indianapolis: Hackett.
Shaw, R. (1962) 'The Length Contraction Paradox', American Journal of Physics 30, 72.

Shimony, Abner (1978) 'Metaphysical Problems in the Foundations of Quantum Mechanics', International Philosophical Quarterly 18, 13.
Sklar, Lawrence (1981) 'Time, Reality and Relativity', in Richard Healey (ed.), Reduction, Time and Reality, Cambridge: Cambridge University Press.
Smith, Quentin (1993) Language and Time, New York: Oxford University Press.
—— (1996) 'Tense and the New B-Theory of Language', Philosophy 71, 5–26.
Smoot, G. F., M. Y. Gorenstein, and R. A. Muller (1977) 'Detection of Anisotropy in the Cosmic Blackbody Radiation', Physical Review Letters 39, 899.
Stapp, Henry P (1988) 'Are Faster-Than-Light Influences Necessary?' in Franco Selleri *(ed.)*, Quantum Mechanics versus Local Realism, Physics of Atoms and Molecules, New York: Plenum Press, pp. 71–72.
Stein, Howard (1968) 'On Einstein-Minkowski Space-Time', Journal of Philosophy 65, 14.
—— (1991) 'On Relativity Theory and the Openness of the Future', Philosophy of Science 58, 147–67.
Taylor, J. G. (1975) Special Relativity, Oxford Physics Series, Oxford: Clarendon Press, preface.
Taylor, Edwin F. and John Archibald Wheeler (1966), Spacetime Physics, San Francisco: W. H. Freeman.
Tittel, W., J. Brendel, N. Gisin, and H. Zbinden (1999) 'Long Distance Bell-type Tests Using Energy-Time Entangled Photons', *Physical Review A* 59/6, 4150–63.
Van Inwagen, Peter (1990) 'Four Dimensional Objects', Noûs 24, 245–55.
Varicak, V (1911) 'Zum Ehrenfestschen Paradoxen', Physikalische Zeitschrift 12, 169.
Wenzl, A. (1949) 'Einstein's Theory of Relativity, Viewed from the Standpoint of Critical Realism, and its Significance for Philosophy', in P. A. Schilpp (ed.), Albert Einstein: Philosopher-Scientist, Open Court: Library of Living Philosophers 7 La Salle, Ill, pp. 583–606.
Winnie, John (1972) 'The Twin-Rod Thought Experiment', American Journal of Physics 40, 1091–94.
Winnie, John (1977) 'The Causal Theory of Spacetime', in John S. Earman, Clark N. Glymour, and John J. Stachel (eds.), Foundations of Space-time Theories, Minnesota Studies in the Philosophy of Science 8, Minneapolis: University of Minnesota Press, pp. 134–205.
Zahar, Elie (1973) 'Why Did Einstein's Programme supersede Lorentz's? (I & II)', British Journal for the Philosophy of Science 24, 95–123; 223–62.
Zeeman, R. C. (1964) 'Causality Implies the Lorentz Group', Journal of Mathematical Physics 5, 490–93.

2 Finding "real" time in quantum mechanics

Craig Callender

Quantum mechanics seemingly offers something to everyone. Some find free will in quantum mechanics. Others discover consciousness and value. Still others locate the hand of God in the quantum wave function. It may come as no surprise, therefore, to hear that many believe quantum mechanics implies or at least makes the world more hospitable to the tensed theory of time.[1] Quantum mechanics rescues the significance of the present moment, the mutability of the future and possibly even the whoosh of time's flow. It allegedly does so in at least two different ways:

A. Quantum non-locality is said to make a preferred foliation of space-time into space and time scientifically respectable again. Tensers need worry no longer about "no-go" theorems proving the incompatibility of the tensed theory with special relativity. Quantum non-locality provides the foliation they need.
B. Wave function collapse injects temporal "becoming" into the world.

The aim of this paper is to show that the kind of reasoning underlying these claims is at least as desperate as that finding freedom, value, the mind and God in quantum mechanics – which is pretty desperate. The bulk of the paper concentrates on A; discussion of B is reserved for the Appendix. After setting things up in the first three sections, the next section develops what I call the "coordination problem" for tensers. The upshot of this problem is that if tensers escape the threat of relativity, they do so only by embracing conflict with the branch of physics they believed saved them, quantum mechanics. The following section entitled "Quantum gravity to the rescue?" briefly considers what lessons we might draw for tenses from quantum gravity. Finally, in the concluding section I step back from the fray and examine some methodological issues, concluding that scientific methodology will always be "against" tenses as they are currently conceived.

Special relativity against tenses

The argument from special relativity against tenses is familiar, so I will be brief. The basic idea begins with the relativity of simultaneity in Minkowski space-time, the space-time appropriate to special relativity. A special relativistic world is a 4-dimensional manifold of space-time events endowed with Minkowski metric and matter fields. A foliation of this manifold carves up space-time into space and time via an equivalence relation, simultaneity, and time is the 1-dimensional linearly ordered quotient set induced by this relation. The famous relativity of simultaneity implies that there are many different foliations of space-time into space and time. Though a tension between relativity and common sense conceptions of time was recognized very early on, Putnam (1967) and Rietdijk (1966) were perhaps the first to set out the argument against tenses from relativity explicitly.

The basic idea is as follows (see Figure 2.1): consider two inertial observers, A and B, traveling in opposite directions but intersecting at some event e, and some distant inertial observer C. Simply put, using the standard Einstein-Poincaré synchronization, A has a different hyperplane of simultaneity than B does. Hence A and B will disagree about what events on C's history are simultaneous with e. A will declare that event C_1 is simultaneous with e whereas B will declare that event C_2 is simultaneous with e. In typical terrestrial situations, C_1 and C_2 may be so close together that their difference is not subsequently noticeable to A or B. For Cs that are very far away

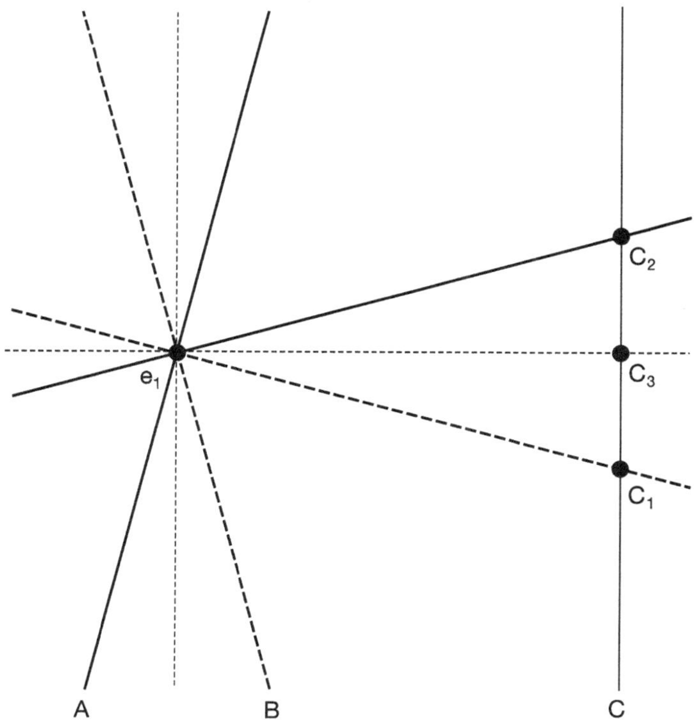

Figure 2.1 Relativity of simultaneity.

(or for As and Bs with very high relative velocity with respect to one another), however, there can be great disagreement. Now take some event C_3 such that $C_1 < C_3 < C_2$. Since one's simultaneity hyperplane divides the world into the future and past – and on any tensed view this has ontological repercussions – C_3 is in A's past but B's future. Furthermore, the so-called "principle of relativity" asserts that neither A nor B is privileged in any way. If the future is ontologically unlike the past or present (i.e. non-existent, indefinite, etc.), as the tensed theory demands, then A judges C_3 ontologically different than B judges C_3. But why should C_3's *ontological* status be relative to one's state of motion?

Scores of papers respond to this argument. I won't summarize them all here, but let me comment on a few strategies of reply (see Savitt 2002). Some tensers have bitten the bullet and suggested relativizing existence to one's state of motion. Others have flatly denounced special relativity as false (which it is, but they mean even if gravitational and quantum effects are negligible). These claims are obviously very radical. Others, like Stein (1991), have claimed that Putnam's argument is wrong, that a "becoming" relation is perfectly well definable on Minkowski space-time. The trouble with this claim is that Stein's "tensed" theory is not remotely close to any tensed theory ever devised and lacks any philosophical virtues apart from having a relation definable on (temporally oriented) Minkowski space-time (see Callender 2000; Saunders 2002). Callender (2000), furthermore, argues that if one uses a relation remotely like those found in tensed theories (that is, where at least two events can be co-present), then one can invert Stein's theorem and prove a "no go" theorem showing that becoming is incompatible with Minkowski space-time (see also Clifton and Hogarth 1995; Dorato 1996; Rakic 1997).

In my opinion, by far the best way for the tenser to respond to Putnam *et al.* is to adopt the Lorentz 1915 interpretation of time dilation and Fitzgerald contraction.[2] Lorentz attributed these effects (and hence the famous null results regarding an aether) to the Lorentz invariance of the dynamical laws governing matter and radiation, not to space-time structure. On this view, Lorentz invariance is not a space-time symmetry but a dynamical symmetry, and the special relativistic effects of dilation and contraction are not purely kinematical. The background space-time is Newtonian or neo-Newtonian, not Minkowskian. Both Newtonian and neo-Newtonian space-time include a global absolute simultaneity among their invariant structures (with Newtonian space-time singling out one of neo-Newtonian space-time's many preferred inertial frames as the rest frame). On this picture, there is no relativity of simultaneity and space-time is uniquely decomposable into space and time. Nonetheless, because matter and radiation transform between different frames via the Lorentz transformations, the theory is empirically adequate. Putnam's argument has no purchase here because Lorentz invariance has no repercussions for the structure of space and time. Moreover, the theory shouldn't be viewed as a desperate attempt to save absolute simultaneity in the face of the phenomena, but it should rather be viewed as a natural extension of the well-known Lorentz invariance of the free Maxwell equations. The reason why some tensers have sought all manner of strange replacements for special relativity when this comparatively elegant theory exists is baffling.

The main concern about the Lorentzian theory is that dynamical symmetries do not mirror space-time symmetries on this view, or as Einstein said, "there are asymmetries in the theory not found in the phenomena" (Janssen 2002). The matter fields are Lorentz invariant but the space-time is not. For this reason, all else being equal, one ought to prefer the Einstein-Minkowski interpretation to the Lorentzian interpretation. Positing otherwise unnecessary unobservable structure – absolute simultaneity – does violence to Ockham's razor. But is all else equal? If the case for tenses is elsewhere strong, that may tip the balance over to the Lorentzian interpretation. The Lorentzian picture is logically consistent and empirically adequate, after all. What are a few lost explanatory virtues in contrast to ________ (fill in the blank with whatever tenses explain)? There are many assumptions in our overall world picture, and we know from Quine-Duhem that there are many ways of organizing them. The no-go theorems focus on only a small piece of this theorizing and are only as good as their assumptions. In particular, what symmetries one takes a space-time to have depends on prior assumptions about what one takes to be in the space-time in the first place. If quantum non-locality spoils the Lorentz invariance of Minkowski space-time, then this would override the explanatory deficit of the Lorentzian view. Does quantum mechanics help tip the balance toward a space-time structure more friendly to tenses?

The quantum challenge

Sir Karl Popper, reflecting on recent experiments violating Bell's inequality, writes:

> It is only now, in the light of the new experiments stemming from Bell's work, that the suggestion of replacing Einstein's interpretation by Lorentz's can be made. If there is action at a distance, then there is something like absolute space. If we now have theoretical reasons from quantum theory for introducing absolute simultaneity, then we would have to go back to Lorentz's interpretation.
>
> (1982: 30)

According to Popper, the underdetermination between Lorentz and Einstein, which had persisted for more than sixty years, was finally broken with an *experimentis crucis.* Quantum non-locality, experimentally vindicated by Aspect's violation of Bell's inequality, demands absolute simultaneity. For the would-be tenser, Popper's reasoning to a physically preferred foliation of space-time is precisely what one wants. But why would Popper, or anyone, think quantum non-locality entails absolute simultaneity?

To answer this question we must take a detour through the philosophical foundations of quantum mechanics. In brief, the idea is as follows. Experiments in the late twentieth century revealed robust correlations between space-like separated events, i.e. events that are not connectable by a light signal. Quantum mechanics must explain these correlations. Different interpretations of quantum mechanics explain the space-like correlations differently, but the thought is that however this is done, it will entail picking out a preferred foliation of space-time. Quantum mechanics, once interpreted plausibly, must posit a mechanism requiring absolute simultaneity if it is to explain the Bell correlations. Let's flesh this out slightly and briefly evaluate the claim. For more details, see Bell (1987) and Maudlin (1994, 1996).

To begin, start with the now canonical spin ½ version of the famous 1935 Einstein-Podolsky-Rosen (EPR) paradox by Bohm. A pair of electrons, 1 and 2, in the spin singlet state:

$$\frac{1}{\sqrt{2}}(|\uparrow_x>^1\ |\downarrow_x>^2 - |\downarrow_x>^1\ |\uparrow_x>^2)$$

emerge from a common source and are sent in opposite directions. Each electron is then measured by a Stern-Gerlach device that sorts spins, up or down, in the x-direction. In the singlet state, thanks to spin conservation, the probability of the measurements on systems 1 and 2 disagreeing is one for any measurement orientation. Upon measuring electron 1 and finding a definitely spin up or down state, therefore, we know with certainty the result on electron 2. Assuming locality, that is, that the measurement of 1 didn't affect the state of 2, EPR reason that it must be that 2 already had a definite spin state – even when it was in the singlet state, which doesn't have a definite spin state. Hence we have EPR's dilemma: either quantum mechanics is non-local or it is incomplete.

Later, Bell derived in 1964 an inequality from the distant correlations encoded by the singlet state and some natural locality assumptions. For various orientations quantum theory predicts the violation of this inequality; and in a host of experiments since systems have vindicated this prediction. Though there remain theoretical and experimental loopholes still to close, these loopholes are increasingly desperate. There is now wide consensus that theory and experiment have discovered space-like correlations not attributable to any local hidden variable theory. Complete or not, quantum mechanics is non-local.

Does this non-locality conflict with the relativity of simultaneity? In the absence of an interpretation of quantum mechanics, it is impossible to answer this question. The mechanism responsible for enforcing the space-like correlations varies with interpretation. Not including the mechanism in the discussion of conflict is like leaving the guest of honor at a party uninvited. We need to know how different interpretations understand the space-like correlations and then we need to ask if the physics posited is Lorentz invariant. To answer these questions, we need a quick survey of the measurement problem and reactions to it. As we will see, Popper's reasoning has some purchase with *some but not all* solutions to the measurement problem.

Here is a quick and simplistic description of the measurement problem (for more details, see Albert 1992.) First, suppose that the governing equation for the quantum state is linear. The equations are in fact linear in both non-relativistic and relativistic quantum theory. Second, suppose we have a reliable measuring device. For our spin ½ system, "reliable" means that if the state of the electron is spin up (down), then the measuring device will register a spin up (down) outcome. Together these assumptions entail that if we let the measuring device M measure an electron in the superposed state $\psi = \frac{1}{\sqrt{2}}(|\uparrow_x> + |\downarrow_x>)$ then we get a macroscopic measuring device in a superposed state too. In other words, the initial state

$$\Psi = |ready>_M \frac{1}{\sqrt{2}}(|\uparrow_x> + |\downarrow_x>) \text{ at time } t_1$$

will evolve into

$$(1) \qquad \Psi = \frac{1}{\sqrt{2}}(|''up''>_M |\uparrow_x> + |''down''>_M |\downarrow_x>) \quad \text{at time } t_2.$$

We then have a macroscopic state that is not reading "up," "down" or anything in between. What does this mean? We never see anything suspended between distinct macroscopic properties. And of course, if *everything* is governed by the linear quantum equation, then we are also in a superposition, suspended between myriad distinct macroscopic states; and so is the Earth, the solar system, and anything else entangled (even to a small degree) with the superposed system.

Since measurements seem to have determinate outcomes, something has gone badly wrong. To have an empirically adequate physics, the measurement problem must be solved. How do we solve it? The lack of determinate outcomes is largely a result of holding two theses:

a. Ψ is representationally complete (i.e. the so-called eigenvalue-eigenstate link holds).
b. Ψ always evolves according to a linear dynamical equation.

To solve the measurement problem, we must deny one of these or explain away the mismatch between the macroscopic superposition and experience. The denial should be part of a full-blown theory that is empirically adequate and logically consistent. These theories are called "interpretations" of quantum mechanics. There are scores of interpretations, but they fall naturally into one of three classes (see Albert 1992 and Barrett 1999 for discussion and references).

The first class is called "collapse" interpretations. What makes a theory a collapse interpretation is its denial of proposition b above. The "standard" Copenhagen interpretation states that upon measurement there is an instantaneous wave function collapse from a superposition to an eigenstate (when the state is expanded in the relevant basis for the observable being measured). Other collapse theories rewrite the dynamical equation so that it is sensitive to the mass density or the particle number; when a certain threshold is reached a collapse is triggered. The most developed theory of this kind is known as GRW, after Ghirardi, Rimini, and Weber (1986).

The second class is sometimes called "hidden variable" interpretations. These theories deny proposition a. In addition to the wave function evolving according to some linear equation, these theories add what Bell (1987) calls "beables" (typically a particle or field ontology) and a separate dynamics for these beables. The ontology is dualistic: interpreted realistically, there are both beables and wave functions in the world. By claiming the Ψ description to be incomplete, they can say, for instance, that a macroscopic pointer is in a definite position because its beables are definitely located – even if the quantum state description is a superposition of distinct positions as in (1). In these theories measurement-like situations stimulate what are sometimes called "effective collapses" – events such that only one component of the superposed wave function becomes non-negligible for the subsequent evolution of the beable. Bohmian mechanics and modal interpretations are the best-known versions of this kind of reaction.

The third class forms a heterogeneous group. These theories neither supplement the wave function description of the world nor interrupt its evolution. What unifies them, if anything, is that they seek to explain away the mismatch between (1) and experience. Advocates of relative-state interpretations claim that (1) does describe our experiences accurately, but that the way our experience supervenes upon (1) is more complicated than one normally thinks. They often speak of (1) as corresponding to different worlds, different branches, or different observers, with one world branch or observer seeing the pointer pointing up and another with the pointer pointing down. Decoherence effects are often invoked as being crucial. Another very different group, which for our purposes we might include in group three because they don't supplement or modify quantum mechanics, is one that treats quantum mechanics instrumentally. These thinkers consider the wave function an epistemic device and see collapse as a kind of Bayesian conditionalization. We learn new information about the system and change our credences accordingly, but collapse is not a real physical process nor is the dynamics supplemented with hidden variables.

We are finally in a position to ask our question, namely, do all interpretations that solve the measurement problem (or purport to) enforce Bell's space-like correlations in a way that conflicts with Lorentz invariance? The answer is straightforwardly "no," for we don't get a conflict on the "epistemic" treatment of the wave function mentioned in class three. Also, depending on how one understands the metaphysics of branches, observers and worlds, it *may* be possible to escape conflict with a relative-state interpretation (Bacciagaluppi 2002). So interpretations in class three hold out hope of not conflicting with Lorentz invariance. It is also possible that hidden variable theories be Lorentz invariant, in the sense that no proof to the contrary has ever stood up. Some hidden variable theorists are also reluctant to posit a dynamics for their beable; not doing so can make it unclear as to whether the theory is Lorentz invariant or not. At any rate, it is clearly contentious whether all interpretations that solve the measurement problem also entail a violation of Lorentz invariance. In addition, there are non-standard ways of understanding Lorentz invariance ("hyperplane dependence"). If successful, this understanding would allow any of our interpretations to be Lorentz invariant. With all these qualifications now in place, we can only say that Popper's conclusion threatens most if one adopts a standard collapse or hidden variable interpretation of quantum mechanics as well as a standard reading of Lorentz invariance.

To get a sense of the trouble, consider real collapses in Minkowski spacetime (for more see Aharanov and Albert 1981). We'll consider no-collapse dynamics later. Consider two spin-1/2 particles in the singlet state. Both particles emerge from a common source, with particle 1 traveling to the left in the diagram and particle 2 traveling right (see Figure 2.2). At event L

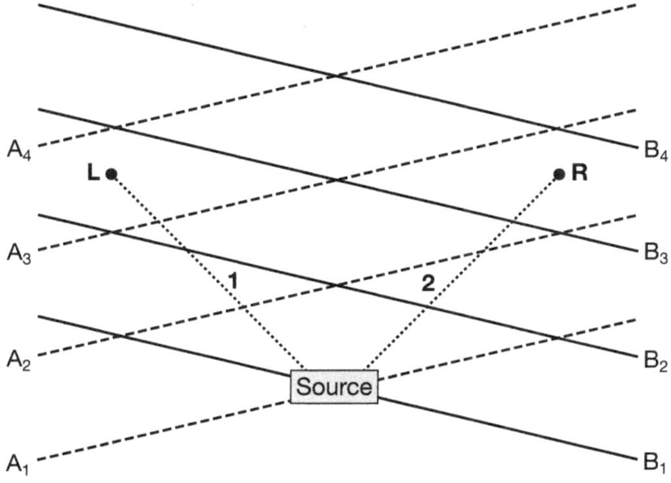

Figure 2.2 Measurement of the singlet state.

particle 1 is measured for its value of x-spin; at event R, which is space-like related to L, particle 2 is measured for its value of z-spin. Now consider two foliations of space-time, A, wherein R happens first, and B, wherein L happens first. What is going on in the world according to inertial observers to whom B is appropriate is as follows.

Start off in the singlet state

$$\frac{1}{\sqrt{2}}(|\uparrow_x>_1 |\downarrow_x>_2 - |\downarrow_x>_1 |\uparrow_x>_2)$$

at time B_1. Then by B_3 an x-spin measurement happens at L. Suppose the result at L is x-spin-up. Then the quantum state instantaneously reduces to:

$$|\uparrow_x>_1 |\downarrow_x>_2 .$$

By time B_4, R has occurred and the scientist at the right finds particle 2 to be z-spin-up:

$$|\uparrow_x>_1 |\uparrow_z>_2 .$$

But from the perspective of an inertial observer to whom A is the appropriate foliation, we instead get the sequence:

A1: $\frac{1}{\sqrt{2}}(|\uparrow_x>^1 |\downarrow_x>^2 - |\downarrow_x>^1 |\uparrow_x>^2)$

A3: $|\downarrow_z>^1 |\uparrow_z>^2$

A4: $|\uparrow_x>^1 |\uparrow_z>^2$

by parallel reasoning. The two histories are very different. History A says R collapsed the singlet state into a factorizable state; history B says L did. History A says that R measured particle 2 to be x-down; history B says it measured the singlet state. History A says that L measured the singlet state; history B says it measured particle 1 to be z-down. History A and B also disagree on which measurements results were determined and which ones were chancy.

If we take the wave function at all seriously – that is, as a real entity in the world rather than a summary of information – disagreements like this will not do. If real wave functions really collapse, then either A's story is right or B's story is right. The same goes for no-collapse hidden variable theories. Hidden variable theories will have the beables behaving one way if A's story is right and behaving another way if B's story is right (see Section 4 below). Again, since the beables represent the fundamental ontological furniture of the world in these theories, disagreements like that between A and B won't do there either.

Hence, on two large and natural classes of interpretation, the mechanism responsible for explaining the Bell correlations does require a preferred frame. Since I am personally partial to these types of interpretation, I have some sympathy with tensers claiming quantum non-locality lends some pressure to believe again in a preferred frame.

Whether a quantum preferred frame would actually push us all the way back to the Newtonian or neo-Newtonian space-time of Lorentz is yet another question. As Maudlin (1996) points out, another option would be to retain Minkowski space-time but add a physically preferred foliation (see also Dürr *et al.* 1999). Like the Newtonian options, it would introduce an asymmetry in the theory not fond in the phenomena. All three space-times, Newtonian, neo-Newtonian and Minkowskian-with-preferred-frame, would be hospitable to tensers hoping to re-introduce a global simultaneity hyperplane.

Quantum preferred frames

To one defending the Putnam argument against tenses, the situation regarding quantum mechanics is nothing less than embarrassing. By adopting the principle of relativity Putnam claims that there can't be anything in the world that doesn't "commute" with the symmetries of Minkowski space-time. But the speculations of the previous section suggest that physics itself – indeed, arguably our best scientific theory ever – violates Putnam's reasoning! Putnam's argument, run on quantum mechanics rather than tenses, would prohibit good interpretations of quantum mechanics from reproducing and enforcing violations of Bell's inequality.

Here is Lucas (1998) savoring the irony:

> But physics goes further. It not only defeats the would-be defeaters of the tense theory, but offers positive support. Quantum mechanics, if it is to be interpreted realistically, distinguishes a probabilistic future of superimposed eigen-states from a definite past in which each dynamical variable is in one definite eigen-state, with the present being the moment at which – to change the metaphor – the indeterminate ripple of multitudinous wave-functions collapses into a single definite wave. Admittedly, many of those who think about quantum mechanics are not realists, and admittedly again, there are horrendous difficulties in the way of giving a coherent account of the collapse of the wave-function. But an obstinate realism, as well as a slight sympathy for our feline friends, precludes my envisaging any long period in which Schrödinger's cat could be half-dead and half alive, and this whether she be in a laboratory in Europe or on some planet circling Betelgeuse. There is a definite fact of the matter, there as much as here, whether or not we are dealing with a superposition of functions or one definite eigen-function. And hence there is a unique hyperplane advancing throughout the whole universe of collapse into eigen-ness.

Lucas finds at least two attractive elements in quantum mechanics. One is the unique hyperplane presumably corresponding to the tensed now advancing through history. Another is that what is to the future of this hyperplane is genuinely open. For many who find tense in quantum mechanics, it is quantum mechanics' alleged probabilistic future that make tense so attractive. One should point out, of course, that the two features, preferred frames and stochastic dynamics, do not go hand in hand. Bohm's theory requires a preferred frame, for instance, but does not need a probabilistic dynamics for its hidden variables (though it usually has such in its field theoretic extensions). So if it's a genuinely "open" future that attracts you, preferred frames don't necessarily get you that. We'll turn to the idea of quantum becoming in the Appendix.

I want to argue that even if quantum mechanics does imply a preferred frame, matters are hardly rosy for tensers. In fact, one can argue that there is a real in principle problem for tensers. What I want to do is consider the foliation from the perspective of Bohmian mechanics and GRW. The former is the best worked out "realistic" no-collapse interpretation of quantum mechanics; the latter perhaps the best worked out "realistic" collapse interpretation. Assuming neither can be modified and made fundamentally Lorentz invariant, then quantum phenomena plus a solution to the measurement problem may demand a preferred frame. I want to concentrate on these theories since, for the tenser seeking to escape Putnam, things will get *no better* than if one or the other of these interpretations is true.

The coordination problem

Quantum mechanics is not the answer to the tenser's prayers. Even if we charitably assume quantum non-locality does require a preferred frame, there is a Putnam-like argument lurking nearby in quantum mechanics. Let's run the argument with the Bohm interpretation and then briefly point out the slight differences that we obtain when working with GRW instead.

The basic idea of non-relativistic Bohmian mechanics is that there is, in addition to the wave function, particles. Relativistic versions of Bohmian mechanics have been developed for some fields and are in the process of being developed for others.[3] The basic structure is the same, however. There is the wave function and there are the beables; the wave function evolves according to the relevant linear dynamical equation (Schrödinger equation, Dirac equation, Klein-Gordon equation, etc.) and the beables have a velocity that takes the wave function as input. In the non-relativistic particle version, the particles evolve according to:

$$\mathbf{p} = \frac{\partial \mathbf{S}}{\partial \mathbf{X}}$$

where **p** is the canonical momentum equal to md**X**/dt, **X** is a point in configuration space, and S is the phase of the wave function (S = ImlnΨ). In the non-relativistic case, the Schrödinger equation and this guidance equation are the two fundamental laws of nature according to the Bohmian. One often makes another assumption, namely, that the initial probability density of particles is given by the absolute value of the initial wave function squared, $|\Psi|^2$. Call this assumption the *distribution postulate*; some Bohmians view it as a law of nature. The amazing thing is that if you assume the distribution postulate and that the beables and wave function evolve according to the above equations, you will find that the particles' predicted statistics precisely match those of "standard" quantum mechanics.

Consider our earlier experiment with two spin ½ particles in the singlet state. Stern-Gerlach devices essentially split the regions of positive wave function support into two disjoint sections, an upper and lower section as measured along the vertical axis of the magnet. In the version of Bohm's theory under consideration, spin is not fundamental. The value of spin is contextual, meaning that it depends on the initial location of the particle in the wave packet and the kind of device it meets. If the Bohm particle starts off in the upper half of the wave packet (along the i-th dimension, where i = x, y, or z), then the dynamics will evolve it to the region we call spin-up in the i-th dimension; if the Bohm particle begins in the lower half of the wave packet, then it will evolve to the position we call spin-down in the i-th dimension. There is *no* possibility of a transition from low to high, or vice versa. This impossibility is due to the fact that the Bohmian dynamics is first-order and deterministic; the combination means that trajectories can't cross in configuration space.

Suppose the wave function is in our singlet state above and that the configuration point representing the two-particle system is located in the upper left half of the wave packet before the system is measured. (See Figure 2.3, fashioned after Barrett, 1999: 142.) Suppose also that the measurements occur at space-like separation, so that the measurement events occur in different orders in at least two different foliations of space-time. According to one foliation, A measures first; according to another foliation, B measures first. Let it be the case that A measures first. Since the particle is in the upper half of the wave packet, A will find the particle to be x-spin up; when the observer corresponding to B's foliation subsequently measures, the observer must find the particle to be x-spin down. That is, A's measurement will have effectively collapsed the state to $|\uparrow_x\rangle^A\ |\downarrow_x\rangle^B$. Now let it be the case that B measures first. If B measures first, B will find the particle to be x-spin up and therefore A will find the particle to be x-spin down, i.e. B's measurement would have effectively collapsed the state to $|\downarrow_x\rangle^A\ |\uparrow_x\rangle^B$. So in Bohm's theory, the *actual outcome* of the measurements depends on who measured first!

But "first" is not a relativistic invariant for space-like separated events. There shouldn't be a fact of the matter about who measured first, yet on

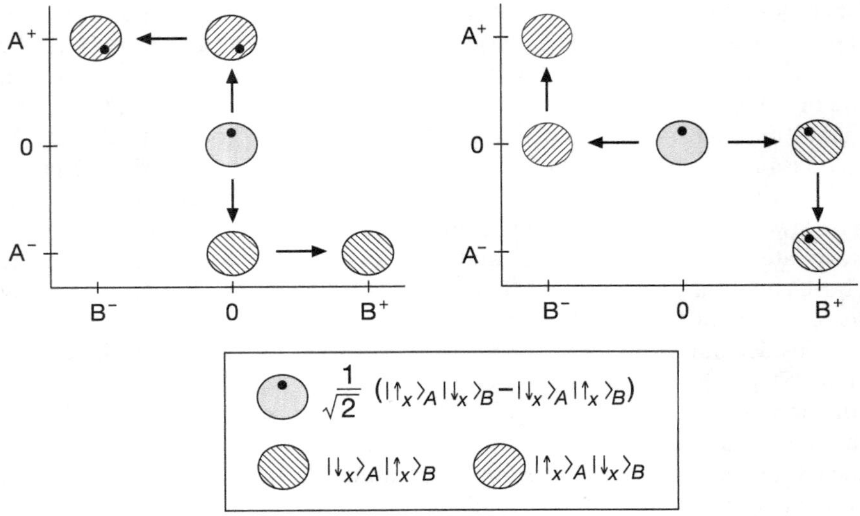

Figure 2.3 The preferred frame in Bohm's theory.

Bohm's theory there is. Since there is a fact of the matter about the outcomes there must be a fact of the matter about who measured first. Hence one foliation must be preferred over the others.

You may be curious why the above reasoning doesn't simply show that Bohm's theory is observationally incompatible with known relativistic facts. Doesn't the above reasoning demonstrate that Bohm's theory is false? Recall the distribution postulate. We don't know where the representative particle point is. If we get spin-down, we know it was in the lower half of the initial wave packet; but there is no way to know this beforehand. Moreover, it can be shown that if the distribution postulate is satisfied, then we can in principle never find out (see Albert 1992). Intuitively, finding out the point is in the upper half in the x-direction doesn't allow a reliable inference about whether it is in the upper or lower half of the y-direction. In any case, because the distribution postulate ensures that Bohmian mechanics reproduces exactly the predictions of ordinary Copenhagen quantum mechanics, the so-called "no signaling theorem" holds in Bohmian mechanics too. That is, one can show that it's impossible in Bohmian mechanics to exploit these space-like correlations for communication. At the statistical level, special relativity holds. It is only at the sub-quantum level that the outcomes pick out a preferred frame. But since we, in principle, don't have access to the initial location of particles in wave packets, we can never tell which of the indefinitely many possible inertial frames is the preferred one.

We now have enough background to make a very simple point. There is an, in principle, irresolvable *coordination problem* between the two preferred foliations, the metaphysically preferred foliation posited by the tenser and the physically preferred one by Bohmian mechanics. There is simply no reason to think the two are the same. Only blind faith leads one to expect that the two are coordinated. In our above experiment, A might measure first and then B measure second according to the Bohm frame, yet according to the temporal becoming frame B measures first and A second. Assuming the becoming frame is primary, we would say B really happened before A; meanwhile fundamental physics would say that A happened before B. Since it would be a miracle if the two frames coincided exactly, with near certainty this will be the case for *some* pairs of events. *Hence the tenser is committed to asserting that with near certainty fundamental physics gets the order of some events the wrong way round.* Far be it from quantum mechanics saving tenses, the tenser merely trades one conflict with fundamental physics for another. And since the distribution postulate prohibits in principle us from finding out the preferred frame, the coordination problem is in principle irresolvable.

This situation also seems to undermine many of the reasons motivating the tenser. For suppose – as an act of blind faith – that the Bohm frame and tense frame were one and the same. Well, it's still true that *you* are a Bohmian system corresponding to a piece of the universal wave function and a bunch of Bohmian particles. So whenever you experience anything or even introspect, *you* are making a Bohmian "measurement". The same general limitations on Bohmian measurements hold for you too. If the distribution postulate is a law of nature, then the laws of nature in a Bohmian world prevent you from having any *reliable* feeling or impression or introspective reflection that could at all indicate which frame is the becoming frame. Unlike in Putnam's argument, there *is* a fact of the matter about which foliation is the foliation needed by physics (i.e. Bohmian mechanics). But Bohmian mechanics makes it in principle impossible to determine via any interaction whatsoever which one this is. *Your intuitions, introspections, etc., all being species of interactions, can be in principle no guide to which foliation is the true foliation or even whether there is one.* If the world becomes or enjoys an objectively privileged present, then it is not something at all connected to experience (assuming physicalism). Since tensers regularly appeal to experience to support their theory (whether they should is another matter, see Callender (ms)), this conclusion cannot be congenial to the tenser. Hence the tenser faces a dilemma: either the becoming frame and preferred quantum frame are one and the same, in which case Bohmian mechanics implies that no physical experience could be a reliable guide to this frame, or they differ, and then the tensed theory conflicts with physics over the order of some events. Since there aren't any good reasons to endorse the former horn of the dilemma, it appears that the tenser is stuck with the latter, as depicted in Figure 2.4.

Matters change only slightly when we switch to the GRW interpretation. Again we will have a preferred foliation (the collapse dynamics is not Lorentz invariant), and again there will be no reason to think the foliation matches the one produced by becoming. So again the tenser will be postulating a

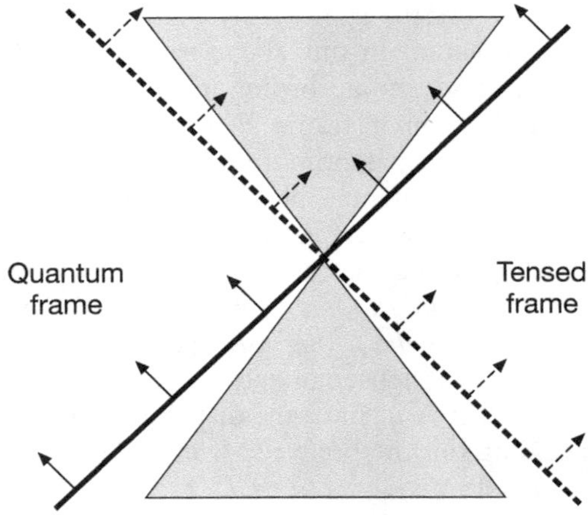

Figure 2.4 The coordination problem.

new coincidence into our world system. The only significant difference between the case of Bohm and of GRW regarding the present discussion is that, unlike in Bohm's theory, a relativistic version of GRW does at least in principle allow one to find the quantum preferred foliation (see Albert 2000). Second quantized GRW may predict slight observable violations of Lorentz invariance, in which case sophisticated experimental investigation of Lorentz invariance may pick out GRW's preferred frame. But since there is absolutely no reason to suspect that our brains are already doing the experiments in Albert (2000), there is no reason to think our experience in any way is a reliable guide to where the preferred frame is.

One might respond to the coordination argument as follows. We sense becoming or flow or what-have-you. Making sense of this experience entails the existence of a preferred frame. But the experience itself doesn't tell us *which* frame is preferred – it just tells us that one is needed. Monton (2005) makes a claim similar to this one. My reply is that our experiences aren't so unselective. What could our experience of becoming be if it is so startlingly insensitive? For all we know, the quantum foliation might be one that treats events in what we call the very early universe as "present". Don't confuse my argument with one merely pointing out that due to our necessarily coarse terrestrial measurements (arising from our being clumsy macroscopic creatures) we'll have trouble finding the preferred frame. If that were the argument, then one could imagine coarser measurements that might give us some estimate of the foliation. But my argument is that no measurement whatsoever will even remotely narrow down the preferred foliation. Contrast this case with the following one in cognitive science. In some circumstances, subjects will report that they experience two events (for instance, two flashes on a computer screen) even though they cannot tell which happened first (for discussion and references, see Callender (ms)). These subjects would be quite right to say that the experience entails various propositions, yet it does not entail knowing which event happened first. Notice, however, that many other experiments will reveal which events happened first. As we separate our events in time, we'll begin to experience one after another. Or if we leave the events the same temporal distance apart, we can devise mechanisms we might experience that tell which came first. Not so in the present case. No experiment whatsoever will even give us a hint about the true location of this foliation.

Far from being out of the woods thanks to a preferred foliation, a quantum foliation only serves to embarrass the tenser.

Quantum gravity to the rescue?

One feature all of the theories we have discussed have in common is the fact that they are false. Special relativity gives way to general relativity, non-relativistic quantum mechanics to relativistic quantum field theory; furthermore, there are reasons for thinking general relativity and quantum field theory are mutually incompatible and must themselves give way to quantum gravity. Perhaps quantum gravity can rescue preferred frames? This is the hope expressed by Monton (2005).

There are already sketches of theories of quantum gravity that yield a preferred foliation, including some to which I'm partial (e.g. Callender and Weingard 1995). But it must immediately be acknowledged that there are also sketches of those that do not require a preferred foliation, e.g. loop quantum gravity, and probably there are more of the latter than the former. None of these theories is at all well developed. Most are constitutionally unable to make predictions. Some, like topological quantum field theory, have no "local physics" whatsoever. Some turn out to be mathematically incoherent. Some are under-funded; others perhaps over-funded. Who should we bet on? At this stage it's way too early to glean anything. The tenser is right: relativity will be superseded. But with what, when, how and why are all up in the air.

Tensers can always cross their fingers and hope. If quantum gravity lets them down, perhaps then the next theory will accommodate. If science is a state of permanent revolution, as Popper thought, there is always hope, if nothing else. In the conclusion we return to this issue, where I hope to dash even this thin reed of hope.

Conclusion

This paper has argued that even if tensers get what they want from quantum mechanics, i.e. a preferred foliation, there is no reason to think it will serve the tensers' purposes. The foliation preferred by quantum mechanics may not be that preferred by metaphysics. Indeed, there is no reason to think it will be.

The same argument can be made with respect to tensers replying to the threat from general relativity. In general relativity, tensers find the cosmic time definable in a certain class of "realistic" solutions to Einstein's field equations congenial to their purposes. These cosmic times are defined in various ways, but usually they hang on various averaging procedures to determine the center of mass frame. The matter distribution picks out a preferred foliation. But why think that the psychological lapse of time or our perceived present marches in step with the foliation dictated by the center of mass frame? There is no reason to link the two.[4]

Why are tensers having such trouble connecting their theories to what they find in physics? I claim that it is no accident. Let's step back and reflect on what is happening. If we believe physics is time translation invariant, then we believe that whether an experiment is done at 2 p.m. or 3 p.m. doesn't matter, so long as the experimental procedure is the same in both cases. If for some reason we thought the property of happening at 2 p.m. were physically relevant, then we wouldn't be inclined to think physics is time translation invariant. If our reasons are good, then we are justified in denying time translation invariance; if our reasons don't stand up, nor do they for any other time, then time translation invariance looks plausible. If, as is in fact the case, its obtaining helps explain other phenomena as well, such as energy and energy conservation, then we have a strong case for time translation invariance. For why should science posit a fundamental property that doesn't do any work and whose existence would adversely affect science's simplicity? What symmetries and laws we take to hold of the world hangs on what things we take to be real. But also, if there is a great deal of motivation for certain symmetries and laws, then what things we take to be real hangs in part on what symmetries and laws we take to hold.

We ought to think of the arguments against the tensed theory of time in this same light.

The argument from special relativity against tenses was never just a matter of one physical theory implying the rejection or acceptance of tenses. Putnam's argument and the other no-go arguments arrange an inconsistent set of propositions so that Lorentz invariance is a premise and the denial of the tensed theory of time is a conclusion. Of course, any inconsistent set can be re-arranged; one might instead take objective tenses (a now, becoming, etc.) as real and turn the argument over, making tense a premise and the violation of Lorentz invariance a conclusion. As we know from Quine-Duhem, how we arrange the premises and conclusions depends upon background assumptions. Putnam's argument assumes that physics gets along perfectly well without tenses, just as our argument for time translation invariance assumes physics can manage without "2 p.m.-ness". This claim is contentious, to be sure, but it is there behind the scenes. The Minkowski space-time structure explains away asymmetries in the theory not found in the phenomena. Giving it up for neo-Newtonian space-time or Minkowski space-time with a preferred foliation introduces otherwise unnecessary unobservable structure to the theory.

From the Minkowskian perspective, it also introduces unexplained coincidences: why do those rods and clocks keep contracting and dilating, respectively? As a kinematical effect in Minkowski space-time, Minkowski space-time is a common cause of this behavior, which is otherwise brute in the Lorentzian framework.

Unobservable theoretical entities and unexplained coincidences are found throughout science, and there is often nothing wrong with that. Quarks are examples of the former. The unexplained equivalences among passive, active and gravitational masses in Newtonian gravitational theory are an example of the latter. Though replete with such entities and coincidences, science accepts them reluctantly: only if their cost is compensated for elsewhere and the alternative is worse. Do the benefits of accepting a coincidence among types of mass outweigh the costs? Given the astounding success of Newtonian gravitational theory, the coincidences more than pay their weight. By contrast, tensers have a stock of arguments about temporal indexicals that also apply to spatial and personal indexicals, dubious appeals to experience, and ordinary language analysis. Since this is not a fair trade for coincidences or extra unobservable structure, the balance tips and we claim special relativity rules out tenses rather than tenses rule out special relativity.

The same goes with my coordination argument against tenses. It desires to eliminate otherwise unexplained coincidences from science. The same goes also with Gödel's famous argument against tenses from general relativity. Belot (forthcoming) faults Gödel for assuming the symmetries of the general relativistic laws are more important than the matters of fact that support a preferred foliation of space-time. However, Gödel's argument assumed, again as a kind of background assumption, the idea that science could operate perfectly well without privileging a particular foliation. Why posit this structure if it's not needed?

Tensers are wasting their time trying to find an image of the tensed theory in physics. Specific physical theories will be more or less hostile to tenses, but in general they will be against tenses so long as there is no clear need for them. Show physics a need for tenses and it will quickly accommodate them. Until then, merely as a by-product of scientific methodology, physics will not accommodate them. Those hoping to rescue tenses will do best by returning to the fundamentals and showing that we can't do without them. But since most of these fundamental reasons arise solely from ordinary language analysis – a mostly bankrupt enterprise in my opinion – I, for one, will not hold my breath. From this perspective, physics – and science itself – will always be against tenses because scientific methodology is always against superfluous pomp.

Appendix: Quantum becoming

Tensers sometimes find vindication or inspiration for their views from objective wave function collapse. We met this view in Lucas (1998) above and it is found in Stapp (1977), Whitrow (1980), Popper (1982), Shimony (1993, 1998), Lucas (1999), and elsewhere. Quantum mechanics is supposed to require wave function collapse, and wave function collapse is supposed not only to pick out a preferred frame but also to make the future open, indeterminate or mutable. Here is Lucas (1999):

> There is a worldwide tide of actualization – collapse into eigenness – constituting a preferred foliation by hyperplanes (not necessarily flat) of co-presentness sweeping through the universe – a tide which determines an absolute present ... Quantum mechanics ... not only insists on the arrow being kept in time, but distinguishes a present as the boundary between an alterable future and an unalterable past.
>
> (1999: 10)

The collapse of the wave function, interpreted realistically, suggests a picture of a fixed past (wave functions collapsed to the eigenstates of the relevant observable) and an open future (wave functions as superpositions of such eigenstates). In fact, the path from objective collapses to tenses is a two-way street. While some reason to tenses from collapses, others reason to collapse interpretations from a prior commitment to tenses. Here is Christian (2001) on his motivation for pursuing Penrose's collapse theory:

> [It] implicitly takes *temporal transience* in the world – the incessant fading away of the specious present into the indubitable past – not as a merely phenomenological appearance, but as a *bona fide* ontological attribute of the world [...] For, clearly, any gravity-induced or other intrinsic mechanism, which purports to actualize – as a *real* physical process – a genuine multiplicity of quantum mechanical potentialities to a specific choice among them, evidently captures transiency, and thereby not only goes beyond the symmetric temporality of quantum theory, but also acknowledges the temporal transience as a fundamental and objective attribute of the physical world
>
> (2001: 308)

Many who crave a tensed time find just what they need in objective wave function collapse and vice versa.

Let us first consider whether quantum mechanics, on a collapse interpretation, supports one's pre-theoretical views of the openness or mutability of the future, as Lucas suggests. I do not believe it does. We can approach this question by asking whether the open/fixed distinction maps at all neatly into the superposition/eigenstate distinction? The answer is "no". To begin with, the symmetry of Hilbert space implies that we can expand our wave function in any of an indefinite number of bases, e.g. position, momentum, spin. A wave function that is a superposition in one basis may not be a superposition in another; for instance, the wave function of x-spin down is a superposition of up and down spins in the z-spin direction. Now, if we believe collapses are real physical mechanisms, then we must decide in what basis they occur. A natural choice might be position basis, as in GRW. A preferred position basis will entail (if it solves the measurement problem) the absence of superpositions of distinct *macroscopic* properties in other bases too, but it does not entail the absence of superpositions in these bases. Eigenstates in position space still correspond to non-eigenstates in momentum space. For the topic at hand, this fact raises the natural question: is the momenta of past objects "still" open? Or do we care only about macroscopic openness? Either view would represent a significant departure from the ordinary conception of an open future. Furthermore, how should we view the future measurement of systems already in eigenstates of the relevant observable? These measurements aren't open in the sense of a superposition collapsing to the eigenstate of the relevant observable. Are the outcomes nonetheless open because of the future or are they fixed because of eigenstates? If the former, then quantum mechanics has little to do with openness; if the latter, then again we have a drastic departure from our pretheoretic intuitions of openness. The actualization Lucas gets from quantum mechanics is more a series of partial drips and splashes than a worldwide surge.

Perhaps the link with openness and transience arises instead from the single-case objective probabilities needed for a collapse theory? Shimony and Popper stress throughout their work the benefits of a truly probabilistic process, seeing in it an open future, the flow of time, and even freedom. The intuitions underlying these links are clear enough. Suppose at time t there is an objective chance of 0.5 that a radium atom will decay tomorrow. For this to be true, some believe, there must not "already" at t be a unique determinate future with (say) a decayed radium atom in it tomorrow. That would entail, in one way of understanding objective chances, an objective probability of 1, not 0.5. Since the tenseless theory of time entails that there is a unique determinate future – in a sense – the existence of non-trivial objective probabilities requires the tensed theory of time (see Shanks 1991).

Here I only want to point out that the inferences in this reasoning are more tenuous than is usually acknowledged. First, if the reasoning goes through at all, it does so only for some interpretations of chance and not others. The reasoning is perhaps most natural on a Popperian propensity interpretation, but there exist other (and I would argue, better) interpretations of objective single-case chances that won't yield the desired conclusion. On Lewis' 1994 theory of chance, for instance, non-trivial chances are compatible with a tenseless theory of time. Lewis views chance as a theoretical entity that increases the overall strength and simplicity of the best systematization of nature. Crucially, on this theory information about whether or not a radium atom decayed after t is "inadmissible" at t and therefore doesn't affect the value of the chance at t. Second, and at least as important, the justification for the line that a "fixed" future implies trivial values of objective chance is similar if not identical to the famous argument for fatalism. The sea battle tomorrow spoils freedom today just as the radium atom's decay tomorrow spoils non-trivial values of chance today. But if one believes, as I do, that the argument for fatalism is flawed (see Sobel 1998 for a penetrating critique) then the existence of the sea battle tomorrow doesn't undermine freedom today; one therefore needn't see the tenseless view and its implications of either a sea battle or not tomorrow as a threat to freedom. Nor need it be a threat to non-trivial chances. The actual world may contain our radium atom in it decayed tomorrow, yet today it still may have a one-half chance of decaying.

Notes

1 Throughout I assume familiarity with tensed theories of time. For clear introductions see Dainton (2001) and Savitt (2002). In what follows I make no attempt to disentangle the various distinct theses classified as "tensed theories". Primarily I have in mind the theories commonly referred to as presentism and possibilism (or the "growing block" view).

2 See Bell (1987: 67–80) and Janssen (2002) for discussion and references. Craig (2001) is the first tenser I know of who adopts a Lorentzian perspective to defend the tensed theory. Arguably, he uses the wrong version of Lorentz's theory (see Balashov and Janssen 2003).

3 For some philosophical discussion of Bohmian field theories, see Callender and Weingard (1997). For the latest Bohmian quantum field theory, see Dürr *et al.* (2003). For an attempt to write a Bohm theory that wouldn't pick out a preferred frame, *in asense*, see Dürr *et al.* (1999).

4 Tensers in general have a problem linking the phenomenon they want to explain to the ontology they believe explains it, even when the ontology is their own metaphysics. See Dainton (2001: 75) for an argument that there is no necessary coordination between the becoming arrow and the memory arrow, and see Callender (2005) for a generalization of this objection.

References

Aharonov, Y. and Albert, D. (1981) 'Can We Make Sense out of the Measurement Process in Relativistic Quantum Mechanics?' *Physical Review D* 24, 359–70.

Albert, D.Z. (2000) 'Special Relativity as an Open Question', in Heinz-Peter Breuer and Francesco Petruccione (eds.) *Relativistic Quantum Measurement and Decoherence*, Berlin: Springer, pp. 3–13.

Albert, D. (1992) *Quantum Mechanics and Experience*, Cambridge: Harvard University Press.

Bacciagaluppi, G. (2002) 'Remarks on Space-time and Locality in Everett's Interpretation', in T. Placek and J. Butterfield (eds.), *Non-Locality and Modality*, Dordrecht: Kluwer Academic, pp. 105–22.

Balashov, Y. and Janssen, M. (2003) 'Presentism and Relativity: A Critical Notice', *British Journal for the Philosophy of Science* 54, 327–46.

Barrett, J.A. (1999) *The Quantum Mechanics of Minds and Worlds*, NY: Oxford University Press.

Bell, J.S. (1987) *Speakable and Unspeakable in Quantum Mechanics*, Cambridge: Cambridge University Press.

Belot, G. (forthcoming) 'Dust, Time and Symmetry', *British Journal for the Philosophy of Science.*

Callender, C. (2000) 'Shedding Light on Time', *Philosophy of Science*, 67, S587–S599.

Callender, C. and Weingard, R. (1997) 'Trouble in Paradise? Problems for Bohm's Theory', *Monist*, 80 (1), 24–43.

—— (1995) 'A Bohmian Model of Quantum Cosmology', *Philosophy of Science (Proceedings)*, 1, 228–37.

Clifton, R. and Hogarth, M. (1995) 'The Definability of Objective Becoming in Minkowski Spacetime', *Synthese*, 103, 355–87.

Christian, J. (2001) 'Why the Quantum Must Yield to Gravity' in Callender, C. and Huggett, N. (eds.) *Physics Meets Philosophy at the Planck Length*, Cambridge: Cambridge University Press, pp. 305–38.

Craig, W.L. (2001) *Time and the Metaphysics of Relativity*, Dordrecht: Kluwer Academic Publishers.

Dainton, B. (2001) *Time and Space*, Montreal: McGill Press.

Dorato, M. (1996) 'On Becoming, Relativity, and Nonseparability', *Philosophy of Science*, 64, 585–604.

Dürr, D., Goldstein, S., Tumulka, R. and Zanghì, N. (2003) 'Bohmian Mechanics and Quantum Field Theory', *Physical Review Letters*, 93, 090402, 1–4.

Dürr, D., Goldstein, S., Münch-Berndl, K. and Zanghì, N. (1999) 'Hypersurface Bohm-Dirac Models', *Physical Review A*, 60, 2729–36.

Ghirardi, G.C., Rimini, A. and Weber, T. (1986) 'Unified Dynamics for Microscopic and Macroscopic Systems', *Physical Review D*, 34, 440–91.

Jansen, M. (2002) 'Reconsidering a Scientific Revolution: The Case of Einstein Versus Lorentz', *Physics in Perspective*, 4, 421–46.

Lewis, D. (1994) 'Humean Supervenience Debugged', *Mind*, 103, 473–90.

Lucas, J.R. (1998) 'Transcendental Tense II', *Aristotelian Society Supplementary Volume*, 72, 29–43.

—— (1999) 'A Century of Time' in Jeremy Butterfield (ed.) *The Arguments of Time*, NY: Oxford University Press, pp. 1–20.

Maudlin, T. (1994) *Quantum Non-Locality and Relativity: Metaphysical Intimations of Modern Physics*, Cambridge, MA: Blackwell.

—— (1996) 'Space-Time in the Quantum World', in J. Cushing, A. Fine and S. Goldstein (eds.) *Bohmian Mechanics and Quantum Theory: An Appraisal*, Dordrecht: Kluwer, pp. 285–307.

Monton, B. (2005) 'Presentism and Quantum Gravity', *The Ontology of Spacetime*, edited by Dennis Dieks, forthcoming.

Popper, K. (1982) *Quantum Theory and the Schism in Physics*, Hutchison: London.

Putnam, H. (1967) 'Time and Physical Geometry', *Journal of Philosophy*, 64, 240–47.

Rakic, N. (1997) 'Past, Present and Future and Special Relativity', *British Journal for the Philosophy of Science*, 48, 257–80.

Rietdijk, C. (1966) 'A Rigorous Proof of Determinism Derived From the Special Theory of Relativity', *Philosophy of Science*, 33, 341–44.

Savitt, S. (2002) 'Being and Becoming in Modern Physics', in Edward N. Zalta (ed.) *The Stanford Encyclopedia of Philosophy (Spring Edition)*, available online at: http://plato.stanford.edu/archives/spr2002/entries/spacetime-bebecome/

Saunders, S. (2002) 'How Relativity Contradicts Presentism', in C. Callender (ed.) *Time, Reality and Experience*, Cambridge: Cambridge University Press.

Shanks, N. (1991) ‘Probabilistic Physics and the Metaphysics of Time,’ *The South African Journal of Philosophy*, 10 (2), 37–43.

Shimony, A. (1993) *Search for a Naturalistic World View*, Vol. II, Cambridge: Cambridge University Press.

—— (1998) ‘Implications of Transience for Spacetime Structure’ in S.A. Huggett, L.J. Mason, K.P. Tod, S.T. Tsou, and N.M.J. Woodhouse (eds.) *The Geometric Universe: Science, Geometry and the Work of Roger Penrose*, Oxford: Oxford University Press, pp. 161–72.

Sobel, J.H. (1998) *Puzzles for the Will*, Toronto: University of Toronto Press.

Stapp, H.P. (1977) ‘Quantum Mechanics, Local Causality, and Process Philosophy’, edited by William B. Jones, *Process Studies* 7/3, pp. 173–82.

Stein, H. (1991) ‘On Relativity Theory and Openness of the Future’, *Philosophy of Science*, 58, 147–67.

Whitrow, G. (1980) *The Natural Philosophy of Time*, Oxford: Oxford University Press.

3 A radical rethinking of quantum gravity: rejecting Einstein's relativity and unifying Bohmian quantum mechanics with a Bell-neo-Lorentzian absolute time, space and gravity

Quentin Smith

Introduction

Replacing Einsteinian relativity with Lorentzian absolutivity in a Bell-Bohm-Lorentz quantum gravity theory

The major problem with contemporary physics is commonly acknowledged to be the incompatibility of Einstein's general theory of relativity (GTR) and quantum mechanics (QM). The prospects of unifying these two theories into a quantum gravity (QG) theory seem insurmountable. Witten's M-theory (and string theory in any form), perturbative QG (e.g. the Hartle-Hawking and Vilenkin proposals), topological QG, loop QG and other ideas about QG are not unifications but are speculations about what might belong to an *approximation* of a "complete quantum gravity theory" that is not yet known to be possible or even conceivable. No resolution of the incompatibility of Einstein's GTR and QM has been achieved and there is none in sight.

I believe that if we ponder the implications of Bell's theorems we should regard GTR the same way that Popper suggests we regard a Special Theory of Reference (STR). Writing about Aspect's confirmation of Bell's theorems, he writes:

> we have to give up Einstein's interpretation of special relativity and return to Lorentz's interpretation and with it to ... absolute space and time. ... The reason for this assertion is that the mere existence of an infinite velocity entails [the existence] of an absolute simultaneity and thereby of an absolute space. Whether or not an infinite velocity can be attained *in the transmission of signals* is irrelevant for this argument: the one inertial system for which Einsteinian simultaneity coincides with absolute simultaneity ... would be the system at absolute rest – whether or not this system of absolute rest can be experimentally identified.
>
> (Popper 1982: xviii, 20)

If Einstein's STR is endangered by Aspect's confirmation of Bell's theorems, it certainly seems to follow, indeed, "follow" in the sense of logical implication, that Einstein's GTR is equally endangered. Does anybody really think that instantaneous, non-local, space-like, universe-wide relations of absolute simultaneity (and EPR causal correlations) are logically, mathematically and ontologically consistent with Einstein's GTR? Of course not. Why is GTR not subjected to the same criticism as is STR? Could it be that people are thinking that we must "make do" with GTR "as approximately predicatively accurate in many cases" until a complete QG theory is developed, where this means a unification of GTR with QM? But what would be the point of unifying it with QM if we know it is disconfirmed even on the large scales where it is supposed to be most successful? The relation of instantaneous, non-local, absolute simultaneity is universe-wide; this is a large scale feature of the universe. It seems that to be consistent, we must treat GTR in the same way that we treat Newton's theory, namely, that it is false, i.e. we know on the basis of observational evidence that it does not describe the nature of physical reality. Physical reality is not a Newtonian reality. Nor is it a general relativistic reality. Both Newton's theory and GTR are useful for making predications within the approximate limited circumstances, but they do not give us a physical ontology. They must be interpreted instrumentally, as instrumentally useful in certain circumstances, but they cannot be given a realist interpretation. What is a scientific realistic to believe about the nature of physical reality?

A major task is to develop what may be called a Lorentzian GTR. Popper suggests we should adopt a Lorentzian theory of inertial motion or reference frames, what might be called a Lorentzian STR, where "STR" now means a Special Theory of Reference frames. It is a special theory since (like Einstein's) it is only about inertial frames. A Lorentzian GTR is a General Theory of Reference frames; it is general since it is about both inertial and non-inertial reference frames. And just as Einstein's GTR included a theory of gravity and a cosmology, so must our Lorentzian GTR.

The surprise is that this paper does not present some magnificent mathematical structure that is a new and closer approximation to "the Master Equation" of QG, such as Witten's M-theory or the Hartle-Hawking theory, but rather that the results we reach show how easy it is to unify a classical (non-quantum) theory of space, time, gravity and cosmology with QM. What's the catch? The key move is to (a) reject Einstein's classical GTR and substitute for it a classical neo-Lorentz GTR; and to (b) select only Bohm's 1952 interpretation of QM, which interprets QM as a supplement to or form of Newtonian mechanics. As physicists use these terms, a semi-classical theory quantizes the matter field in the space and time. QED and QCD are semi-classical theories. Einstein's GTR, Lorentz's theory and Newton's theory are classical, in that they do not quantize space and time or matter fields. QG theory quantitates space and time as well as the matter fields. I believe quantitizing space, time and gravity can be achieved if we choose a neo-Lorentzian theory of space, time and gravity and unify it with Bohm's 1952 interpretation of QM, which is classical in its approach and is "a *form of classical mechanics.*" Specifically, Bohm accepted Halpern's (1952: 389) characterization of Bohm's "*quantum mechanics as a form of classical mechanics involving special quantum forces.*" By classical mechanics Bohm here means Newtonian mechanics, including Euclidean space, absolute Newtonian time, and two of Newton's three laws of motion and either including or being consistent with gravitational non-inertial motion. If we give these ideas a neo-Lorentzian formulation, we have a QG theory, and the main difficulty no longer is the QM-GTR unification, but developing a Lorentz GTR that can reproduce the accurate predications of Einstein's GTR.

The reasons QG has seemed so difficult to Bohmians is the same sort of reason other physicists have. The problem with Bohmians is that they have been trying to unify de Broglie-Bohm theory with Einstein's GTR, which is impossible, since the de Broglie-Bohm theory has a non-local, space-like, instantaneous, universe-wide, EPR causal correlation among events and this is logically incompatible with Friedman GTR's basic laws that causal correlations are time-like, local, propagated at a finite velocity (not exceeding that of light) and are non-instantaneous. However, once we recognize that Aspect's confirmation of Bell's theorems disconfirms Einstein's GTR no less than they do his STR, then we should look to the theory that has traditionally been regarded as observationally equivalent to Einstein's STR, namely, Lorentz's STR, and see if we can generalize this theory. Once we do, we will find that Bohm's 1952 interpretation includes part of a Lorentzian GTR.

A Bohm-Lorentz QG consists of ideas and equations scattered throughout the physics literature and which have not been conjoined and organized and presented as a theory. An introductory outline of how these ideas can be unified is the project of this paper. A central claim is that a neo-Lorentzian GTR can incorporate the results of the neo-Newtonian cosmology and gravity theory presented by McCrea and Milne (1934) and developed by Bondi (1960), North (1965), Sciama (1971) Peebles (1993, esp. p. 48), Harrison (2000: esp. pp. 323–338) and others. They show that a neo-Newtonian cosmology is observationally equivalent to Friedman's cosmology on a cosmological scale. Harrison further develops this neo-Newtonian theory into a more comprehensive theory of gravity and shows it is observationally equivalent to Einstein's GTR at smaller scales (Harrison, 2000: 334). But these physicists discuss the neo-Newtonian theory merely as a heuristic device for understanding Einstein's GTR and do not discuss the new relevance it has given that this Neo-Newtonian theory, but not Einstein's gravity theory, is consistent with the non-local EPR correlations. Furthermore, some of the problems with a more comprehensive neo-Newtonian gravity theory, such as Harrison's theory, can be resolved if the neo-Newtonian equations are modified to become neo-Lorentzian equations. Bohm's 1952 interpretation of QM contains laws of motion that are part of the neo-Lorentzian equations (the part that needs to be transformed by the Lorentz transformations). The mentioned parts of Bohm's equations become complete Lorentzian equations through undergoing a Lorentzian transformation. Bohm's 1952 "formally classical" was of interpreting of QM (where the quantum force **Q** is an additional force to the classical forces) either implies or is already unified or consistent with Newton's first two laws of motion and his universal law of gravitation.

The phrase "quantum gravity" has been used to refer to a unification of Einstein's GTR with QM. But that is not the meaning of this phrase, or at least it should not have such a limited meaning. I take it to mean a unification of a classical (non-quantum) theory of space, time, gravity and the universe with QM. QM has (supposedly) been unified with a classical theory of space and time (Minkowski space-time) in quantum electrodynamics and quantum chromodynamics. The classical (non-quantum) theory is Minkowski's theory of space and time. Minkowski space-time is a flat space-time. Recall that Lorentz's theory has traditionally been regarded as observationally equivalent to Einstein's special theory of relativity (1905) or its space-time formulation in Minkowski (1908). If we substitute a flat Euclidean space and an absolute time (required by Aspect's confirmation of Bell's theorem) for Minkowski space-time, we have a theory that is observationally equivalent to Minkowski QED and QCD and also consistent with the absolute, instantaneous simultaneity that was observationally confirmed by Aspect's confirmation of Bell's theorems. This substitution makes all coordinate systems but one effective (merely apparent).

The further unification of the "relativist effects" (described by Lorentz transformation equations) with Newtonian mechanics and gravity shows us that a unification of the relativistic, Newtonian and "quantum effects" are already built into Bohm's equations of motion. The classical potential V and quantum potential **Q** belong to the same equation of motion for particles, $\boldsymbol{m}\mathbf{dx/dt}=-\nabla(\mathbf{V})-\nabla(\mathbf{Q})$, which provides a "built in" criterion for determining the strength of the quantum potential **Q**. The Newtonian laws in Bohm's equations of motion are given a neo-Lorentzian formulation and these can provide classical predictions of observational data. The measurement of the degree of inaccuracy of the classical equations and the observational data is a measurement of the strength of the quantum potential's **Q** contribution to the total causal force. However, some classical laws of motion are not implicit in Bohm's theory, such as the law for gravitational acceleration, and these need to be built from the simpler laws that are implisit in Bohm's equations of motion. An example is a non-gravitational equation for acceleration which I first formulate as a kinematic equation and then as a dynamical equation, which includes the mass and force. Bohm's equations of motion, in their limit, include the kinematics of Newton's second law of motion, i.e. $\mathbf{d(v)/dt}$. A neo-Lorentzian formulation that modifies this Newtonian law of acceleration **A** is:

Neo-Lorentz kinematic equation of acceleration

$$\mathbf{A} = \mathbf{dv/dt}$$
$$\sqrt{1 - \mathbf{v}^2/\mathbf{c}^2}$$

v is velocity, **v** = **dx/dt**, the rate of change of position **x** with time. With Newton, we have **d(v)/dt**, but now we introduce into the kinematic equation the acceleration **dv/dt** divided by a Lorentzian factor. **A** is either the acceleration force or the acceleration field. But accurate predictions require that this be formulated in an equation with Bohm's quantum potential **Q** which causally affects acceleration, as in the double slit experiment. A more comprehensive, dynamic acceleration equation will be given in Part Two.

To differentiate between the classical and quantum potentials, a sufficiently broad neo-Lorentzian theory of motion, gravity and cosmology needs to be developed.

Part one: Relativity is the omission of the Galilean transformation

The "advant garde" discussion in the last third of the nineteenth century and the early twentieth century centered around Maxwell's laws, even though many universities still did not teach Maxwell's (1865) theory because it was "too radical and new". What were these laws? In Maxwell's equations, **D** is the electric displacement, **H** the magnetic field, **p** the charge density, **μ** the charge velocity and **c** the speed of light. **∇** is the vector operator and ∂ the partial differential sign in (∂/∂**x** ∂/∂**y** ∂/∂**z**), where **x**, **y** and **z** are the three spatial variables.

What we shall focus upon is the implied role of space and time in these laws, i.e. the temporal variable **t** and that vector operator ∇ is a partial differential of *the three spatial coordinates **x**, **y**, **z***

$$\text{(1)} \quad \nabla \mathbf{D} = \mathbf{p} \qquad \text{[Eq. 1]}$$
$$\text{(2)} \quad \nabla \mathbf{H} = 0$$
$$\text{(3)} \quad \nabla \times \mathbf{D} = -1/4\pi \mathbf{c}^2 \partial \mathbf{H}/\partial \mathbf{t}$$
$$\text{(4)} \quad \nabla \times \mathbf{H} = 4\pi(\partial \mathbf{D}/\partial t) + \mathbf{p}\boldsymbol{\mu}$$

The electric field due to a point charge can be determined at *an absolute location **x**, **y**, **z**, and time **t*** by Coulomb's force law.

Where does this law obtain? According to Lorentz, it obtains in the absolute rest frame, where the relations of absolute simultaneity obtain. At the absolute time **t**, anything simultaneous with **t** is absolutely simultaneous. In the rest frame, whatever is at rest in the absolute rest frame denoted by ***x**, **y**, **z*** in Maxwell's equations, is at rest relative to absolute space. This implies they are absolutely at rest. (a) Space is "absolute" in the sense of being substantival rather than relational ("relational" meaning reducible to relations among events or sets of events or possible events). (b) It is also absolute in the sense that all physical laws are instantiated *only* by bodies that are at rest relative to absolute space. The absolute rest space (as distinct from the many "relative rest" positions) is where spatial and temporal coordinates measure the *real* (not merely apparent) spatial and temporal values. Maxwell, Lorentz, Fitzgerald and others we discuss posit a flat, non-dynamical space with an independent time variable. We shall see that this understanding of time and space corresponds to Bell's and Bohm's.

By "absolute simultaneity" I mean (a) a two-termed relation, such that even **e** can be simultaneous with event e′ without being also related by a relativity relation of "relative to the reference frame F". (b) If **e** and **e′** sustain a two-term relation of simultaneity, then anything **x** simultaneous with events **e** and **e′** is simultaneous with everything else **y** in the universe that is simultaneous with **e** and **e′**. Each relation of simultaneity can be decomposed into a two-term simultaneity. If event **x** is simultaneous with **e** and **e′** this is equivalent to **x** being simultaneous with **e**, and with **x** being simultaneous with **e′**, and with **e** being simultaneous with **e′**. In addition (c) a maximal relation; there is nothing at all that is neither *simultaneous* nor *not simultaneous* with **e** and **e′**. It is a universe-wide simultaneity relation. Also (d) a relation that connects some space-like separated events and is non-local. If the line of absolute simultaneous events **e**, **e′**, etc. are space-like related to some light comes, the reason is that the distinction between "time-like" and "space-like" in relativity theory is about certain types of causal activities and their relation to light; it has nothing to do with time unless one stipulates that some mechanism on a world-line is going to be pragmatically used as a conventional "clock". It also means (e) an instantaneous relation; such that if **e** and **e′** are absolutely simultaneous, they both are temporally located at the instant **t**, which is an instant and thus has zero duration. Anything **x** or **y** absolutely simultaneous with **e** and **e′** is instantaneously occurring with everything else, across the universe, that is occurring at the same instant **t** as **e** and **e′**. Bohm, Lorentz, Aspect, Bell, etc. regard absolute simultaneously as possessing all these features. (f) Time is substantival, composed of instants and intervals, and not of physical or mental events. It can exist without events or matter.

Distinct from absolute simultaneously is a non-instantaneous "preferred simultaneity plane" or the time of a "privileged frame". For example, a Friedman universe is alleged to have a "preferred time" in this sense, but it needs some argument to show that two events simultaneous in this preferred time frame are absolutely simultaneous in my sense, or even that the imperfect homogeneity and isotropy of our universe makes a "privileged frame" an averaged out frame, and thus a conventional construct rather than an ontological reality. It is interesting that many defenders of a preferred time assume that this preferred time is not an "averaging" among imperfectly homogenous frames and imperfectly isotropic frames and is instead an ontological reality; but it would be real only if the universe had a perfectly homogeneous and isotropic frame.

Bell's EPR correlations occur at the absolute time **t**. It is difficult in any theory, where non-local instantaneous causal relations or correlations obtain, to try and admit an absolute time but adopt a reductivist and relational theory of space. The correlation is causal, given suitable definitions of causality, such as David Lewis' type counterfactual definitions (for example, see Maudlin 1994).

For two distant events to be EPR correlated, the correlating relation must extend from one event to the other and this requires a pathway in substantival space. There are no candidate "events" or "bodies" to try and construct a reductivist theory of space. One would need continuum-many points extending from one place to another – the pathway of the EPR correlation. Positing such points is another way of saying one is positing substantival space. The idea that a place could be a "possibility" for something to be located is not obviously intelligible. For example, what could it mean that an EPR correlation instantaneously and physically transmits information through continuum-many possibilities? Possibilities are abstract objects and cannot constitute something concrete such as space. With non-local instantaneous relations in the absolute frame, we require substantival space. Bohm's 1952 space is also Euclidean, flat, non-dynamical and infinite. It has an absolute frame of simultaneity, detectable by observing EPR correlations.

According to Lorentz and Maxwell, absolute space is flat and non-dynamical. Considering some concepts pertinent to later twentieth century observations and theories, we may say that the Lorentz-Maxwell absolute space is not expanding or contracting and is not identical (even partially) with curves, dips, waves of space, holes and the like. The alleged "observational evidence" for such gravitational effects are the behaviors of large bodies, or, more precisely, large aggregations of fermions and bosons (such as the earth and its moon); the Einstein GTR postulate of curved space is observationally equivalent to a theory that what curves, dips, expands, etc. is not *space* but certain *types of movements of aggregations of bosons* (photons, gluons, weak vector bosons, etc.) or *fermions* (electrons, quarks, etc.). These trajectories are describable (perhaps) by a Bohm-Lorentz QG theory, where gravity is a "force" on bodies analogous to the "force" of the quantum potential **Q** (Bohm 1952).

According to Lorentz, any reference body *moving in absolute space* has its real spatial and temporal variables revealed through either being at rest in absolute space or through being related by a Galilean transformation equation to the rest system. The absolute frame and the Galilean transformation contains the real, absolute (Galilean) coordinates, which absolutely relate the coordinates of an inertial frame at rest in absolute space to another inertial frame that is uniformly moving in absolute space. As physicists now commonly state, the reintroduction of instantaneous, non-local absolute simultaneity and absolute space brings back the Galilean transformations, which, given recent evidence (mainly Aspect's), makes these transformations "new" in the sense that they are needed in theories that accommodate Bell-Aspect-Bohm absolute simultaneity.

The real, absolute coordinates, i.e. the Galilean coordinates, are stated below in bold face. They state the spatial coordinates, ***x***, ***y***, ***z***, of a body at rest in absolute space and they state the absolute time **t**, which is the same for resting or moving bodies. The regular type-face states the coordinates of a system moving in absolute space. The Galilean transformation equation that relates the moving body's coordinates to the rest frame's coordinates is:

$$\mathbf{x}' = \mathbf{x} - \mathbf{vt}, \quad \mathbf{y}' = \mathbf{y}, \quad \mathbf{z}' = \mathbf{z}, \quad \mathbf{t}' = \mathbf{t} \qquad \text{[Eq. 2]}$$

The Galilean transformation, Lorentz, Fitzgerald, Lamor, Poincaré and others noted, did not allow Maxwell's equations to be expressed similarly in all frames, but only in the frame absolutely at rest. The magnetic field **H**, the electric displacement **D**, the charge density ***p*** and the charge velocity ***u*** and the speed of light **c** obey Maxwell's equations in the absolute rest frame, but different and more complicated equations are needed if we are to express something analogous to them in moving frames. For example, **c+v** or **c−v** would need to be substituted for **c** in Maxwell's equations, since a moving body's velocity would need to be included in the number for the velocity of light. The law of the addition of velocities really obtains only in the absolute rest frame, but the causal effects of moving through absolute space have as one of their consequences, that it could appear to a hypothetical observer on the moving body that the law of the addition of velocities, in particular regarding light, does not obtain.

If we want to retain the form of expression of Maxwell's less complicated equations, we need to use, not the Galilean transformation, but the Fitzgerald-Lorentz transformation equations. The need for these new equations arose consequently upon the realization that the absolute rest frame could not be observationally distinguished from absolutely moving frames. What could be the cause of the observational indistinguishability of resting and moving frames?

The evidence offered for a causal explanation comes from the various and unsuccessful attempts by such physicists as Bradley (1728), Young (1804), Arago (1810), Fresnell (1810, 1811, 1815, 1818), Fizeu (1851), Hoek (1868), Michelson (1881), Michelson-Morley (1887) and others to discover the motion of the earth relative to absolute space or an aether at rest in absolute space. There is also "evidence of a kind" from some thought experiments by Einstein (see the letters interchanged between Alfred Einstein and Marvaria before 1905, where Einstein, assuming there is an aether absolutely at rest, thought about new ways to try and detect the earth's motion relative to it). Fitzgerald (1889) was the first to develop the generally accepted and natural causal explanation, that the spatial length of an object is shortened in the direction of its motion by the factor $(1 - \mathbf{v}^2/\mathbf{c}^2)^{1/2}$, due to the fact (then assumed by many) that since matter is electrical, or is bound together by electric forces, it will contract in the direction of motion as it moves through the electromagnetic field or aether or field. The usual histories of physics mistakenly say Fitzgerald (1889) presented an equation but no causal explanation. This is false. Furthermore, standard histories are not entirely correct when they say that Lorentz (1892) or Lamor (1900) first discovered the clock retardation equation. Since a light clock tipped in the direction of motion reveals the same contraction, the factor for clock retardation is easily obtained from Fitzgerald's equation: $(1 - \mathbf{v}^2/\mathbf{c}^2)^{-1/2}$

The "one mathematical step" that led twentieth century physicists along a relativistic path rather than a Fitzgerald-Lorentzian path, is that *Einstein simply omitted the Galilean transformation equation in his 1905 essay*, stated above, from Lorentz's full transformation equation. (Poincaré had done the same.) Let us see how this one omission could lead to a relativism about space and time.

The Galilean variables for space and time in the absolute rest frame are in bold face, **x**, **y**, **z**, **t**.

The variables for a reference frame $\mathbf{R}^*$ that is moving in absolute space are starred: $\mathbf{x}^*$, $\mathbf{y}^*$, $\mathbf{z}^*$, $\mathbf{t}^*$.

The variables for a second reference frame **R**, which is moving in absolute space in such a way that it is also moving relative to $\mathbf{R}^*$, are primed variables $\mathbf{x}'$, $\mathbf{y}'$, $\mathbf{z}'$, $\mathbf{t}'$. The full Lorentz transformation is:

$$\mathbf{x}' = \left[(1 - \mathbf{v}^2/\mathbf{c}^2)^{-1/2}\right]\mathbf{x}^* = \left[(1 - \mathbf{v}^2/\mathbf{c}^2)^{1/2}\right](\mathbf{x} - \mathbf{vt});$$

$$\mathbf{y}' = \mathbf{y}^* = \mathbf{y};\ \mathbf{z}' = \mathbf{z}^* =;\ \mathbf{t}' = \mathbf{t}^* - \mathbf{v}$$

$$\left[(1 - \mathbf{v}^2/\mathbf{c}^2)^{-1/2}\right]^2/\mathbf{c}^2\mathbf{x} = \mathbf{t} - \mathbf{v}\left[(1 - \mathbf{v}^2/\mathbf{c}^2)^{-1/2}\right]^2/\mathbf{c}^2(\mathbf{x} - \mathbf{vt}) \qquad \text{[Eq. 3]}$$

$$= \left[(1 - \mathbf{v}^2/\mathbf{c}^2)^{-1/2}\right]^2(\mathbf{t} - \mathbf{v}/\mathbf{c}^2\mathbf{x})$$

Lorentz shows by these equations several things. First, that since the electric displacement **D** of the electric field, and the magnetic field **H**, at rest in absolute space, are certain functions of the spatial and temporal variables, then to obtain a similar form to Maxwell's laws in a system moving in absolute space, the electrical displacement and magnetic field of these moving systems should be a function of their spatial and temporal variables. We recall that the vector operator **∇** is a partial differential of *the three spatial coordinates* **x**, **y**, **z**. If our equations that hold in the absolute rest frame are a function of the coordinates, the same should be true for a moving system if Maxwell's laws are to have the same form, which would allow observers on the two frames to each view themselves as at absolute rest and the other frame as moving.

But this sounds more like Poincaré (1902, 1905) and not like Lorentz, at least until Lorentz (1909, 1915). Before this, Lorentz emphasized that the only spatial and temporal coordinates are the Galilean ones, and that the ones for moving bodies are not coordinates of space or time, since space and time exist in their real nature only in the absolute rest frame. In moving frames, the coordinates record merely the contraction effects caused by motion through space (or, as Lorentz would prefer to say, the aether). How could $\mathbf{t}^*$, for example, be considered a temporal coordinate if it at best measured merely a contracted and thereby distorted clock reading? How could $\mathbf{t}^*$ be a time? More basically, Lorentz viewed the non-Galilean coordinates as merely "effective variables," as physically uninterpreted mathematical devices that were aids to computation.

Further, the Galilean transformations were not assigned values, since they were undetectable, but were left in to indicate that it is real time and space that we are talking about, even though only its phenomenal effects are perceivable.

We see here the basis for the absolutist theory that can be developed, added to and modified to produce the absolutist theory that can replace GTR. The Lorentzian approach fits in well with later observational evidence, e.g. Aspect's experiments and with the theories of absolute time and space (of course, Lorentz's aether, his electron and Maxwell's equations need to be considerably updated, but here I am talking about Lorentz's basic ideas). Lorentz developed transformation equations relating phenomenal frames to the absolute inertial frame, where physical systems have their real, absolute, temporal and spatial coordinates. Ignoring for the moment the adjustments required by non-inertial movement, this absolute rest frame is the only frame where absolute time and undeformed spatial shapes and sizes exist. Again, ignoring the adjustments required by non-inertial movement, this absolute frame is the absolute frame disclosed in the Aspect experiments, i.e. the frame wherein there exists the instantaneous, non-local EPR *causal correlations.* This absolute EPR frame, as shown by Bohm and Hiley, Valentini and other Bohmians, is the absolute frame along which the quantum potential **Q** (Bohm and Hiley 1993; Callender and Weingard 1994) or the phase **S** of the universal wave function (Valentini 2002a; Dürr *et al.* 2003) instantaneously acts at arbitrarily far, non-local, space-like distances. In fact, in de Broglie-Bohm theory, there is only one wave function, the universal wave function, and effective wave functions for particles or "beables" (to use Bell's term) are obtained by abstracting from the universal wave function. The Bohmian causal **Q**-potential is universe-wide and any local application requires "mentally abstracting" an inseparable part of this universal, instantaneous causal relation. We also could follow Valentini and others and omit the **Q**-potential, identifying the causal element directly with the Bohmian causal phase S of the universal wave equation, but Bohm's 1952 is more near the neo-Lorentzian approach of our other theories.

It is usually said that Einstein's (1905) theory is at least equal in predictive power to Lorentz's theory, since they are observationally equivalent.

Many writers in the 1950s, 1960s and 1970s were fond of saying that Einstein's (1905) theory was the most well-confirmed scientific theory in history (the same has also been said of QM and QFT). Given the observational equivalence and predictive power equivalence is acknowledged, it is at best misleading to say that Einstein's (1905) is the most experimentally well-confirmed theory. One should say that Einstein's (1905) theory, Lorentz's, Poincaré's, Fitzgerald's, Lamor's and others are tied for the most well-confirmed, since they are all observationally equivalent; whatever experiment result confirms any one of these theories, it confirms the other.

An important issue in Lorentz's full equation lies in the type of causal explanation it gives to the various values of the spatial and temporal variables of the phenomenal frames. By regarding the absolute rest frame as "superfluous" in developing phenomenal physics, Einstein deprived his theory of virtually any explanatory causal power; it is a theory filled with "brute facts". In Fitzgerald's and Lorentz's theory, the rest frame with Galilean coordinates is interpreted as containing a force field that causally acts on moving frames and shortens their length $\mathbf{x}'$ and $\mathbf{x}^*$, in the direction in which they are moving and, as a part of this contraction effect, slowing the clock-motion processes $\mathbf{t}'$ and $\mathbf{t}^*$. The causal activity explains why lengths contract and physical processes are retarded. It also explains why the velocity of light $\mathbf{c}$ appears constant in every frame, enabling Maxwell's laws to appear to be the same in any frame, absolutely at rest or absolutely moving. By characterizing the variables for moving frames as effective variables, Lorentz maintained that real spatial lengths, heights and widths, and real time, existed only in the absolute rest frame. But omitting the Galilean equation, Einstein's theory lost this causal explanatory power. *The coordinates (and all properties) of his inertial frames become "brute facts."* If one adopts the positivist interpretation of 1905 as a theory that the theory of phenomena is a theory of reality, and as not including the absolute frame as part of reality, the result is that there are *different* Maxwell equations held relative to *each* different co-moving inertial frame or reference frame. Lorentz's parsimony is lost, for Lorentz posited only one real Maxwellian law and equations, those obtaining in the one, mind-independent absolute frame, with the phenomenal frames being mere appearances. Appearances, constructed by observers misinterpreting the distorted shapes, sizes and the slowed rate of physical activities of their moving system as being *measurements* of real sizes and real times.

If Lorentz's theory has greater causal explanatory power then perhaps "Einstein's" theory wins on grounds of "symmetry," which has been a traditional view. But Lorentz's theory in fact had greater symmetry. This is obscured by the metaphorical charge (the "derision" fallacy, as I call it) that Lorentz's theory implied that nature was intrinsically paranoid that there was a "conspiracy of silence". This metaphorical derision is still widely believed, but what does it amount to beyond a "derision" attempt to try and make Lorentz's theory sound ridiculous?

Examined precisely, it is seen that Lorentz's theory meets none of the conditions of this "derision" fallacy of the "conspiracy of nature". The literal facts that lie behind this metaphor are in fact opposite to those one would expect if one assumed this to be an accurate way of metaphorically stating a charge against Lorentz's theory. In Lorentz's theory in the 1890s and 1900s, there is no conspiracy but a clear, well-defined, causal explanation of the observable phenomena and *a greater symmetry than Einstein's 1905 theory.*

A typical statement of this "derision" informal logical fallacy, which tries to avoid mere metaphors more than most statements of this criticism, reads as follows: note that this charge is stated in a way that Lorentz's theory superior symmetry is "twisted and ridiculed" so to speak, into some sort of strange type of asymmetry (even Einstein did not say anything like this; the asymmetry he and others were criticizing was Faraday's, not Lorentz's).

> It is unlikely that Nature contains both deep asymmetries and compensatory factors which exactly nullify these asymmetries. Such a state of affairs is not logically impossible and to envisage it is not meaningless; but it is unlikely, or improbable, in the same intuitive sense in which a series of coincidences and accidents having a single global effect are improbable.
>
> (Zahar 1973: 39)

What does this mean? What is a deep asymmetry that is exactly nullified by a compensatory factor? Does the "problem" consist merely in the type of emotive or rhetorical words used in stating the issue? I think so. What if the relevant clause in the first sentence is rephrased as:

> Nature's fundamental symmetry is manifest in the fact that phenomena are caused to possess the same sort of symmetrical structure that the one uncaused frame, the one at absolute rest, possesses. This symmetry makes Lorentz's theory such an elegant and natural theory that Einstein's elimination of the main symmetrical element results in an unacceptably less symmetrical theory?

But we get a clue about why a emotively negative "derision" is used to describe this exceptional symmetry. We can see where the basic misunderstanding of Lorentz's theory lies, in a clause in the next sentence: "it is unlikely, or improbable, in the same intuitive sense in which a series of coincidences and accidents having a single global effect are improbable" (Zahar 1973: 39). If it can be seen that Lorentz's theory is the opposite of something like a series of coincidences and accidents, this may cause critics to think exactly how the symmetry in Lorentz's is somehow problematic, and not rest content with burying his symmetry beneath the officially accepted "derision" argument.

A series of coincidences or accidents having a single global effect are compounds of *nomologically independent* events that have a single global effect. A coincidence is inexplicable and the compound of events constituting the coincidence has necessary and/or sufficient conditions (or probabilistic conditions) that are independent of one another. (See David Owens, *Causes and Coincidences.*) However, in a Lorentzian theory, the so-called "compensatory factors which exactly nullify these asymmetries" *are* nomologically explained and have the same sufficient and necessary conditions (or have conditions that are not independent or nomically independent of one another). There is an aether in absolute space that transmits both the electromagnetic forces and molecular forces. This transmission causally affects (is a causally sufficient condition of) contracting the moving body in the direction of its motion. Lorentz deduces the Lorentz-Fitzgerald contraction hypothesis and the phenomenal constancy of the velocity of light from this causal hypothesis. The "ad hoc" derision fallacy is quickly dismissed. Compare Fitzgerald and Einstein. Fitzgerald looked at the Michelson-Morley experiment and developed a mathematically accurate causal explanation of it. Einstein looked at it and adopted it as a brute fact. Whether Zahar is right or not about the Molecular Forces Hypothesis, Lorentz's theory is a good one in the respect in which it is similar to Fitzgerald's. A new piece of evidence is discovered. One person looks at it, goes home and develops a theory that causally explains its existence. A second person looks at it and say "its just another brute fact. I will adopt it as an unexplained assumption". Whose theory will have greater theoretical virtue?

The Galilean transformation gives real temporal and spatial coordinates which, due to bodies departing from their natural equilibrium state (being absolutely at rest) are caused to be in the non-equilibrium state of being contracted. But the remarkable symmetry of Lorentz's theory, borne out by the phenomena, is that the non-equilibrium state the moving body is caused to possess is caused to be symmetrical to the body's rest state, the state the body would have if it were absolutely at rest and in its equilibrium state. Nature is symmetrical at such a basic level that the causal effect produced by motion is such that the cause produces an effect in the moving bodies that transfers to them a structure that makes them symmetrical to their rest states. If what occurs in the absolute frame is considered as the causal field, and the phenomenal arena as the arena where the resultant effects appear, then we can say this theory satisfies the old philosophical maximum, "causes produce effects that are alike the causes"; "like causes, like effects". In fact, Pierre Currie (1894) can speak for us: "when certain causes produce certain effects the symmetry elements of the causes must be present in the effects which they reproduce". There is no clause better than this to capture a theoretical virtue of Lorentz's theory. This is a marvelously symmetric theory. Einstein did not see it since he was interested in a phenomenal relation between two phenomena, the magnetic and conductor, which he wanted to explain in purely phenomenal terms, without recourse to the absolute frame. One need merely consider the phenomenal motion that the magnetic and conductor have relative to each other. Unfortunately, he did not realize that he could have retained his phenomenal symmetry, attained by explaining the experiment in terms of their relative motion, and at the same time retained the remarkable deeper symmetry of nature that Fitzgerald, Lorentz, Poincaré and others discovered.

We have explored some of the respects in which Lorentz's theory had greater theoretical virtues than Einstein's 1905 theory. And we know we need to develop an outline of neo-Lorentz-Bohm quantum field theory, which will not directly (by "third quantization") lead to a discussion of quantum gravity; rather, QG requires returning to Bohmian mechanics and its implicit Newtonian mechanics and some gravitational acceleration equations. This will lead to an outline of a neo-Newtonian gravitational field equation that can be unified with Bohmian QFT.

Part two: Lorentzian-Bohm quantum field theory and QG

I hope the burden of proof for justifying the unification of an absolutist theory with STR has by now shifted to some defenders of absolute simultaneity, such as some Bohmians. Pitowsky writes that "Bohm's theory, as it stands, has no satisfactory special relativistic formulation" (1991: 343). I hope not! If it did, it would be falsified by Aspect's observational data, assuming STR could even survive the criticisms made in the last section.

The unification of QM with STR is a desirable goal for most Bohmians and one wonders where the problem lies. STR and Bohmian QM are manifestly inconsistent and one must be eliminated if a "unification" of any sort is to take place. The underlying space-time of an adequate quantum field theory (QFT) cannot be Minkowskian, for that prohibits absolute simultaneity. The spatial-temporal structure of a 3+1 Lorentz space and time needs to be the flat space and absolute time on which quantum electrodynamics, quantum chromodynamics and quantum electroweak theory are formulated. This will be observationally equivalent to existent QED and QCD, except for the respects in which Minkowskian relativity renders the formulations inconsistent with the observational data, includes Aspect's observations, and any changes required will be changes motivated by a desire to accommodate Bohm's QM, not STR.

I wonder why Bohmians' approach to quantum electrodynamics and quantum chromodynamics reveals a fundamental ambivalence. Maudlin expresses the typical ambivalence of the Bohmian to QED and QCD: " ... the most straightforward route to putting quantum theory into a relativistic space-time is to add something to the space-time, and, hence, at least implicitly, to reject Relativity as the complete story." (Maudlin 1996: 295). The "added something" is "something like the classical notion of absolute simultaneity; a fundamental physical relation between events at spacelike separation" (Maudlin 1996: 295). It seems to me there should be no ambivalence and that another route (I believe hinted at by Maudlin) is the preferable one. This is best done through examining some ideas of Callender-Weingard (1997) about QFT, who adopt their own equations and make the attendant philosophical issues especially clear.

Note that Callender and Weingard (1997) write the following (the separation into numbered sentences is mine):

(1) According to contemporary physics, the quantum mechanics of non-relativistic particles is not fundamental.
(2) Instead, what we call particles are manifestations of relativistic quantum fields.
(3) Relativistic quantum field theory, then, is a more fundamental theory than elementary quantum mechanics, and it follows that believers in Bohm's theory should apply it to relativist quantum fields.

(1996: 28)

The inference of (3) from (1) and (2) is invalid, since all that follows from (1) and (2) is

(4) According to contemporary physicists quantum mechanics is not fundamental unless it is unified with STR.

Perhaps "according to contemporary physicists" should read "according to the great majority of contemporary physicists". What Callender and Weingard could be discussing at this point is why (1) is considered by most physics to be true, even though it is false. Further, there are two distinct issues involved. One concerns the impossible task of unifying Bohmian QM with STR. The other concerns whether field theory is more fundamental than particle theory. Shall bosons be given a field theory ontology and fermions a particle ontology? Bohm and Hiley (1989, 1993) believe so. Or should both be given a field ontology (Cushing 1994; Valentini 1996)?

I suggest that the reasons are that the absolutists' theories have been regulated to specialist branches. For example, one's field of specialization could be the Aspect experiments. Or it could be Bohmian mechanics. But the great majority of physicists do not specialize in this area. This fact, along with the "officially sanctioned status" that STR and especially GTR has in the physics community, a specialist may say that the Aspect experiments do not lie in his own of specialization and with that remark proceed blithely with the assumptions of STR and GTR.

There are some good features to the compartmentalized, specialized, slow, cautious, hesitant, resistant way of dealing with fundamental changes in physics, specifically, rejecting GTR and replacing it with a neo-Lorentz gravity-Bohmian QM theory, i.e. a QG theory. I emphasized at the beginning of this essay that one part of the problem is that there is no picture of what physics will be left if GTR is jettisoned. They can't change their beliefs if they don't know what to change their beliefs to. I suggest we *already* have a replacement for Einstein's GTR on hand, if we combine several difference sources, theories, ideas into one unitary framework.

One thing could change now regarding QFT. Consider Callender's and Weingard's guidance equation for a scalar field. The field values are indexed by the spatial position **x**. **S** is the phase of the wave function. $\boldsymbol{\delta\varphi(x^{>})}$ represents the differences or "displacement" of the field at successive times in the time evolution of the field $\boldsymbol{\varphi(x^{>})}$. In the equation of motion for the time evolution of the field, one reads $\boldsymbol{\varphi(x^{>})}$ as a function that assigns the value of the field to each point **x** of 3-dimensional space.

$$\boldsymbol{\delta\varphi(x)/\delta t = \delta S/\delta\varphi(x)}$$

This is the equation of motion for the time evolution of the scalar field $\boldsymbol{\varphi(x)}$. Recalling that **S** is the phase of the Schrodinger wave function on a *Bohmian* interpretation of QM, it is immediately apparent that there is absolute simultaneity in an absolute frame along with non-local, space-like, instantaneous causal relations are propagated and that the temporal evolving displacements in the field $\boldsymbol{\delta\varphi(x)/\delta t}$ from one time to another are instantaneous and due to non-local, space-like causal correlations. In brief, there is a violation of Lorentz invariance of the field dynamics. But it will not be observationally detected. Callender and Weingard acknowledge this and say it is not "that bad" since "we will not detect a violation of Lorentz invariance, even though it is violated by the field dynamics" (1996: 30). In other words, relativist quantum field theory is ontologically inconsistent with Bohmian field theory, but at the observational level "we will not detect this".

Callender and Weingard add that in QFD we do not have

> to give up the view that there is an underlying flat spacetime structure. What we have to accept is that there is additional, physically significant spacetime structure, namely, that there is a preferred frame. We do not. . . . postulate that Lorentz [Minkowski] metric breaks down at very small length scales.
>
> (1996: 30)

I beg to differ. They adopt the standard convention in QFT that the "Lorentz metric" refers to Poincaré-Minkowski space-time. But "Minkowski spacetime" does not permit a privileged frame, let alone non-local, instantaneous, space-like causal connections and absolute simultaneity. "Adding spacetime structure" in the present instance is like adding volume to a point. The result is what should have been announced as their goal in the first place: to reject Minkowski space-time, replace it by a Lorentzian flat space, (not "Lorentzian" in the sense in which this word is used in Einsteinian relativity theory, but in the sense of a Heinrich Lorentz's (updated) ontology), a flat space with an independent absolute time and an absolute frame along which there are transmitted non-local causal correlations. In particular, the absolute frame alone is the one in which the law $\delta\varphi(\mathbf{x})/\delta \mathbf{t} = \delta \mathbf{S}/\delta\varphi(\mathbf{x})$ obtains and (the neo-Lorentzian idea here should be added that the movement relative to this absolute frame distorts the other frames and that the "relativistic effects" observed are causal effects of this movement). The "relativistic effects" are symmetrical to their absolute cause, have a sensory observable nature that enables the physicist to imagine (without conflict from observations except for Aspect's) that he is on a Minkowski frame in a closed phenomenal system (i.e. with no absolute causal frame), or, more nearly on the right track, that his frame is the absolute rest frame and the others are the moving and distorted one or (exactly on the right tract) that he is on a frame and he cannot distinguish between the absolute rest frame and the distorted moving frame and that for pragmatic purposes he could adopt the phenomenal part of a Lorentzian absolute/phenomenal ontology and develop a theory, useful for pragmatic purposes, that the phenomenal part is a closed system (i.e. "all of reality, there being no absolute frame") whose spatial-temporal laws are Minkowskian. Exactly what justification do many Bohmians (I discussed Callender-Weingard merely as a representative example; their position is more or less the standard Bohmian approach to QFT) have for not adopting this neo-Lorentzian ontology that seems to me the fairly obvious approach that a Bohmian should take to QED and QCD? The "success" of these Bohmians in developing a theory that is "observationally Lorentz invariant" or that the absolute frame is "observationally undectactable" is not a success, for it requires physicists to misinterpret the sensible phenomena as a closed system that obeys Minkowski or Poincaré-Minkowski laws. This is not a "success," for the theoretical requirement that the observations be misinterpreted is a requirement for a "failure" and for there to be a "success" there must be a theoretical requirement that the observations be correctly interpreted as mere phenomenal effects, symmetrical distortions caused by the absolute frame, with one of the frames being absolute by being observationally indistinguishable from the rest. But even this is wrong. Electrodynamics and chromodynamics are at the atomic and subatomic level at which Aspect's EPR correlations were observed, which detected the absolute frame, and thus QED and QCD should be interpreted as capable of being observationally distinguished into the absolute frame and the "merely relativistically appearing" frames. In each case, the absolute frame is (if one performs an Aspect experiment) detectable, so we have to disagree with Callender-Weingard that Minkowski-QED and Bohm-Bell-Aspect QED are "observationally equivalent" in that the "lack of Lorentz invariance will not be observationally detectable".

This way of formulating QFT should be natural to the Bohmian. Bohmian QM implies the apparent "quantum effects" are mere appearances due to our ignorance of the unobserved causes and this ontology (they should acknowledge) carries over into QFT where the apparent "relativistic effects" are mere appearances due to our ignorance of the unobserved causes. One could say that QFT is about unifying one set of misinterpretations or "mere appearances" (Minkowski "relativistic" ones) with another set of misinterpretations or "mere appearances" (quantum ones). There is something real: the unobserved, absolute causation of the contractions/retardations in the one case and the unobserved trajectories guided by a **Q**-potential or phase **S** of the wave function in the other. The question I am ultimately aiming to answer is how to integrate the absolute classical frame with the absolute quantum frame (QG, as it appears below).

Changes such as this to QED and QCD is how progress should be made towards a new basic physics. An objection, that QED is the most well-confirmed theory physicists have ever constructed, ought to be met with the same attitude as we now meet the frequent claims from 30 to 60 years ago that STR is the most well-confirmed theory physicists had ever constructed. In fact, the so-called confirmations of (Poincaré-Minkowski) STR were confirmations of a neo-Lorentz theory, which physicists have acknowledged is observationally equivalent to STR (as long as Aspect's experiments are ignored). Aspect's experiments pose no problem for a Lorentzian theory, but we know the situation is different with regard to STR. Bohmian QED on an absolute, neo-Lorentz space and absolute time is observationally equivalent to non-Bohmian QED on a flat Minkowski space-time, apart from the fact that the retention of a relativistic ontology renders QED falsified on the basis of the Aspect experiments.

What really is the problem with the Bohmians (virtually all Bohmians) who still cling to STR when doing QED and QCD? Surely, it is not an intellectual confusion, since they themselves note the contradictions. Is it due to the fact that Lorentzian physics has disappeared from the horizon of mainstream physics, rhetorically dismissed with the emotive rhetoric of "ad hoc" and "conspiracy of silence" informal fallacies of "derision" phrases? Insofar as "ad hoc" means something substantive, it is reducible to the theoretical virtue or "vices" I mentioned: lack of sufficient causal explanatory power, parsimony, symmetry, predictive power, conservativeness, mathematically simplicity (in the sense of the fewest independent mathematical equations used as axioms), etc. Neo-Lorentzian theory as it has been developed throughout the twentieth and early twenty-first century is both untaught, unknown to the point where virtually no one has even heard of the physicists' names, and, even if one desires to learn it, one finds it is virtually inaccessible. The undergraduate and graduate training in physics does not include neo-Lorentzians such as Builder, McCrea, Ives, Stillwell, Procoviknic, Bastin, Jannossy, Sherwin, Keswani, and others. However, these thinkers developed only a "neoLorentz special theory of relativity," adopting the positivist, and operationalist approach of Einstein in his 1905 STR. Furthermore, by "absolute simultaneity" they mean a privileged frame of reference, whose plane of simultaneity is the fundamental one. But they reject instantaneous, non-local simultaneity of the sort Lorentz, Bell, Bohm and others adopt. Further, their main focus is on defending the specific physical ideas of an aether that causes the contraction of moving bodies. However, I believe that absolute simultaneity is not a privileged plane of simultaneity, e.g. in Friedman's or Prokhovnik's sense, but a non-local instantaneous universe-wide relation. Further, I postulate no aether. It is the movement of a body in a flat Euclidean rest space that causes the contraction. No additional cause is needed, especially when the theory of its nature is varied constantly, depending on the latest advances in electrodynamics, elementary particular theory, and their attempts to accommodate Einstein's GTR. Thus I am not a "neo-Lorentzian" in this "old sense". I am neo-Lorentzian in the "new" sense in which Bell's chapter can be considered as presenting a neo-Lorentzian theory (1987). However, I do not agree with the specifics of Bell's ideas, e.g. the manner in which he incorporates QM into neo-Lorentzian non-gravitational acceleration equations or his unwillingness to reject an "aether" that causes the contraction. But I agree with the basics; absolute simultaneity is instantaneous and non-local, and that bodies that move in absolute rest space have their lengths contracted and clock movements retarded. Unfortunately, Bell never connected his neo-Lorentzian theory to Bell's theorems, to de Broglie-Bohm QM, to Einstein's General Theory of Relativity, or to QG. In a certain sense, this essay may be seen as "going beyond Bell, but in the same direction" to develop the various new theories that he could have, but did not, develop or even hint at. Naturally, I could say the same of de Broglie-Bohm quantum physicists who integrated Bell's theorems into their results. Both Valentini (1996) and Bohm and Hiley (1993) appear to have gone far in this direction, but Valentini still regards QG as a unification of Einstein's GTR with Bohm's QM and the Bell-Aspect EPR correlations. Bohm and Hiley (1993) become mystical and posit a "subquantum realm" at which an entirely new and unknown physics reigns, without the slightest justification for their assertion. Valentini perhaps has gone the furthest, especially with QED and QCD, but does not address the issue that Friedman's universe is a solution of Einstein's GTR and has a metric and Robertson-Walker line element that is logically, mathematically, and physically inconsistent with non-local and space-like, instantaneous, EPR causal correlations. York slicing does not help since it is also inconsistent with absolute simultaneity of this sort. And Valentini does not seem to address that the Friedman privileged plan of simultaneity is an "averaged out" concept of various non-connecting planes of simultaneity, thereby preventing it from being something unique and real, a real feature of the universe, rather than a concept constructed by averaging out the small divergences from perfect homogeneity and isotropy. I recommend to Valentini that he pursue a QG where space has a Euclidean metric, not a GTR metric (not even the "flat" metric of a Minkowski space-time), which automatically makes Bohm's 1952 interpretation consistent with a non-Einsteinian and neo-Lorentzian general theory of inertial and non-inertial motion. This unifies classical space and time with quantum space and time, the main goal of "quantum gravity" as it is usually understood. It also shows how QED and QCD are consistent with gravitational acceleration, which is based on neo-Lorentzian equations. This achieves the goal usually associated with "quantum gravity" and make it a Bohm-neo-Lorentz-Valentini research program. It is indeed welcome news that Valentini, even if only in a brief paragraph about Bell (1987), indicates that Bell showed him how Lorentzian physics could be used instead of Einstein's relativist physics. Valentini states that in QED

> what emerges in equilibrium is not Einstein's special relativity [Minkowski] but Lorentz's earlier interpretation of the Lorentz transformations. The view of Lorentz – that one particular frame is at absolute rest – is perfectly consistently classically. A single frame with coordinates (**x**,**t**) is sufficient to describe all of physics – including the physical response of the moving equipment (See for example Bell 1987). ... Relativist spacetime is a construct of equilibrium observers who are unable to see nonlocality directly; it is, like 'quantum ontology, a misguided projection, a misguided projection of the contingent human limitations onto Nature herself.
>
> (1996: 56)

Valentini, indicating his influence by Bell, takes Bell's ideas one step further in QED and QCD. Valentini seems to be instantiating (at least in outline form) the particular ontological approach to QFT that belongs to a Bohm-neo-Lorentz QFT and the rarity of its instantiation is explained by how Valentini words his passage, namely, that he learned about a Lorentzian theory from reading Bell (1987).

Bell's own approach to QFT involves the same problem of "adding structure" to a Minkowski space-time that we previously discussed:

> The aspects of quantum mechanics demanding non-locality remain in relativistic quantum mechanics. It may well be that a relativistic version of the theory, while Lorentz invariant [i.e. having Minkowski metric] and local at the observational level, may be necessarily be non-local and with a preferred frame (or aether) at the fundamental level. Could we not then just omit this fundamental level and restrict the classical variables to some "observable" macroscopic level? The problem then would be to do this with clean mathematics, and not just talk.
>
> (Bell 1987: 132–133)

The fundamental level that Bell asks if we could "omit" pertains to the level that Maudlin and Callender and Weingard discuss in terms of the "added structure." In all three cases we have a Minkowski space-time and an absolute, non-local spatial-temporal structure added to it as the preferred frame on the non-observational, fundamental level

We are aiming to provide a space-time metric to QFT that does not involve adding a preferred frame to Minkowski space-time.

This can be most simply achieved by using some of the many Riemannian geometric and generally covariant formulations of both Minkowski and neo-Newtonian space-time. On the geometric theory, it is possible to obtain a Euclidean space and absolute time by subtracting most of the structure of Minkowski space-time.

In a Minkowski space-time, there is a light cone at each point p. Each world-line through p is an inertial frame and for each of these world-lines there is a distinct plane of simultaneity. Each of these planes of simultaneity (considered in abstraction from the other frames of Minkowski space-time) has a Euclidean spatial metric, the same as a plane in an absolute frame in Lorentz's space and time. Each world-line through p has as a set of successive planes of simultaneity and each of these world-lines has its own temporal metric, measuring the temporal distance or duration between its successive planes of simultaneity (e.g. see M. Friedman 1983: 126–127).

If we eliminate all but one of the world-lines through point p, the one remaining world-line will have a Euclidean spatial metric and an absolute time **t**, which is the one remaining frame. We eliminate its light cone structure (the conformal structure, a coefficient of the Minkowski metric $\mathbf{g}_{ij}$)) and extend the space-like region of the (former) light cone to a maximal Euclidean space. But now it is not space-like related to the point. There is no structure that precludes a non-local, instantaneous causal correlation from connecting any point on the plane of simultaneity to any other point on the plane of simultaneity. There is a 3 + 1 Euclidean space and absolute time whose Lie group is O^3 plus **T** (for an explanation, see Friedman 1983), which is the absolute frame of neo-Lorentzian space and time. The Minkowski metric $\boldsymbol{g}$ is dependent on the existence of the other inertial frames and the Minkowski metric cannot exist if there is only one inertial frame, the absolute Euclidean 3D space and the independent Newtonian time **T**.

But this not yet a complete description of a neo-Lorentzian absolute space and time. A neo-Lorentzian space and time contains moving bodies whose nature is described in terms of the absolute rest frame coordinate system. The laws describe in part how moving bodies have lengths, retardations, etc in the absolute frame; it describes the kinds of changes that are associated with kinds of movements.

But how could we describe the real laws if we have no local experimental evidence about whether we are on the absolute rest frame? It is true that we *may be* on the absolute, neo-Lorentz frame. Accordingly, if the observer assumes she is on the absolute frame and records moving systems as absolutely distorted and as bodies moving in the absolute frame. This neo-Lorentzian formulation is based on the symmetry of cause to effect described in Part 1 in Lorentz's theory, that what the observer is assuming to be the absolute frame is similar in respect of its laws to the absolute frame, so she can grasp the nature of the real absolute laws, the nature of the laws stating the kinds of *proportionate dependence* of the length, rate of processes and amount of mass of a moving body to the absolute velocity of the body, its velocity relative to the absolute rest frame.

Bohm and Hiley (1993) mention a neo-Lorentz QFT in general terms but it is hard to avoid the impression that their theory is indefensible; they claim the microwave background radiation is the "ether" that causes contraction and retardation. Their theory may seem hard to evaluate since no details or equations are offered, but one does not need these since the presented statements are sufficient for an evaluation. The astronomical estimates tend to vary, but the figure of 340,000 years after the big bang is often identified as the time when the conditions of expansion of the material universe were suitable for the formation, "release" or "beginning to exist" of the background microwave radiation. The problems with this theory are many; for example, many elementary particles moved prior to this time and yet the (effective) light cone structure existed prior to this, which, even if "effective," requires the contraction and retardation effects (relative to the absolute frame). More importantly, the background microwave radiation is not perfectly homogeneous and isotropic and this leads to the testable prediction that similar, moving bodies have degrees of contraction and retardation that vary proportionally to the variation in homogeneity of the radiation, a prediction falsified by the numerous tests that have been said to "confirm STR," which we take to mean "confirm Lorentzian STR." Third, the background radiation is more intense the in the more distant past, and if Bohm and Hiley's theory is correct, what we should observe is that this radiation puts up a correspondingly stronger resistance to movement and consequently causes greater contraction and retardation, a prediction falsified by our astronomical observations of distant stellar and galactic systems.

Furthermore, the ether is the medium of electromagnetic propogation, which includes microwave propogation, and the medium of microwave propogation cannot be microwave propogation, since the medium through which some **x** travels is different than **x**. If microwave radiation is the medium of microwave propogation, this implies both **x** = **x** and the negation of **x** = **x**. The microwave radiation, further, is said to be at rest with respect to the expansion of space. This seems to suggest something analogous to Lorentz's thesis that the ether is at rest with respect to space. But Bohm and Hiley cannot both accept and reject GTR; their classical forces V are Newtonian, not general relativist. On neo-Newtonian cosmology, microwave radiation moves through space; space does not expand. Rather, basic large-scale clusters of galaxies recede from each other within this stationary Euclidean space. One of the neo-Lorentzians, Builder (1958), seems to put forward a more plausible thesis when he says the neo-Lorentzian ether should be identified with space. But this is more ambiguous than might appear; there is not merely the question of whether space causes the contraction or whether it is a body's state of movement that causes its other state of contraction (and the other effects, such as retardation) Rather, we have multiple possible positions available corresponding to the many definitions of causality in the philosophical literature.

Bohmian QFT should not accept Bohm's view of the Heisenberg uncertainty principles. We must reject Bohm's realist interpretation of the Heisenberg uncertainty principle and his groundless claim that there is an entirely new and unknown physics governing the behavior of processes that occur at the Planck dimensions. However, I think we should go beyond merely reporting our disagreement; these two theses are logically invalid in the context of the Bohm-de Broglie interpretation. A de Broglie-Bohm type of interpretation of QM and QGT must give the uncertainty principles a merely epistemic interpretation; a realist interpretation results in an inconsistent theory. For example, one of the inconsistencies that would arise concerns the fact that the Q-potential acts instantaneously, which imply the laws governing the Q-potential apply at times shorter than the Planck time 10^{-43} second, and, second, this implies that there is a time shorter than the Planck time. Our quantum laws (e.g. the Shrödinger equation) and the Q-potential govern quantum processes at each instant, where an instant has zero duration, ruling out as logically inconsistent the Bohm-Hiley interpretation that QM does not apply but that, instead, "new unknown physics" apply, to processes shorter than 10^{-43} second.

We need a neo-Lorentz acceleration equation, since our aim is to explain both non-gravitational non-inertial motions and gravitational acceleration. A significant early example is Lorentz's acceleration equations (1904), whose equations continue to be born out by experimental evidence (Bell 2000). Lorentz developed an equation for *smoothly accelerating* moving particles in an electric field, in particular for any electron ***e***. Lorentz offers the equation $\mathbf{dr}/\mathbf{dt}\boldsymbol{e} = \mathbf{p}/\sqrt{\boldsymbol{m}^2} + \mathbf{p}^2\mathbf{c}^{-2}$. Here **r*e*** is the position of the electron, ***m*** is matter and **p** is momentum. Lorentz definitely did not consider this as gravitational acceleration (Lorentz 1910), but as electromotor acceleration, but his notion of acceleration is verified (Bell 1987). Other work on the acceleration of electrons was done by Lienard (1898), "Champ Electrique et Magnetique produit par une change concentree en un point et animee d'un movement qualconque," Eclairage Electrique, Vol. 16, pp. 5, 53, 106 and Wichert (1900) "Elektrodynamische Elementargesetz," Arch. Neerll. 5 (2), 1900, p. 549. A. Lienard's and Wichert's works are discussed at length in W.K.H. Pnanofsky and M. Phillips, *Classical Electricity and Magnetism*, Addison-Wesley, 1964, eqs. 20–13, 2–15. J.S. Bell (1987) summarized some of Lienard's, Wichert's and Lorentz's acceleration equations and developed his own neo-Lorentzian laws for acceleration. Bell briefly stated these laws or their more pertinent aspects (1987, 2004) and indicated they are highly confirmed by the observational evidence of contraction-retardation effects of smooth accelerations (Bell 2004: 78–79).

Bell presents an equation, using r and **r** as different symbols. The distance r of an accelerated particle from its rest source **r** can be calculated if we assume the source **r** has been at rest for some time and that r is initially just the instantaneous distance to the source **r**. Here r is the distance of the electron from the rest position **r**. **N** is the neutron. The electron ***e*** is gently accelerating away from the rest position **r**. Initially r is the instantaneous distance to the position **r** but the equations tell us how to keep track of it as it changes. The distance r subsequently changes in a law-like way, which can be seen by integrating the differential equation

$$\mathbf{dr/dt} = \mathbf{s}^{-1}\mathbf{1r}(\mathbf{dr}_e/\mathbf{dt} - [\boldsymbol{v}]) \qquad \text{(Eq.1)}$$

which follows from

$$\mathrm{r}^2 = (\mathbf{r}_\mathrm{e} - [\mathbf{r}_\mathrm{N}])(\mathbf{r}_e - [\mathbf{r}_\mathrm{N}]) \qquad \text{(Eq. 2)}$$

on differentiating with respect to time,

$$\mathbf{d/dt}[\mathbf{r}_\mathrm{N}] = [\mathbf{v}](1 - \mathbf{dr/cdt}) \qquad \text{(Eq. 3)}$$

For much larger accelerations, there is different mathematics. Bell verifies that the acceleration transformations work with an example of a high-velocity acceleration. Bell incorporates some quantum mechanical ideas (which in my case will naturally follow from a version of Bohmian QM, which I explain in a later section, so I avoid QM in this section apart from explaining Bell's equation). Bell uses as an example a hydrogen atom (**Z** = 1) with a Bohr radius:

$$\boldsymbol{h}\left(\boldsymbol{mcZa}^{-1}\sqrt{1} - (\boldsymbol{Za})^2\right) \qquad [9]$$

Here ***a*** is the fine structure constant, 1/137, of photons, which (being the bosons constitutive of light) travel at the velocity ***c***. The fraction 1/137 is the strength of the electromagnetic force relative to the strong force, set at 1. Bell notes that the hydrogen atom **Z** can have much higher values, e.g. **Z** = 70, and the Lorentzian contraction and retardation affects can be obtained.

These equations for the smooth acceleration of the orbits of an electron (Lorentz's equation) and Bell's more complete and general equation for smooth orbital acceleration, and Bell's equation for larger orbital accelerations where **Z** = 70, are verifiable for some accelerations, as Bell indicates. But we need something more general for a general theory of acceleration and for a neo-Lorentz theory of gravitational acceleration. It will appear in the next section that a classical theory of gravity is most easily quantized by a Bohmian QM. Avoiding for now QM notions, let us see what we can do to obtain a general neo-Lorentz theory of gravity.

I do not think we can just say "we will add some sort of gravity in the quantum potential **Q**" since the question at issue is "what gravity"? I do not follow the path of Pitowsky (1991) in his discussion of quantizing GTR in "Bohm's Quantum Potentials and Quantum Gravity". He describes a situation where the only classical potential **V** present is that of gravity, where **G** is the gravitational constant:

$$\mathbf{V} = \mathbf{G}m/\left[\mathbf{x}_1 - \mathbf{x}_{2,}\right]$$

Without the quantum potential **Q** present, the two particles will move towards each other. But with the addition of **Q**,

> the sum of the effective gravitational potential [is] the sum of the classical potential in the equation and the quantum potential **Q**. We see that effective gravity is no longer attractive. Unlike classical gravity, the effective force depends not just on masses and distances, but also on the quantum state. ... This means it might be possible to absorb the effective gravitational force in the general relativist framework"
>
> (Itamar Pitowsky (1991) "Bohm's Quantum Potentials and Quantum Gravity," *Foundations of Physics*, 21 (3), 345)

But this is a suggestion of what is epistemologically possible in a QG theory; it is not itself such a theory. We do not even learn the nature of the "effective gravitational potential". Nor do we need the sort of effective gravitational potential that Pitowsky has in mind, since we are jettisoning GTR.

Both Pitowsky (1991) and Callender and Weingard (1994, 1997), among many others, believe a path integral over GTR spaces and matter fields is desirable for a unified GTR and Bohmian QM. This cannot possibly work, since the GTR spaces and matter fields do not permit non-local, space-like, instantaneous, causal connections. Regarding Callender's and Weingard's (1994, 1997), I would note that Einstein's GTR 3-spaces or histories of ordered 3-spaces are defined in terms of their light cones, their relativistic metric, time-like curves and causal lines. The non-local, space-like causal correlations and the continuum-many, temporally successive Bohmian instantaneous simultaneity relations that relate instantaneously non-local, space-like, distant spatial regions completely shatter the Einstein GTR structure of each and every 3-space. Everything (or almost everything) in Einstein's GTR is destroyed (or, less metaphorically) is mathematically, logically and physically inconsistent with these Bell-Bohm relations. Einstein says each GTR 3-space (we are dealing with those that are like or approximately like Friedman's 3-spaces) has a light cone structure that sets the conditions upon causality, space and time. The limit upon causality is that it has to be time-like or null curve (if "time-like" is no longer the right word, say the inside of the light cone that is bound by the light cone). There cannot be a causal relation where the cause is space-like related to the effect. The limit upon space is that no spatial region can have a causal relation to another spatial region that is non-local. The limit upon "time" is that a hypothetical observer's light clock determines the measured simultaneity relations. With Bell-Bohm, causality is frequently not time-like or null-like; it is space-like. With Bell-Bohm, each spatial region is instantaneously and non-locally causally connected to every other spatial region on that instantaneous 3-space. With Bell and Bohm, time is not measured by light signals, but is an absolute, instantaneous "equatable flow" on the absolute plane that is undetectable by methods used in GTR. Furthermore, the GTR metric is a metric for *space-time*, not for a 3-space or an order of 3-spaces. Time is the radius of the successive 3-spaces, or the density of matter in these 3-spaces, or a component of the curvature of space-time. Callender and Weingard talk about *curved spaces*. But there is no such thing in GTR. What is curved is space-time, not space.

I conclude that it is clear that Bell-Bohm theory cannot be made consistent or unified with Einstein's GTR. Bohmians are looking in the wrong direction. They should be looking at a non-dynamical empty Euclidean space into which a material universe expands (see the last section on cosmology), and they should have an absolute time and space that coincides with a Bell-Bohm-neo-Lorentz theory. The major task is to fill out the multiple details and refinements of the Bohm-neo-Lorentz QG theory I am outlining in this essay. I believe the essential ideas are here. There is here a QG theory that has overcome any major obstacles, that is, "major obstacles" in the sense of those confronting Einstein GTR's unification with QM, which no one yet has any idea of how to overcome. There are no such obstacles in a Bohm-neo-Lorentz QG theory. In a manner of speaking, its accomplishment is just as easy as the various attempts at Einstein-QM unification are hard (or, rather, impossible). What remains are the numerous minor obstacles that consist in developing this theory further, refining it, and working out ever more precise, physically interpreted mathematical equations.

We need some more general ideas about neo-Lorentzian acceleration than the ones obtained from Bell. Instead of using some GTR and some Newtonian gravity theory as source material for mathematics, could we somehow combine GTR and Newtonian gravity theory? Bondi thinks this can be done. There is a theory of gravity that is both non-Newtonian and non-Einsteinian if a gravitational field is defined in terms of the *relative acceleration* of neighboring particles. Newton's theory is that a gravitational field is the gradient of the gravitational potential V. Suppose we adopt some suggestions made by Bondi (1979) for a theory of gravity that differs from both Newton's and Einstein's, although involving some mathematics used in both theories. There is a basic idea of Bondi's that we will be able to use, namely that the gravitational field is the relative acceleration of two particles. But it remains to be seen if Bondi's equations allow for an acceptable neo-Lorentz theory of gravity. Bondi uses Newton's and Einstein's mathematics, but not anything Lorentzian in nature. Using a tensor formulation, Bondi identifies the tensor that represents the gravitational field as the tensor linking the relative acceleration vector of neighboring particles to their separation vector. He introduces a relation between the relative acceleration vector $\boldsymbol{\delta f}^{\mathrm{i}}$ and the relative displacement vector $\boldsymbol{\delta}\mathbf{x}^{\mathrm{j}}$

$$\boldsymbol{\delta f}^{\mathrm{i}} = \mathbf{S}^{\mathrm{i}}_{\mathrm{j}}\boldsymbol{\delta}\mathbf{x}^{\mathrm{j}} \qquad [12]$$

Here $\boldsymbol{f}$ is a function of force, accelerating particles. The tensor of rank 2 $\mathbf{S}^{\mathrm{i}}_{\mathrm{j}}$ wholly characterizes the gravitational field. This enables the gravitational field to be characterized differently than in Newton's equations, where $\partial\mathbf{V}/\partial\mathbf{x}^{\mathrm{i}}$ represents the gravitational field. Bondi notes that two conditions need to be applied to **S**. **S** is the second derivative of a scalar ($-\mathbf{V}$), establishing the conservative character of the field,

$$\mathbf{S}_{\mathrm{ij}} = \partial^2\mathbf{V}/\partial\mathbf{x}^{\mathrm{i}}\partial\mathbf{x}^{\mathrm{j}} \qquad [13]$$

The source of the field is the density $\boldsymbol{p}$ of matter, given by the familiar Poisson's equation, where **G** is the constant of gravitation.

$$\mathbf{S}^{\mathrm{i}}_{\mathrm{j}} = 4\pi\mathbf{G}\boldsymbol{p} \qquad [14]$$

Now Bondi faces the problem that restrictions must be added to provide the Lorentzian transformational equations (but without having Lorentzian theory or ontology in mind). He tries to solve this problem in terms of the curvature tensor R, regarded as symmetric in (**i**, **j**) and (**k**, **l**) and anti-symmetric in (**i**, **k**) to ensure that the force function $\boldsymbol{f}$ cannot accelerate particles through the velocity of light and also to manifest the Lorentzian contraction and retardation effects. But the problem with introducing the curvature tensor **R** in this manner is that it eliminates absolute (instantaneous, space-like) simultaneity; the space-time becomes relativistic. What we want is an absolute space and time where the "relativistic effects" are due to the movement of bodies in the absolute rest frame. Bondi's gravity theory is *really* relativistic, as distinct from *being really absolute and only apparently relativistic due to the symmetrical contraction effects caused by movement in absolute space.* A neo-Lorentzian gravity theory should give us merely apparent relativist effects. Bondi's space-time is inconsistent with the absolute frame revealed in the Aspect experiments.

Further, Bondi's neo-Lorentzian theory does not distinguish between gravitational acceleration and non-gravitational acceleration; in fact, he explicitly identifies gravity with acceleration. This is false since a body passing through a magnetic field accelerates towards the magnet and this is not gravitational acceleration. The acceleration is due to the interchange of photons between the magnet and body; it is not due to the gravitons emitted from the magnetic. We recall that in the standard model photons are said to "carry the electromagnetic force"; weak vector bosons "carry the weak force," gluons "carry the strong force" (and in an extension of the standard model, even if only based on an argument from analogy, gravitons "carry the gravitational force"). It is already known that a "gravitons" vs. "curvature" interpretation of GTR are mathematically and observationally equivalent (but ontologically different), but gravitons can be used in a neo-Lorentzian gravity theory, but not "spacetime curves," which presupposes the falsity of a neo-Lorentzian theory of space and time.

Nonetheless, we can develop a neo-Lorentzian gravitational theory where the gravitational field is the relative gravitational acceleration of two particles and not the gradient of Newton's gravitational potential **V**. In fact, one of Bondi's problems was to introduce GTR formulae into a reformulation of Newton's theory. It seems to me the clearest path is to do the minimum that is necessary to generalize Lorentz's transformation equation for inertial motion into an equation for acceleration and then proceed to add the conditions necessary to make it a neo-Lorentzian gravitational acceleration.

Let us begin with the familiar **x**, **y**, **z**, **t** as the absolute spatial and temporal coordinates of an event at the position vector P, where P = **x**, **y**, **z**, which is an event's spatial coordinates at some time **t**. The coordinates **x**, **y** and **z** are the three components of the position vector P. We are describing the coordinates composing P on a body that is moving uniformly in the absolute rest frame. The velocity of an body is a velocity vector V, a function whose components are **x**, **y**, **z**, **t**, $\left[\sqrt{\ } 1 - (\mathbf{dx/dt}, \mathbf{dy/dt}, \mathbf{dz/dt})^2/\mathbf{c}^2\right]$.

Consider the variable **x**, the variable for spatial coordinates on the **x**-axis. The rate of change of a body's **x** coordinate, its change from one value on the **x**-axis to another, is mathematically represented by a fraction whose numerator is the differentiation of **x** with respect to time, **dx/dt**. The denominator of the fraction gives the object's velocity along the **x**-axis,

$$\frac{\mathbf{V} = \mathbf{dx/dt}}{\left[\sqrt{\ } 1 - (\mathbf{dx/dt})^2/\mathbf{c}^2\right]}. \qquad \text{(Eq. 15)}$$

The velocity vector is a function which orders its components in the following way:

$$\frac{\mathbf{u} = \mathbf{dx/dt}, \mathbf{v} = \mathbf{dy/dt}, \mathbf{w} = \mathbf{dz/dt}.}{\sqrt{\ } 1 - (\mathbf{dx/dt},)^2/\mathbf{c}^2 \left[\sqrt{\ } 1 - (\mathbf{dx/dt})^2/\mathbf{c}^2\right] \left[\sqrt{\ } 1 - (\mathbf{dw/dt})^2/\mathbf{c}^2\right]} \qquad \text{(Eq. 16)}$$

This equation can be simplified as

$$\frac{V = \mathbf{dP/dt}}{\left[\sqrt{}1-(\mathbf{dP/dt})^2/\mathbf{c}^2\right]} \qquad \text{(Eq. 15)}$$

The rate of change of an object's velocity **V** is the object's acceleration. An accelerating (or decelerating) object is moving at a velocity that is increasing or decreasing. These basics enable us to formulate an acceleration function for the numerator of the fraction, the function being **A (P, V, t)**, where **A** is a function of the position vector **P**, the velocity vector **V** and the scalar **t**. The acceleration function **A(X, V, t)** has the components

$$\mathrm{du}/\mathbf{dt}, \mathrm{dv}/\mathbf{dt}, \mathrm{dw}/\mathbf{dt}.$$

These components are the number of the fraction representing the acceleration of the body's velocity along the **x** axis, namely, the numerator (which is itself a fraction) **du/dt**, which differentiates velocity component u with respect to time. Since $\mathbf{u} = \mathbf{dx/dt}$, the numerator of the fraction representing the acceleration along the **x** axis is $\mathbf{A} = \mathrm{d}^2\mathbf{x}/\mathbf{dt}^2\mathrm{d}$, that is, the position variable **x** differentiated twice with respect to time. This information about position, velocity and time, enables us to formulate a numerator of a kinematic acceleration equation, i.e. an equation that does not take into account the dynamical aspects of acceleration, such as its cause. The numerator of the kinematic acceleration equation is:

$$\mathbf{A} = \mathbf{dV/dt} = \mathbf{du/dt}, \mathbf{dv/dt}, \mathbf{dw/dt}.$$

Neo-Lorentzian physics requires that the acceleration of a body in the absolute frame includes a transformation equation relating the absolute accelerating body to this body considered as at rest in the absolute frame, where the body's length is not contracted and its clock's motions and other physical activities are not slowed down. This compares the absolute rest coordinates of the body to its contracted and retarded coordinates in acceleration motion. Since a body considered at rest has no absolute velocity, $\mathbf{V} = 0$, and thus no absolute acceleration, $\mathbf{dV/dt} = 0$. Bringing in the neo-Lorentzian transformation equation, we have the complete fraction,

$$\frac{0}{\left[\sqrt{}1-(\mathbf{dV/dt})^2/\mathbf{c}^2\right]}.$$

But is not a fraction but is an undefined construction, since 0 divided by any number is 0.

The only components left are its position coordinates **x**, **y**, **z**, at an absolute time **t**. The real, absolute rest spatial and temporal coordinates of the resting body **S′** can be obtained from its spatial and temporal coordinates in its acceleration function **A(P**, **V**, **t**, **L)**, where **L** is one of our neo-Lorentzian transformation equations. (Recall that by a "neo-Lorentzian transformation equation" I do not mean one of "Lorentz's transformation equations". I am not here doing a "history of physics" that recounts what Lorentz wrote a hundred years ago or so. Rather, I am developing, modifying, adding to his theory, as well as retaining some parts of it. Accordingly "one of our neo-Lorentz equations" does not mean "one of Lorentz's equations".) The acceleration function is related to the body's "absolute rest" position function P' (**x′**, **y′**, **z′**) at time **t′**, when the body is at, or at least is hypothetically considered to be. The "approximately equal" symbol ≈ will be used to relate the absolute rest coordinates **x′**, **y′**, **z′**, **t′** to the coordinates as they are contracted or retarded by the body's acceleration. Since there is only one real coordinate system, the one that obtains in absolute rest space, the description of the resting and accelerating bodies will use the same coordinate system variables.

Since we need to construct some sort of neo-Lorentz transformation equation, we put

$$\frac{(\mathrm{RD})\left[(\mathbf{x}', \mathbf{y}', \mathbf{z}'), \text{at time } \mathbf{t}')\right] \approx (\mathrm{d}^2\mathbf{x}/\mathbf{dt}^2, \mathrm{d}^2\mathbf{y}/\mathbf{dt}^2, \mathrm{d}^2\mathbf{y}/\mathbf{dt}^2,)}{\sqrt{}1-(\mathrm{d}^2\mathbf{x}/\mathbf{dt}^2, \mathrm{d}^2\mathbf{y}/\mathbf{dt}, \mathrm{d}^2\mathbf{z}/\mathbf{dt}^2)^2/\mathbf{c}^2}$$

The approximate equality ≈ measures the degree to which the real rest coordinates **x′**, **y′**, **z′**, **t′** are distorted by the acceleration. An alternate distortion equation is

$$\frac{(\mathrm{RD}^*)(\mathbf{x}', \mathbf{y}', \mathbf{z}', \mathbf{t}') \gg \mathrm{d}/\mathbf{dt}(\mathbf{dx}/\mathbf{dt}, \mathbf{dy}/\mathbf{dt}, \mathbf{dz}/\mathbf{dt})}{\sqrt{}1-(\mathbf{dx}/\mathbf{dt}, \mathbf{dy}/\mathbf{dt}, \mathbf{dz}/\mathbf{dt})^2/\mathbf{c}^2}$$

In this equation as well, the right hand side of the approximation equation measures the magnitude of the distortion of the real, rest coordinates of the body. The acceleration equation can be abbreviated as

$$\frac{\mathrm{A} = \mathbf{dV}/\mathbf{dt}}{\sqrt{}1-\mathbf{V}^2/\mathbf{c}^2}$$

and if this is substituted in the right hand side of (RD) or (RD*), it will be the right hand side of the acceleration distortion equation. Note that these equations, the various (RD) equations, are not transformation equations; nor are they equations that Lorentz or neo-Lorentzians used. Lorentz used, as I explain in the first part of the paper, the "Fitzgerald-Lorentz-Lamor equations," with the Galilean equation as an intermediary.

We can derive phenomenal distortion equations if we imagine there to be a hypothetical observer on each system or body, e.g. one on the body **S** that is absolutely at rest and one on the accelerating body **S′**, and further imagine that the observer **O′** is misinterpreting her contracted and retarded system as the system that is absolutely at rest. Suppose also that we have an observer **O** on **S** who cannot distinguish that his absolute rest frame, which is appearing to him veridically from a moving frame that is imagined to be at absolute rest. In this case we have phenomenal acceleration equations. The observer **S** on the absolutely accelerating frame imagines himself as being on the rest frame described by the left hand side of the distortion equation and imagines the absolutely resting observer to be really accelerating, thereby satisfying the right hand of the equation. Since movement in absolute rest space causes contraction and "clock" retardation, the velocity of light will appear to be **c** on absolutely moving bodies, insofar as there is a real or hypothetical observer who measures the velocity of light. Observer **S** will interpret **S′** as accelerating relative to himself, and **S** imagines himself to be at rest, so these equations are imagined to be satisfied:

O imagines his own system **S** as absolute rest: (**x**, **y**, **z**, **t**),

And **O** imagines that the system **S′** will have the coordinates of an absolutely accelerating system:

$$\frac{\mathrm{d}V'/\mathbf{dt}'}{\sqrt{\ }1-\mathbf{V}'^2/\mathbf{c}^2}$$

Bohm's equation of motion implies Newton's first and second laws of motion. To unify our classical (non-quantum) neo-Lorentzian theory with Bohm's QM, we can formulate a neo-Lorentzian law, a *dynamical* law of acceleration that is modeled on Newton's. This dynamical law includes both mass and the force producing the acceleration. By formulating a neo-Lorentzian equation, we can make the Bohm-Lorentz QG theory more adequate, although they will not be adequate until we add the quantum mechanical laws for Bohm's quantum potential **Q**. For now, we will formulate a classical dynamical law of acceleration and a law of gravity, both being neo-Lorentzian versions of Newton's laws, in conformity with the requirements of Bohm's 1952 interpretation of the de Broglie-Bohm theory. The "***m***" stands for mass and "$-\nabla(V_{A'})$" stands for the negative gradient of the acceleration potential (***V*a**). The "***V***" is used to make it correspond with Bohm's equation, where "***V***" is the classical potential variable and "**Q**" the quantum potential variable.

$$\frac{\mathrm{d}(\mathbf{m}'\mathbf{V}')/\mathbf{dt}' = -\nabla(\mathbf{V}_{\mathbf{A}'})}{\sqrt{\ }1-\mathbf{V}'^2/\mathbf{v}'+\mathbf{c}^2} \qquad \text{Eq. 16}$$

QG theory requires a specific form of acceleration, gravitational acceleration, and the addition of new gravitational variables into the acceleration equation will provide us with a classical gravity equation that appears in the classical part of Bohm's equations of motion, namely, gravitational motion. I am including some neo-Lorentzian equations to the classical part of Bohm's law of motion.

Is gravity a force or a field or a boson or something else? Whether gravity is a force or field or values of a field, and whether or not field values are identical with a system of gravitons (the fourth type of boson, in addition to photons, weak vector bosons, and gluons) moving at the speed of light c, and whether or not a gravitational force is identical with the force carried by gravitons, is an issue parallel to a similar issue about the electromagnetic field. Is electromagnetism an "electromagnetic *field*," *high values* of an electromagnetic field, a configuration of photons, or a configuration of photons and electrons, or is it an "electromagnetic *force*," or "*photons that carry the electromagnetic force*, for example, the force carried by the photons that are emitted from an electron and bind the electron to the protons in the nucleus of an atom, such that the interchange of photons between the electron and protons carry the electron in an orbit around the nucleus of the atom. These are issues that need not be decided here. The physical ontology is underdetermined by the conjunction of the observational evidence and the equations. Some physicists, such as Valentini, and philosophers of physics, such as Callender-Weingard, believe the ontology is not underdetermined and they argue for varying sorts of ontologies, but I will not enter that debate here and I will assume an ontological underdetermination. I use "gravitational field," "values of a gravitational field," "graviton configuration" and "gravitons carrying a gravitational force," and "gravity force **F**" as different ways of referring to gravity and I do not commit myself to an ontology that might seem to be associated with some of these expressions.

A gravitational acceleration is distinguished by the source of the gravitational force or field values and by the nature of the force or values that cause the acceleration.

Let "gravity" mean the force exerted on the mass $\boldsymbol{m}$ by the distribution of other masses (we use only masses and their distribution for the sake of simplification) $\boldsymbol{m_i}$ at their various positions $\mathbf{X}_i$, such that in our gravitational acceleration equation we are considering $\boldsymbol{m}$ as subject only to the gravitational force. Let F be a gravity force acting on a mass $\boldsymbol{m}$ that is moving in absolute space. Since the gravity force is acting on mass $\boldsymbol{m}$, and since its source is the distributions or positions of the other masses, we need to indicate this in our gravitational acceleration equation. The acceleration of any mass is gravitational if the force **F** or the gravitational potential of the field is what is stated on the right hand side of the equation; we will assume that $\boldsymbol{m_1}$ is the mass of a body at rest relative to absolute space and that $\boldsymbol{m'_2}$ is the mass of a body that is moving towards $\boldsymbol{m_1}$. The unprimed $\boldsymbol{m_2}$ indicates that it is the rest mass of the body and the primed $\boldsymbol{m_2}$ is the mass distorted by its movement. The neo-Lorentz law of gravity then takes this form, where V_g denotes the classical gravitational potential, and $\mathbf{L}^2$ is the distance or "Length" between the two masses.

$$\frac{\mathbf{G}m_1m_2'}{\mathbf{L}^2}\frac{(\sqrt{}1+\mathbf{V}/\mathbf{c})}{1-\mathbf{V}'/\mathbf{c}} = -\nabla(V_g) \qquad \text{Eq. 19}$$

If two bodies with mass m_1' and m_2' are absolutely gravitational accelerating in the same direction and at the same rate $\mathbf{dV}'$ and time $\mathbf{t}'$ our neo-Lorentz law of gravity takes the form, where distance between the two masses is written in a different form,

$$\frac{\mathbf{G}(\mathbf{m}_1'(\mathbf{dV}'/\mathbf{dt}'))(\mathbf{m}_2'(\mathbf{dV}'/\mathbf{dt}'))}{[\mathbf{P}_1^{\not c}(\mathbf{dV}'/\mathbf{dt}')\mathbf{P}_2^{\not c}(\mathbf{dV}'/\mathbf{d}')]} = -\nabla\mathbf{S}\nabla(V_{g'}) \qquad \text{Eq. 20}$$

It is clear that these are gravitational laws based on a neo-Lorentzian transformation of a Newtonian rather than an Einsteinian theory of gravity. One finds no indication here of Einstein's equation, which in a simplified form reads $\mathbf{G}_{\mathbf{uv}} = -(8\pi\mathbf{G}/\mathbf{c}^2)\mathbf{T}_{\mathbf{uv}}$. One reason is that Newton's first and second laws of motion are contained in Bohm's equation of motion (Bohm 1952) and a second reason is that Einstein's theory is disconfirmed by the Bell/Aspect experiments but a neo-Lorentzian-Newtonian theory is not.

In Bohm's 1952 theory, the wave function is governed by the equation of motion, $\Psi = \mathbf{A}\exp[\mathbf{iS}(\mathbf{x})]$. The Schrödinger equation is rewritten as a modified Hamilton-Jacobi equation, namely,

$$\mathbf{dS}/\mathbf{dt} + (\nabla\mathbf{S})/2m + V + \mathbf{Q} = 0(1) \qquad \text{Eq. 21}$$

Here ∇ usually pertains to the gradient of something but in the present case it means, ***mv***, i.e., mass times velocity, i.e. momentum, ***p***. This is the equation of motion for the wave function. The quantum potential **Q** is a "quantum force" additional to the usual classical forces. There is an equation for **Q** and an equation for the particles (called beables by Bell since they exist even when not observed – Bell's substitute for observables).

The equation governing the particles is

$$m\mathbf{dx}/\mathbf{dt} = -\nabla(V) - \nabla(\mathbf{Q}) \qquad (2)$$

The quantum potential **Q** is

$$\mathbf{Q} = -\mathbf{h}^2\nabla^2\mathbf{R}/2m\mathbf{R} \qquad (3)$$

Bohm's 1952 QM is unified with neo-Lorentzian classical gravity in the following way.

The part of (2), $m\mathbf{dx/dt} = -\nabla(\mathbf{V})$ is a modification of the Newtonian second law of motion ($\mathbf{ma} = -\nabla(\mathbf{V})$) or, written otherwise, ($\mathbf{dp/dt} = -\nabla(\mathbf{V})$). Here m is mass, **a** is acceleration (in Newton's sense), **p** is momentum and $= -\nabla(\mathbf{V})$ is a classical force. Newton's second law of motion is second order and the Bohmian guidance equation is first order, and adds a quantum potential **Q** to the classical force **V**. The Bohmian equation of motion contains Newton's second law, since if $\mathbf{Q} = 0$, then Newton's law governs the particles, $\mathbf{dp/dt} = -\nabla(\mathbf{V})$, or, if you prefer, $\mathbf{ma} = -\nabla(\mathbf{V})$.

We know that Newton's second law is disconfirmed by observations of bodies moving at velocities close to that of light. The so-called "relativistic effects" are relative to absolute rest space and pertain to the contraction of the length, retardation of clock, increase in mass, etc., which are causal consequences of moving in absolute rest space. Since Bohmian theory has a distinction between the absolute frame, where non-local, space-like, instantaneous causal EPR correlations are propagated, and relative phenomenal frames, where we cannot normally observe EPR correlations (unless we are performing an Aspect experiment) we normally do not know which frame is absolute. This parallels neo-Lorentzian ontology, with Bohm adding the **Q**-potential to the absolute frame and Lorentz adding the contraction of bodies that move relative to the absolute rest frame.

Bohm's equations of motion imply Newton's law of acceleration and we can deal with its failure to make correct predictions of "relativistic effects" either by rewriting it as a relativist equation or by substituting for it our neo-Lorentzian acceleration equation. Weinberg (1972: 31) formulates a relativist version of Newton's second law but he does so in terms of the mathematics of a Minkowski solution to Einstein's equation. But this leads us back into Einstein's GTR and an inconsistency with QM. We can write a classical acceleration law that predicts the relativistic effects and yet is not a relativity law but a neo-Lorentzian law, which implies an absolute frame and the sort of absolute simultaneity that is consistent with Bohm's theory. The neo-Lorentzian law of acceleration would have greater influence on particles' behavior when the value of **Q** decreases and less effect when the value of **Q** is high.

Bohm's laws of motion include both a classical law of inertial motion and a classical law of acceleration motion. Since our classical laws of motion are neo-Lorentzian, they already put Bohm's theory in a form where it can be unified with a classical theory that predicts relativistic effects. The Bohmian equation of motion (in its Newtonian rather than Lorentzian form) would be reformulated in a Lorentzian manner and this would put Bohmian theory in a form where its fundamental level would be the form it has in Bohmian quantum field theory (with Euclidean space and an absolute time). In Bohmian quantum electrodynamics, electrons and photons are unified with neo-Lorentzian theory, in the way I discussed regarding Callender-Weingard's and Valentini's approach to Bohmian quantum electrodynamics. As I have indicated, I believe talk of "fields" can be translated into talk of fermions and bosons, e.g. electrons and photons. We can say that the laws of motion in QED are laws for changes in field values over time, or we can say they are laws for changes of electrons and photons, suggesting electrons and photons are individual particles, fermions and bosons. Quantum field theory is about a subset of the particles, the subset (electrons, photons) in the case of QED and (quarks, gluons) in the case of QCD. In Bohmian QG, the subject is values of the gravitational field or gravitons. Gravitons are bosons, just as are photons (in QED) and gluons (in QCD). The fermions in QED are electrons and the fermions in QCD are quarks and the protons and neutrons that quarks compose. QG is about gravitons and not merely the fermions of QED and QCD; gravitons affect photons and gluons as well. The observational effects that gravitons cause are most clearly noticeable in large aggregations of fermions and bosons, such as planets, galaxies, clusters of galaxies and superclusters of clusters of galaxies. Each of these large aggregations is composed of atoms and each atom (except for the lightest, hydrogen) is composed of electrons, protons and neutrons; the aggregations are also composed of the photons that travel between the electrons and protons and bind the electrons in their orbits around the nucleus of the atom. The nucleus consists of protons and neutrons and these consist of quarks; quarks are bound together by gluons and (derivatively) protons and neutrons are bound together by gluons. It would be accurate to say that QG is about the gravitons that are interchanged by large organized composites of electrons, photons, quarks and gluons. The gravitational acceleration law is about the manner in which the gravitons interact with the aggregations of electrons, photons, quarks and gluons. This is oversimplifying of course and not merely for the reason that it would take several books to characterize Bohmian QED, let alone the other matters. There are also other bosons and fermions, such as weak vector bosons, mesons, and there are photons that do not remain inside an atom but travel across large spaces, such as from the sun to the earth.

Many details are left out of this outline, for example, how the neo-Lorentz law of gravitational acceleration accurately predicts the orbit of Mercury (but see Jannossy 1971 for a neo-Lorentzian equation that makes this prediction) and predicts that the paths of photons will curve towards stars as they pass them, due to gravitational effects (also see Jannossy 1971 for a neo-Lorentzian equation that makes this prediction). But we understand the basic idea about how Bohmian QM can be unified with a neo-Lorentzian theory of gravity and a neo-Lorentz GTR (General Theory of Reference frames, inertial and non-inertial). There are no mathematical, logical or ontological inconsistencies within this Bohmian-neo-Lorentzian QG theory. For example, we do not need to worry about reconciling the causal determinism and the space-time continuum of Einstein's GTR with the probabilistic and discrete nature of reality implied by a realist (rather than an epistemic) interpretation of Heisenberg's uncertainty principles or the non-Bohmian interpretations of QM that have no causally determinist laws of motion in addition to the Schrodinger equation. In Bohm's theory, the probabilities obtained from the Schrodinger equation are epistemic and do not belong to physical reality; there are no "probabilistic causes," but only the deterministic causes implied by both Bohm's equations of motion and the new Lorentzian equations of motion. Probabilities are merely a measure of our ignorance of the causally deterministic behavior that satisfies the Bohmian-Lorentzian equations of motion. The classical theory that is contained in Bohm's equations of motion imply an absolute time and space and the classical equations, besides being modified by the Q-potential, are reformulated as neo-Lorentzian classical equations, which both imply there is an absolute time and space and the relativistic effects. Nor do we have to worry about the perturbative non-renormalizability of solutions to the Wheeler-deWitt equation, e.g., the Hartle-Hawking solution and the Vilenkin solution (or, rather, approximations to a solution). The Hartle-Hawking and Vilenkin theories are about Einsteinian GTR space-times and the Wheeler-DeWitt equation describes GTR 3-spaces. There is also no need for string theory or M-theory, since there is no inconsistency that needs to be resolved, namely, the inconsistency between QM and Einstein's GTR. Einstein's GTR has been replaced by a neo-Lorentzian GTR and, if we unify this with Bohm's 1952 formulation of QM, the fundamental problems are not "resolved" but "dissolved".

The issue of whether or not a QG theory can make predictions is answered by the nature of the classical theory and quantum theory that are being unified. This QG theory makes approximately the same predictions as the classical theory does on large scales and the same predictions that quantum field theory does on subatomic scales. If there are new predictions, they are that the Planck dimensions, the Planck era, the Planck length, etc., are merely epistemic nature and that there is Bohmian-neo-Lorentzian physical reality smaller than these dimensions, all the way down to the spatial points and temporal instants in a spatial continuum and temporal continuum. The Planck dimensions are the limitations of our measuring equipment and prevent us from measuring these dimensions (at least until significant advancements are made in the area of measuring instruments utilizing EPR correlations).

Part three: Bohm-neo-Lorentz cosmology, present time, and modal necessetarionism

It is very often argued and noted that on the broadest cosmic level, Newton's laws are observationally equivalent to Friedman's solution to the Einstein equation in GTR. I will discuss this view later, but argue that neo-Newtonian cosmology (originated in 1931 by Lemaitre (1931) and developed by many others up to the present day) needs to be fundamentally changed to become a neo-Lorentz-Bohm cosmology, and that these changes render the cosmology observationally equivalent (in the mentioned respects) to Friedman cosmology.

In fact, these changes are required to have a mathematically consistent, realistic and observationally confirmed cosmology. But the substance of this section is on Bohm-neo-Lorentz cosmology, which is where I begin.

On the cosmological scale, a static homogeneous universe has been found to be either impossible or extremely problematic, given either Newton's or Einstein's theory. On Newton's static, infinite homogeneous universe, any motion brings matter collapsing towards the moving point. On Einstein's static homogeneous universe model (1917), motion also brinks about a collapse towards the moving point. A static infinitely old universe would also result in light from stars filling everyplace, e.g. "night" on earth would be impossible, given the amount of starlight that would have arrived (Olber's paradox). Prior to learning about the observational evidence (which began to be noted by Slipher in 1916 and was made more widely known by Hubble's publications in the 1930s), several thinkers had realized that the nature of gravity, light and the distribution of matter possess intrinsic characteristics that put meta-scientific constraints on any cosmology, be it Newtonian or Einsteinian. An inhomogeneous and anisotropic universe would, as a theoretical virtue be relevant to prior probabilities, without (yet) considering any observational evidence, would make any such cosmological they has much lower prior probability that a cosmological theory that postulated large scale isotropy and homogeneity. Anisotropy and inhomogeneity is highly non-symmetric and violates would the equivocally used word "simplicity" is sometimes used to express, namely, lack of arbitrariness. Determining prior probabilities, a cosmology that has the theoretical virtue of postulating a non-arbitrary ("simple" in this sense) structure of the universe has a greater prior probability than a theory that (for no reason or justification) postulates a universe that is highly arbitrary in its structure, e.g. inhomogeneous and anisotropic matter distributions. Given these several facts, there a cosmologist theory that postulates a universe that is homogenous, but non-static (due to the above-mentioned problems about static Einstein or Newtonian universes) would have much greater prior probability than a theory that postulated either an inhomogeneous universe (static or non-static) or a static and homogenous universe. If the matter in the universe is expanding or contracting on a large scale, it could be homogeneous and resolve the problems of a static universe, be the universe finite or infinite.

For various reasons, some thinkers began developing models of a universe in this sort prior to the observational evidence discovered by Slipher and Hubble. Some physicists realized that the assumption that the universe is static, an assumption adopted or "taken for granted" both by Newton in his cosmological thinking and in Einstein's model universe of (1917), should be rejected. A dynamic universe would solve the host of problems that bedeviled the static universe theories. If we call "the universe" the largest aggregate of matter, e.g. the whole composed of either billions, trillions or of an infinite number of superclusters of galaxies, then the universe must be either expanding or contracting. It appears the first person to recognize this is Friedman (1922, 1924), the second Lemaintre (1931), the third Milne (1933), the fourth McCrea (1934) and after this time the results of Hubble's observations were so well-known that the "Hubble Law," as it came to be regarded, as an observational law of our universe. But this is mistaken; it is a fundamental theoretical hypothesis, a hypothesized Cosmological Law in cosmology, developed by Friedman, Lemaitre, Milne and McCrea; Slipher and Hubble merely confirmed this hypothesized cosmological law. The basic Cosmological Law is that the universe is homogeneous and isotropic on large scales and is either expanding or contracting. The Law states that if the universe is expanding, then it obeys a certain distance-velocity law, that the apparent recession velocity of a large-scale system (e.g. a supercluster of cluster of galaxies) is proportionate to its distance.

Hubble's Law is

$$\boldsymbol{L}/\mathbf{d}(\mathbf{t}) = \boldsymbol{L}(\mathbf{t})/\mathbf{T}$$

Here $\boldsymbol{L}$ is the present distance of between the current large scale plane of homogeneity, or some large system, e.g. a supercluster of clusters of galaxies, on this plane, to some spatially distant (of distance **L**) and long-past large system on the plane of homogeneity that used to exist, say, 5 billion years ago. This distant system would have the spatial distance, $\boldsymbol{L}$ = 5 billion light years, from the present plane of homogeneity or a supercluster on this plane. In Hubble's Law, $\mathbf{L}/\mathbf{d}(\mathbf{t}) = \mathbf{L}(\mathbf{t})/\mathbf{T}$, the symbol $\mathbf{L}/\mathbf{d}(\mathbf{t})$ is the velocity of recession of the spatially distant supercluster, which is presently at distance **L** from the present plane of homogeneity, i.e. the plane at time **T**. Hubble's constant is 1/**T**, where **T** is the present time.

Hubble's constant 1/**T** specifies the present rate of mutual recession of the large scale homogenous and isotropic mass content, i.e., the how fast the superclusters *are* (present tense) receding from each other. The present value of the expansion rate (or of the Hubble constant) is approximately 75 km/sec/Mps; that is, 75,000 kilometers per second per megaparsec (where a megaparsec may be translated as "20 quintillion miles" (where quintillion, comes after quadrillion, and quadrillion comes after trillion, and trillion comes after billion, etc.)). In everyday language, the present rate of mutual recession of superclusters is approximately 50,000 miles per second per 20 quintillion miles.

Since we "observe the past," the plane at distance **L** is (we shall say) the plane of homogeneity that was present one billion years ago (with only the photons, etc., emitted from the superclusters still remaining in existence (existing in the present tense sense)). A photograph of this plane, which is 6 trillion miles away, and one billion years in the past, can be found in *Powers of Ten*, 1982, P. and P. Morrison, Freeman and Co. San Francisco. Such photographs provide an accurate image of a plane of isotropy and homogeneity as one reads about them in a text with no photographs.

The small **d** in Hubble's Law, $\boldsymbol{L}/\mathbf{d}(\mathbf{t}) = \mathbf{L}(\mathbf{t})/\mathbf{T}$, is the familiar differentiation symbol. ***L*** is not a "tenseless distance," but the present distance of a large-scale particle, a past supercluster, from a large-scale particle located on the present plane of homogeneity, the plane of homogeneity at the present time **T**. The semantic content of "**T**" in Hubble's law is the present time. Recall that this velocity-distance law is a basic law of neo-Lorentz cosmology and is not an observational law or statement. The earlier argument that the velocity-distance law is a basic cosmological law was an argument for this thesis. The semantic content of **T** is "whatever time is present" or "whatever time is the present time" and so no particular present time, such as 15 billion years after the big bang, is picked out as the present time. **T** ranges over all possible present times and picks out the one that is actually present.

T can be defined in past-tensed terms which also specifies the present time in relation to which some time is past (e.g. one billion years ago, i.e. one billion years earlier than whatever time is the present time). **T** can be defined in terms of a dating system as *how long ago* (tensed) the large scale particles (superclusters of clusters of galaxies) *have been* mutually receding from the big bang to the present.

In every physics article or book I have read, **T** is defined in presented tensed terms and is the only law in theoretical physics that is explicitly tensed. I will not quote one or two dozen passages from numerous sources as evidence of this, since it is obvious to physicists and philosophers can leaf through any book on cosmology or astronomy, or any article, and find "Hubble's Law" and see that it is defined in present tensed terms.

The fact that Hubble's Law is everywhere formulated in tensed terms is very strong *primae facie* evidence that there is an accepted (by consensus or virtual unanimity) a physically interpreted equation that is *about whatever is the present state of the universe* and that contains terms that refer to an *objective, mind-nondependent present time.* Stated in other's words, Hubble's law is accepted by all or virtually all physicists and in every case it is treated as the only law whose terms refer to the **A**-*time* in which the universe exists, or (stated in still other terms) that refers (by its sense; or, if you prefer, by the definition offered by physicists) to the objective, mind-nondependent present, past and future of the universe. Hubble's law entails *the tensed theory of time* (the **A**-theory of time, presentism, Newton's "flow" of time.) But more than prima facie evidence is needed to state that the tensed theory of time is posited by theoretical physics. I present some arguments for the irreducible tensed nature of Hubble's Law.

The tensed formulation of Hubble's Law is no "accident" that can easily be changed into a tenseless law by rewriting it. "**T**" denotes whatever time is present and this cannot be translated into tenseless language. The Hubble time **T** is a primitive or irreducibly tensed variable, since it cannot be stated in a tenseless language without rendering the Hubble law falsified. For example, on a date-version of the tenseless theory of time, "the present value of the Hubble time **T** is 15 billion years" would have for its truth conditions, truth-maker, (or be translatable by) a statement such as *The value of **T** at 15 billion years later the big bang is 15 billion years later then the big bang*, which renders Hubble's Law useless because its output is merely tautologies.

But if a tenseless time theorist tried to used a different version of the tenseless theory of time, the so-called "token-reflexive version," then the ontological significance of "the present value of the Hubble time, time **T**, is 15 billion years" would be *the value of **T** simultaneous with the sentence-token **S** is 15 billion years.* But then "the present value of the Hubble time T" becomes a temporal constant fixed at the time the sentence-token **S** occurs, and cannot be used to measure earlier or later values of **T**. More importantly, it provides no information about when the sentence-token **S** occurs, so astronomers at any time reading old astronomy texts would read, "the sentence-token **S** is simultaneous with the time that is 15 billion years after the big bang" and have no idea when that statement was made and would have no idea of whether or not the present time T is 15 billion years after the big bang. A series of such arguments showing that this tensed statement and similar ones cannot have their ontological significance stated or explained in tenseless terms can be found in Smith (1993), Craig (2000a, 2000b) and other works and I refer the reader to these texts.

Given the primitive, irreducible present tense in "the present value of **T**," the result is that the basic law of cosmology is an irreducibly present tensed law. Since this law is instantiated by the universe, the present time that is denoted by the present tense is intrinsic to the universe. This argument can reach the conclusion about tensed Bohmian QM that is my view and that was formulated (but not endorsed or argued for) by Maudlin:

> It is perhaps interesting to note that Bohm's theory [in addition to Lorentz's and Hubble's] is deeply congenial to an ontology which maintains that all which exists is that which exists now, i.e., at a point in time classically conceived. Instantaneous velocities can of course be defined, but only as the limit of average velocities over finite periods of time. Those puzzled about the status of velocities in an ontology in which only an instant of time exists can happily adopt a Bohmian ontology of particles (with positions) and the wave function.
>
> (Maudlin (1994) *Quantum Non-Locality and Relativity*, Oxford: Blackwell Publishers, p. 124)

The view that time is tenseless was adopted (as Minkowski explicitly stated) for mathematical convenience, not ontological purposes, in Minkowski (1908). Minkowski's assertion that he was assuming a tenseless time model for pragmatic purposes, not alethic purposes, later was ignored and his pragmatic model came to be misunderstood as "ontological model" that had alethic intention. The pragmatic and non-alethic model of a "tenseless spacetime view of the universe" in Minkowski's mathematical model (and its use in Einstein's GTR) led philosophers such as Russell, Smart, Quine and others to think that "the scientific view" is that time is tenseless. As this misunderstanding gradually became habitual, even physicists thought that "the scientific view of the universe is that it has tenseless time" was part of the definition of science, (forgetting that on a Newtonian or Lorentzian physics, time "flows" and time is not tenseless). Tenseless time is a peculiar artifact of the pragmatic models used in Minkowski-Einstein relativity theories, theories which have now been disconfirmed, by the Bell-Aspect experiments in favor of a Lorentzian science and its view of a "flow" of absolute time.

The basic cosmological law of distance/velocity, with its present value of Hubble's constant as an explicitly present-tense and present-time *defined* part of the law, shows the burden of proof is on the tenseless theorist to show why physical cosmology is mistaken and to engage in a (I believe) futile task of trying to show that the scientific used tensed expressions either can be "translated" into tenseless ones, or have tenseless truth conditions, or that the deep structure of all possible natural and artificial languages is tenseless and is isomorphic to the ontological structure of reality.

The velocity/distance basic cosmological law is part of neo-Lorentzian cosmology and neo-Lorentzian GTR (general theory of reference frames, both inertial and non-inertial). It describes the gravitational behavior of the universe at its largest scales. Lorentz posited an infinite, flat, non-dynamical, substantival, absolute, Euclidean rest space. We have retained this since it provides the best explanation of contraction and dilation (namely movement in this space causes contraction and dilation and other Lorentzian effects, e.g. increase in mass as velocity increases). We do not need an ether. The material universe is the whole system of the expanding superclusters of clusters of galaxies, and this system is growing into an arbitrarily large system of matter and energy, a system that is accelerating (or decelerating) relative to absolute rest space. The inertial frame is at rest relative to absolute space or, more precisely, *is* absolute space itself. Infinite Euclidean space is homogeneous and isotropic. This implies it is symmetrical to the large-scale structures of the expanding universe, in respect of homogeneity and isotropy. But Euclidean space is not expanding. Nor is there a non-Euclidean space that is expanding. There is no observational evidence for a "space expansion" hypothesis. What is observed are superclusters of clusters of galaxies receding from each other with a velocity that is proportional to its distance.

In order to "expand" in a Euclidean non-dynamical space, the material universe must expand into an empty part of Euclidean space. Euclidean space is what set theorists or philosophers of mathematics call an "actual infinite" (an idea derived, of course, from Cantor's transfinite mathematics) and the material universe is what Cantor calls a "variable finite" whose values can take on any arbitrarily high finite number. To take just one spatial dimension, e.g. the **x** axis, no matter how high one reaches on the real number line, a spatial size of the material universe is either below, at, or above that number. If the universe expands for ever, as present observational evidence (since the 1998 observations of two supernovae) suggests, then it is true that for any time **t** in the future, and any "size" or number on the real number line, the size of the universe will become larger than that size. This is symbolized by physicists by "$+\infty$", or "$-\infty$" which they call "infinite", which can be misleading by philosophers, who standardly use "infinite" in the sense of set-theorists, who use "infinite" to mean the number of all rational numbers, aleph-zero, $\aleph_0$, and continuum-many to mean the number of all real numbers. When physicists say the future is open and thus infinite, they do not mean the universe is infinite in the sense of having omega, ω, number of years later than the present year. Omega is the ordinal number of a set with $\aleph_0$ or aleph-zero members, ordered with the natural order of all positive whole numbers $\{1, 2, 3, 4, \ldots n, \ldots\}$; the transfinite ordinal ω is the order type of this set. Physicists mean the future is infinite in the senses that it has (in Cantor's terminology) a variable finite number of years, $+\infty$, meaning that for any finite number 10^{20} on the positive real line, the future is finitely longer than 10^{20} years. William Craig's use of the phrase "potentially infinite" is much closer in meaning to "infinite" than what his philosophical critics mean by "infinite", which in set-theory means a set with at least an aleph-zero number of years..

This is not the Milne-Mcrea (1934) neo-Newtonian cosmology, which has been further developed and modified by Bondi (1960), North (1965), Sciama (1971), Peebles (1993), Harrison (2000), Hartle (2003) and a number of others. Milne and Mcrea (1934) derived from Newtonian physics a cosmology quite different than Newton's own cosmology. Milne and Mcrea's neo-Newtonian cosmology implied there is an infinite Euclidean space that was filled homogeneously with matter, with the density of the matter changing with time. Parts of this universe of the appropriate size can be considered as spheres, which are expanding in the Euclidean space. The Euclidean space is considered to be non-dynamical (not expanding or curved) and at rest, with all matter in motion relative to it. A hypothetical observer in this sphere on a large "particle" that is moving with radial motion will observe that large systems of matter (e.g. superclusters) are receding from himself and their velocity of recession increases proportionately with the distance of the superclusters from himself. The sphere obeys Hubble's law of velocity-distance proportionality. If $\mathbf{M(r)}$ is the mass in the sphere of radius r and v the escape velocity from that mass, then the velocity needed for the "partice" to escape the sphere of mass $\mathbf{M(r)}$ is

$$(1/2)\mathbf{v}^2 = \mathbf{GM}(\mathbf{r})/\mathbf{r}$$

where $\mathbf{G}$ is the gravitational constant. The behavior of the sphere is described by an equation,

$$(\mathbf{dR/dt})^2 = 8\pi\mathbf{GpR}^2/3 - \mathbf{k})$$

where **p** is density and **k** the constant, where $\mathbf{k} = 0$ corresponds to a parabolic path of the particle, +1 to an elliptical path and −1 to a hyperbolic path. This is Friedman's equation, derived from Newtonian principles. Although this reflects the observational equivalence, the meaning of the terms is different in the Newtonian and Einsteinian interpretations. The constant **k** in the Newtonian interpretation refers to the total energy of the particles and in the Einsteinian interpretation it refers to curvature of space. The temporal variable refers to absolute time in the neo-Newtonian theory, but to Friedman's "cosmic time" in the Einsteinian interpretation. Harrison (1981: 238), a proponent of Friedman-Einstein cosmology, remarks about this observational equivalence: "In all its applications, Newtonian theory is only approximately true, and yet in this most unlikely of all cases it yields the correct answer." However, there are at least two insurmountable obstacles to considering this to be a realistic and consistent model. The neo-Newtonians, following Newtonian principles, estimate the gravitational potential in the sphere by ignoring the infinite amount of matter, the infinite number of superclusters, outside this sphere. Only in this way can the velocities of escape be calculated and can the sphere be considered to obey the neo-Newtonian interpretation of Friedman's equation. Secondly, each sphere consists of particles that do not obey the Newtonian laws of motion; they obey neo-Lorentzian laws or (as a proponent of general relativity would say) relativistic laws.

These and other problems are overcome on our neo-Lorentzian cosmology. Our infinite Euclidean space is not completely filled with matter. Rather, part of it is filled with matter, and this part, the material universe, is expanding into an empty Euclidean space. The material universe may be open or closed, as in Friedman's case. If it is closed, the expanding superclusters will eventually begin contracting and contract at an ever increasing rate until they "crash into each other" in a "big crunch". The metaphorical talk is deliberate since there are no observations or known laws that explain what occurs at either the instant of the big crunch or the instant $\mathbf{t} = 0$ at which the expansion began. There are also no observations or confirmed laws that enable us to say anything about the material universe before the beginning of its expansion. It is conceivable that some justified hypotheses can be made, especially once we take into account Bohmian QM, but an attempt to formulate any such hypotheses is beyond the scope of this paper.

If the material universe is open, as observational evidence suggests, it will expand forever and will expand to arbitrarily large distances; that is, for every distance, say, 15 billion light years, there is a greater distance to which the universe will eventually expand. The material universe will become less and less dense the further it expands, the superclusters becoming ever further apart, until they disintegrate into elementary particles or radiation, which themselves will keep receding from each other with no end to their mutual recession and with no end to future time.

The second problem is that the material universe is not composed of particles that obey Newton's laws of motion; rather, they obey neo-Lorentzian laws. More exactly, they obey Bohmian-neo-Lorentz cosmological laws, which in effect reduces to neo-Lorentz cosmology, since there is no observational evidence that the quantum potential **Q** is greater than zero at cosmological scales. Some might think the presence of Planck's constant in the equations imply that it cannot be the case that at cosmic scales **Q** = 0, but this is assuming a realist rather than an epistemic interpretation of Planck's constant. Furthermore, the equations can be formulated without Planck's constant, as Dürr, Goldstein and Zhanghi have shown (1996: 26).

There is a phenomenal "cosmic time," which is the perspective from which we say the material universe began to expand 15 billion years ago. But this does not mean "cosmic time" in the sense of Friedman's universe, where there is no absolute time in the sense of non-local, space-like, instantaneous, universe-wide relations of simultaneity.

The so-called "clocks" introduced by Poincaré (1905) and Einstein (1905) are moving masses or light. Since movement in absolute space contracts the moving body in the direction of length, it "slows down time" in the unrealistic sense that, if a light clock is placed so that its top is in the direction of the motion, the light clock contracts. This has little if anything to do with time, especially since in the absolute, real frame, light has the velocity c and moving systems have a velocity of light $\mathbf{c} + \mathbf{v}$ or $\mathbf{c} - \mathbf{v}$. The law of the addition of velocities is not in reality violated, even though it appears to be in the phenomenal realm. The fact that absolute moving systems have a light velocity of $\mathbf{c} + \mathbf{v}$ or $\mathbf{c} - \mathbf{v}$ is part of Lorentz's theory and his retention of absolute simultaneity and Galilean transformation equations for the absolute time and absolute rest frame.

The nearest to an accurate clock could be any ideally periodic motion or simple harmonic oscillation on a body, a clock, at rest which is relative to absolute space. But in (1998) I argued there is no justification to think that it is necessarily true that the metric of time is either identical with or isomorphic with a light clock or harmonic oscillators or other such "clock" devices. We are immediately acquainted with presence and this is how we identify the absolute frame and absolute time. It cannot be the case, as a critic of the tensed theory of time may suggest that we falsely belief a hyperplane 15 billion years or so later than the big bang is present, whereas in reality a hyperplane near the big bang is present. If a hyperplane near the big bang is present, that is the only hyperplane that exists, since "is present" means "exists" in the present tense sense of "exists". Accordingly, we would not exist and have a false belief about what is present. This argument against the tensed theory seems to be presupposing the "Moving Now" and the **A**-series formulated by McTaggart (and which he later rejected), which posits that each event has a primitive tenseless existence and that the

A-properties of presentness, pastness and futurity have no existential import but are possessed successively by beings that have a primitive and irreducible tenseless existence. It is very difficult to find a tensed theorist who holds this theory. Schlesinger (1981) and Gale (1968) might seem to, but a closer analysis makes it rather ambiguous; in Schlesinger's case, it is largely due to the vagueness and ambiguity of his theoretical formulations. Tensed theorists, as far as I know, believe that what is present exists, what is no longer present existed (i.e. no longer exists) and that what is not yet present will exist (i.e. does not yet exist). For our critic's scenario to work, he has to reject this and identify existence with an irreducible, primitive tenseless existence. But then he is no longer talking about theories held by contemporary (or former) tensed theorists of time. Contemporary philosophers of time do not engage in ordinary language analysis, e.g. of the sort popular in the 1940s and 1950s in England and no one would infer an ontological conclusion about the nature of time from "how we ordinarily talk". The issue is of the isomorphism between the deep structure of all possible natural and artificial languages (including the language of the sciences) and reality; the isomorphism is that the deep structure of language is irreducibly tensed and is isomorphic to the tensed structure of reality (Smith 1993: 215–24; Ludlow 1999). Equally important arguments are from experience of presence (e.g. Craig 2000a; 2000b), the physical sciences (Smith 1985, Craig 2001a; Smith, 1993: 18–24, 225–250), and the formal logical invalidity that is implicit in a tenseless theory of time.

My so-called "watch" (movements of bosons and fermions, or large aggregates of them moving relative to another large aggregate that is relatively and phenomenally at rest to the moving bosons and fermions) may not be similar to somebody's who is moving at a high velocity relative to me, but what has that to do with time? Both of us perceive presence or the present. This is absolute presentness. If we want to measure accurately intervals of time, we can wait until a "Valentini clock" (2005) is invented, one that uses instantaneous signals, so that the signals and the absolutely present instant **t** are all simultaneous. An accurate clock is not only physically possible but technologically possible, even though we are a long way away from making this "technological possibility" into a probability or actuality.

Our Bohmian-neo-Lorentz cosmology suggests some implications regarding metaphysical modalities. If and only if the initial conditions of the universe satisfy the precise probability density, $\uparrow \Psi \uparrow^2$, are the predictions of Bohmian QM accurate; without them, they are widely inaccurate if Bohm QM or some other theory is possible at all. In other words, Bohmian QM implies the initial conditions that satisfy this probability density. This is another way of saying that the probability density, $\uparrow \Psi \uparrow^2$ is not a brute or contingent fact and that Bohmian QM could be true if there were some other initial conditions. Rather it is a way of saying that the probability density is nomically necessary. There is another fundamental nomic necessity about the initial conditions of the universe, this one following from the Bohm-neo-Lorentz cosmology. The basic cosmological law is the distance/velocity law and this law requires precise initial conditions at the origin $\mathbf{t} = 0$ of the distance scale. We can say that at the big bang or infinitesimally close to the big bang the initial conditions had to be exactly as they are to produce a material universe that satisfied the velocity-distance law. They are nomically necessary and since this initial distribution is also necessary for the probability density, $\uparrow \Psi \uparrow^2$ this probability density and the distance/velocity initial conditions law indicate there is a very strong form of necessitarionism in Bohm-neo-Lorentz QG. It seems to me very hard to avoid the conclusion that some other philosophers have reached for other reasons, namely, that metaphysical possibility and necessity (or "broadly logical possibility and necessity") is identical with Bohm-neo-Lorentz nomic possibility and necessity. It seems to me that this suggests that the only modal logic system that is a valid *simpliciter*, i.e. that describes the modal logical form of reality, is **Triv** (Hughes and Cresswell). This modal system implies that what actually exists necessarily exists, and it implies that what possibly exists actually exists.

Clearly a new theory of counterfactuals is needed, especially since laws are distinguished from accidental generalizations by virtue of entailing counterfactuals and modal necessitarionism implies counterfactuals are all necessarily false. There are various ways to re-interpret counterfactuals, probabilities, possibilities and the like. For example, counterfactuals about the particulars will be epistemic in nature, consistent with the Bohmian thesis that we are ignorant of the particular trajectories. "Possibility" in modal talk can be replaced by "conceivable" which can obey some modal system other than Triv, e.g. the modal system T or S5 (except now we should call these systems of epistemic modal logic or, better, systems of the logic of conceivability relations). A number of ways of dealing with this issue are available, but these issues should be discussed in a different work on the metaphysics of modality.

The debate about whether metaphysical necessity (what can and cannot exist) is identical with physical or nomological necessity is tilted towards an affirmative answer if two of our scientific theories, independent of each other, imply that the initial conditions of the universe, and not merely its laws, are nomically necessary. We see that both the Density Postulate of Bohm's QM and the initial conditions required by the basic cosmological law of our neo-Lorentz cosmology, the distance-velocity relation, require that the number, distribution, state of motion or rest, and kind of particle, of the initial conditions, necessarily be what they actually are, licensing an inference from "x is actual" to "x is necessary". Our postulate is that if inferences from actuality to necessity, regarding all of the precise initial conditions of the universe, and all the precise and completely described careers or histories of the particles that form the material universe, are logically entailed by two logically independent theories (Bohmian QG and the velocity-distance cosmology, their unification belonging to our Bohm-neo-Lorentz QG), both theories of which are observationally confirmed, then the modal system *Triv* is the one we are most justified in believing is *validsimpliciter* (and not merely *valid-in-Triv*) and that in an appropriate sense the modal logical laws of *Triv* formulate the "the one and only modal form of the universe. The only valid modal logic system states that, for any true proposition **p**, actually **p** implies possibly **p**, and possibly **p** implies necessarily **p**, and necessarily **p** implies actually **p**. Equivalently, possibly **p** implies **p**, and **p** implies necessarily **p**. There is only one metaphysically possible world, the actual world.

We know the answer to these ultimate why-questions: "Why is this world actual rather than some other possible world?" and "Why is there something rather than nothing?" The answer to the two questions is that it is that there is only one possibility, and nothing at all could have been otherwise. The only possible world is our world, the actual world. It is metaphysically impossible for there to be nothing and it is metaphysically impossible that there be a possible world or any possibility that is not actualized.

The belief that our imagination and our modal intuitions show there are other possible worlds is false. This is even the case for conceiving a Friedman world to be possible. A Friedman universe differs from our Bohm-Lorentz world but only in observationally insignificant ways. But a Friedmann universe is mathematically impossible. Consider that Friedman space-time is a continuum of space-time points, and its radius and density are described in Roberston-Walker metric, which includes two equations, one of which is

$$3(\mathbf{da}/\mathbf{dt})^2 = 8\pi\mathbf{Gpa}^2 - 3\mathbf{kc}^2.$$

Here $3(\mathbf{da/dt})^2$ measures the rate of change of the universe's radius **a** with time, **G** is the gravitational constant and it is multiplied first by 8π, and its product is multiplied is proportional to the density **p**. **p**'s magnitude is proportionate to the radius squared $\mathbf{a}^2$ of the universe, and to the velocity of light squared, and to **k**, the cosmological constant.

Friedman's universe seems nomically possible, does it not? In fact, it is mathematically and logically impossible. Friedman's time, the **t** that appears in the equation, has values that are proportionate to the radius **a** of the universe. In an expanding universe, the numerical values for the **t** variable attain higher and higher values as the radius of the universe continues to expand, e.g., if $\mathbf{k} = 0$. There is a possible light clock, where there is a mirror for the top of the cylinder and another mirror for the bottom of the mirror. This clock measures the Friedman time **t** in his equation our Bohm-Lorentz material universe on sufficiently large scales. But Einstein has identified time with measured values of his clocks. The light-clock has a mirror at its top and a mirror at its bottom; a light pulse is bouncing back and forth between the top and bottom mirrors. Heisenberg's uncertainly principle shows that the light clock cannot get extremely small, for at 10^{-43} s and briefer the gravitational force or field becomes so strong that it makes the light pulse move in a zigzag and non-uniform and non-repeating motion. If all physically possible clocks, within Friedman's laws, are broken apart at this level, it is no longer possible to measure time by light-clocks and therefore, it follows from Einstein's GTR, there is no time.

But this contradicts the fact that Friedman's equations posit a space-time continuum, that there is time "all the way down" to temporal instants that have zero duration. Thus, Friedman's theory entails there is both time and there is no time, and thus is a self-contradictory theory. I believe other alleged "possible worlds" can be also shown to be inconsistent when examined closely. Humans construct vague images and think that is evidence for other possible words. They are not evidence. On the Bohm-neo-Lorentz theory, the Heisenberg uncertainty principle is merely epistemic. Further, absolute time is neither a clock, a measured time, or anything involving bodies and movements. It exists and "flows equably" regardless of whether or not it is measured.

Bibliography

Albert, D. (1996) 'Elementary Quantum Metaphysics' in J. Cushing et al (eds.) *Bohmian Mechanics and Quantum Theory: An Appraisal.* Dordrecht: Kluwer Academic Publishers, pp. 277–84.

Anderson, J. (1967) *Principles of General Relativity.* New York: Academic Press.

Aspect, A., P. Grangier, and G. Roger (1981) 'Experimental Tests of Local Realist Theories via Bell's Theorem', *Physical Review Letters* 47, 460–63.

Aspect, A., P. Grangier, and G. Roger (1982) 'Experimental Realisation of Einstein-Podolsky-Rosen-Bohm Gedanken experiment: A New Violation of Bell's Inequalities', *Physical Review Letters* 47, 91–94.

Aspect, A, J. Dalibard and G. Roger (1982) 'Experimental Test of Bell's Inequalities Using Time-Varying Detectors', *Physical Review Letters* 49, 1804–7.

Bell, J. S. (1987) *Speakable and Unspeakable in Quantum Mechanics,* Cambridge: Cambridge University Press.

Butterfield, J. (1989) 'A Space Time Appoach to the Bell Inequality', in Cushing, J., Fine, A., and Goldstein, S. (eds.). *Bohmian Mechanics and Quantum Theory.* Dordrecht: Kluwer Academic. pp. 114–44.

Bohm, D. (1952) 'A suggested interpretation of the quantum theory in terms of "hidden" variables. I', *Physical Review D* 85, 166–79. (For part II, see pp. 180–93.)

Bohm, D and B. Hiley (1993), *The Undivided Universe.* London. Routledge.

Bondi, H. (1960) *Cosmology.* 2nd. ed. Cambridge: Cambridge University Press.

Bondi, H. (1979) 'Is "General Relativity" Necessary for Einstein's Theory of Gravitation?', in (eds De Finis, F) *Relativity, Quanta and Cosmology in the Development of the Scientific Thought of Albert Einstein*, New York: Johnson Reprint Corp., pp. 179–86.

Builder. G. (1958) 'Ether and Relativity', *Australian Journal of Physics* 11, 279–97.

Callender, C. and R. Weingard (1994) 'The Bohmian Model of Quantum Cosmology', in Hull, D. et al. (eds) *PSA 1994*, Vol. 1, pp. 218–27.

Callender, C. and R. Weingard (1997) 'Trouble in Paradise? Problems with Bohm's Theory' *Monist*, 80(1), 24–43.

Callender, C. and R. Weingard 1995. 'A Bohmian Model of Quantum Cosmology' *Philosophy of Science (Proceedings)* 1, 228–37.

Colladay, D. and V. Kosteleck (1998) 'Lorentz-violating extension of the standard model,' *Physical Review D* 58, 116002.

Coleman, S. and S. L. Glashow (1999) 'High-energy tests of Lorentz invariance', *Physical Review*, D 59, 116008..

Craig, William Lane (2000a) *The Tensed Theory of Time.* Dordrecht: Kluwer Academic Publishers.

Craig, William Lane (2000b) *The Tenseless Theory of Time.* Doredrecht: Kluwer Academic Publishers.

Craig, William Lane (2001a) *Time and the Metaphysics of Relativity.* Dordrecht: Kluwer Academic Publishers.

Craig, William Lane (2001b) *God, Time and Eternity.* Dordrecht: Kluwer Academic.

Cresswell, J. and G. Hughes, (1996) *An Introduction to Modal Logic.* London: Routledge Publishers.

Cushing, J. T. (1994) *Quantum Mechanics: Historical Contingency and the Copenhagen Hegemony.* Chicago: University of Chicago Press

Cushing, J. *et a*l (eds.) (1989) *Philosophical Consequences of Quantum Theory: Reflections on Bell's Theorems.* Notre Dame: Notre Dame University Press

Dampier, W. (1948) *A History of Science.* Cambridge: Cambridge University Press.

De Broglie, L. (1928) 'La Nouvelle Dynamique des Quanta', in *Électrons et Photons: Rapports et Discussions du Cinquième Conseil de Physique*, pp. 105–32, ed. J. Bordet. Gauthier-Villars, Paris..(G. Bacciagaluppi and A. Valentini, translate this essay in *Electrons and Photons: the Proceedings of the* Fifth Solvay Congress*), Cambridge University Press, Cambridge, forthcoming.*

De Broglie, L. (1960) *Non-Linear Wave Mechanics: A Causal Interpretation.* Amsterdam, Elsevier Publishers.

Dewdney, C. and G. Horton (2002) 'Relativistically invariant extension of the de Broglie—Bohm theory of quantum mechanics', *Journal of Physics.* A 35, 10117–27.

Dürr, D., S. Goldstein and N. Zanghí (1992) 'Quantum Equilibrium and the Origin of Absolute Uncertainty'. *Journal of Statistical Physics*, 67, 843–907.

Dürr, D. S. Goldstein, and N. Zanghi (1996) 'Bohmian Mechanics and the Foundation of Quantum Mechanics' in Cushing, J., Fine, A., and Goldstein, S. (eds.). *Bohmian Mechanics and Quantum Theory.* Dordrecht: Kluwer Academic.

Dürr, D., S. Goldstein, R. Tumulka and N. Zanghì (2003) 'Bohmian Mechanics and Quantum Field Theory', *Physical Review Letters* 93, 090402, 1–4.

Dürr, D., S. Goldstein, K. Münch-Berndl and N. Zanghì (1999) "Hypersurface Bohm-Dirac Models", *Physical Review A* 60, 2729–36.

Einstein, A. (1905) 'On the Electrodynamics of Moving Bodies' in A. Miller (1971), op. cit.

Einstein, A. (1924) *Nat. Gen.* 105, Teil II: 85–93.

Fizea (1850) *Annal. D. Phys.* Vol 79, pp. 167ff.

Friedman, A. (1922) 'Uber die Krummung des Raumes', *Zeitschrift fur Physik* 10, 377–86.

Friedman, M. (1983) *Foundation of Spacetime Theories.* Princeton: Princeton University Press.

Grunbaum, Adolf (1973) *Philosophical Problems of Space and Time.* New York: Alfred Knopf.

Fitzgerald, (1894) *Scientific Writings*, p. 313, quoted from page 407 of W. Dampier's 1948 book, *A History of Science*, Cambridge University Press.

Hardy, L. and E. Squires. (1992) 'On the Violation of Lorentz-Invariance in Deterministic Hidden Variable Interpretations of Quantum Theory', *Physics Letters* A 168, pp. 169–73.

Harrison, (2000) Cosmology. Cambridge: Cambridge University Press.

Hartle, J. (2004) 'The Physics of "Now"'; see *arXivigr-qc*/0403001 v. 2 no. 21.

Hartle, J and S. W. Hawking (1983) 'The Wave Function of the Universe', *Physical Review D.*

Hawking, S. W. and Don Page (1986) 'Operator Ordering and the Flatness of the Universe', *Nuclear Physics B264*, 185–96

Holland, Peter (1996) 'Is Quantum Mechanics Universals?' in Cushing, J., Fine, A., and Goldstein, S. (eds.). *Bohmian Mechanics and Quantum Theory.* Dordrecht: Kluwer Academic.

Holland, P. (1993) *The Quantum Theory of Motion.* Cambridge: Cambridge University Press.

Holton, G. (1973) *Thematic Origins of Scientific Thought.* Cambridge: Cambridge University Press.

Howard, Don (1989) 'Holism, Seperability, and the Metaphysical Implications of the Bell Experiments', in J. Cushing et al (eds.) *Philosophical Consequences of Quantum Theory: Reflections on Bell's Theorems.* Notre Dame: Notre Dame University Press, pp. 224–53.

Howard, Don (1992) 'Locality, Separability, amd the Physical Implications of the Bell Experiments', in *Bell's Theorem and the Foundation of Modern Physics*, eds van der Merwe, et al., 306–14.

Janossy, L. (1971) *Theory of Relativity Based on Physical Reality.* Budapest: Akademiai Kiado.

Jacobson, T. and D. Mattingly (2001) 'Gravity with a dynamical preferred frame', *Physical Review D* 64.

Lamore, J. (1900) *Aether and Matter.* Cambridge: Cambridge University Press.

Layzer, D. (1954) 'The Significance of Newtonian Cosmology', *Astronomical Journal* 59, 168.

Lemaitre, G. (1931) *La Revue des Questions Scientific*, 4th series, Vol. 20, pp. 397ff. See Peebles (1993, p. 198) for discussion.

Lorentz, H. *Collected Works.* Dordrecht: D. Reidel.

Lorentz, H. (1895) A translation of a small part of this work is in Einstein, Lorentz, Minkowski *et al. The Principle of Relativity* (1952) New York: Dover Publications.

Lorentz, H. (1904) is in Einstein, Lorentz, Minkowski *et al. The Principle of Relativity* (1952) New York: Dover Publications

Lorentz, H. *On the Theory of Electrons* (1909; 1915). New York. Dover Publications.

Maudlin, T. (1994) *Quantum Non-Locality and Relativity: Metaphysical Intimations of Modern Physics*, Cambridge, MA: Blackwell.

Maudlin, T. (1996) 'Space-Time in the Quantum World', in *Bohmian Mechanics and Quantum Theory: An Appraisal*, J. Cushing, A. Fine and S. Goldstein eds. (Dordrecht: Kluwer) 285–307.

McCrea, W. H. (1955) 'On the Significance of Newtonian Cosmology', *Astronomical Journal* 60: 271–74.

McCrea, W., and E. Milne (1934) *Quarterly Journal of Mathematics*, Vol. 5, No. 73, Oxford.
Miller, Arthur. I. (1981) *Einstein's Special Theory of Relativity.* Reading, MA: Addison-Wesley Publishing Co.
Milne, E. A. (1934) 'A Newtonian Expanding Universe', *Quarterly Journal of Mathematics* 5, 64–72.
Moffat, J. (2003) 'Spontaneous violation of Lorentz invariance and ultra-high energy cosmic rays', *International Journal of Modern Physics.* D 12, 1279–87.
North, J. (1965) *The Measure of the Universe.* New York: Oxford University Press.
Owens, David (1992) *Causes and Coincidences.* Cambridge: Cambridge University Press.
Page, Don (1996) 'Attaching Theories of Quantum Mechanics to Bohmian Quantum Mechanics', in Cushing, J., Fine, A., and Goldstein, S. (eds.). *Bohmian Mechanics and Quantum Theory.* Dordrecht: Kluwer Academic.
Peebles, (1993) *Principles of Physical Cosmology.* Cambridge: Cambridge University Press.
Pitowsky, Itamar. (1991) 'Bohm's Quantum Potentials and Quantum Gravity', *Foundations of Physics*, Vol. 21, No. 3.
Poincaré, H. (1905a) La Valeur de la Science (Flammarion, Paris, 1905). (English translation: The Value of Science: Essential Writings of Henri Poincaré).
Poincaré, H. (1905b) 'Sur la dynamique de l'électron' *C. R. Ac. Sci. Paris* 140, 1504–8.
Poincaré, H. (1906) 'Sur la dynamique de l'électron', *Rend. Circ. Matem. Palermo* 21, 129–75.
Popper, K. (1982) *Quantum Theory and the Schism in Physics* (Unwin Hyman), London.
Prokhovnik, S. J. (1985) *Light in Einstein's University.* Dordrecht: D. Reidel.
Prokhovnik, S. J. (1956) *The Special Theory of Relativity.* Dordrecht: D. Reidel.
Redhead, M. (1987) *Incompleteness, Nonlocality, and Realism.* Oxford: Oxford University Press.
Sciama, D. (1971) Modern Cosmology. Cambridge: Cambridge University Press
Shimony, A. (1990) "An Exposition of Bell's Theorem", in A. Miller (ed.) *Sixty-Two-Years of Uncertainty.* New York: Plenum, pp. 33–43.
Smith, Quentin, (1988) "The Uncaused Beginning of the Universe", *Philosophy of Science*, 55, 39–57.
Smith, Quentin. (1993) *Language and Time.* New York: Oxford University Press. See Ch. 7.
Smith, Quentin, (1994) 'Did the Big Bang Have a Cause?' *British Journal of the Philosophy of Science*, 45, 649–68.
Smith, Quentin (1997) 'The Ontological Interpretation of the Universe' *The Monist*, 80(1), 160–185.
Smith, Quentin, (1998) 'Absolute Simultaneity and the Infinity of Time' in Robin Le Poidevin (ed.), *Questions of Time and Tense.* Oxford: Oxford University Press, pp. 135–184.
Smith, Quentin (2000) 'Problems with John Earman's Attempt to Reconcile General Relativity with Theism', *Erkentniss* 52, 1–27.
Smith, Quentin. (2002) 'Time was Created by a Timeless Point', in Gegansall and Woodruff (eds.), *God and Time.* New York: Oxford University Press, pp. 95–128.
Smith, Quentin and William Lane Craig (1993a). *Theism, Atheism and Big Bang Cosmology.* Oxford: Oxford University Press. (I have in mind the distinct chapters whose author is Smith.)

Smith, Quentin (2003) 'Why Cognitive Scientists Cannot Ignore Quantum Mechanics', in Jokic and Smith (eds.), *Consciousness: New Perspectives.* Oxford: Oxford University Press.

Smith (2007) 'Kalam Cosmological Arguments for Atheism', in M. Martin (ed.), The Cambridge Companion to Atheism. Cambridge: Cambridge University Press.

Squires, E. (1996) 'Local Bohmian Mechanics' in Cushing, J., Fine, A., and Goldstein, S. (eds.). *Bohmian Mechanics and Quantum Theory.* Dordrecht: Kluwer Academic.

Tooley, Michael (1997) *Time, Tense, and Causation*. Oxford: Oxford University Press. (see especially chapter 11 for an original physical argument for substantival space and instantaneous simultaneity).

Valentini, A. (1992) 'On the pilot-wave theory of classical, quantum and subquantum physics'. PhD thesis, International School for Advanced Studies, Trieste, Italy.

Valentini, A.(1996) 'Pilot-wave theory of fields, gravitation and cosmology', in Cushing, J., Fine, A., and Goldstein, S. (eds.). *Bohmian Mechanics and Quantum Theory.* Dordrecht: Kluwer Academic.

Valentini, A. (2001) 'Hidden variables, statistical mechanics and the early universe', in J. Bricmont et al. (eds.) *Chance in Physics: Foundations and Perspectives*, Berlin: Springer, pp. 165–81.

Valentini, A. (2002a) 'Signal-locality and subquantum information in deterministic hidden-variables theories', in T. Placek and J. Butterfield (eds.) *Non-locality and Modality*, Dordrecht: Kluwer, pp. 81–103

Valentini, A. (2002b) 'Subquantum information and computation', *Journal of Physics* 59, 269–77.

Valentini, A. (2002c) 'Signal-locality in hidden-variables theories', *Physics Letters.* A 297, 273–78.

Valentini, A. and Westman, H. (2005) 'Dynamic origin of quantum probabilities', *Proc. Roy. Soc.*. A 461, 253–72.

Vilenkin, A. (1983). 'Birth of Inflationary Universes', *Physical Review* D27: 2848–55.

Weinberg, S. (1972). *Gravitation and Cosmology* New York: Wiley.

Yurke, B. and Stoler, D. (1992). 'Bell's Inequality Experiments Using Independent-Particle Sources'. *Physical Review A* 46, 2229–34.

Zahar, E. (1989) *Einstein's Revolution: a Study in Heuristic* La Salle: Open Court.

Zahar, E. (1973), 'Why did Einstein's Program SuperSucceed Lorentz's? I and II' *British Journal for the Philosophy of Science* 24, 95–123; 223–62.

4 Hidden variables and the large-scale structure of space-time

Antony Valentini

We discuss how to embed quantum non-locality in an approximately classical space-time background, a question which must be answered irrespective of any underlying microscopic theory of space-time. We argue that, in deterministic hidden-variables theories, the choice of space-time kinematics should be dictated by the properties of generic non-equilibrium states, which allow non-local signaling. Such signaling provides an operational definition of absolute simultaneity, which may naturally be associated with a preferred foliation of classical space-time. The argument applies to any deterministic hidden-variables theory, and to both flat and curved space-time backgrounds. We include some critical discussion of Einstein's 1905 "operational" approach to relativity, and compare it with that of Poincaré.

Introduction

> The simultaneity of two events, or the order of their succession, the equality of two durations, are to be so defined that the enunciation of the natural laws may be as simple as possible.
>
> (Henri Poincaré 1905a)

> What really matters is not merely the greatest possible simplicity of the geometry alone, but rather the greatest possible simplicity of all of physics (inclusive of geometry).
>
> (Albert Einstein 1949a)

This article concerns the structure of space-time on large scales, in the context of hidden-variables interpretations of quantum theory. In particular, we shall be addressing the question of how macroscopic quantum non-locality can be embedded in an approximately classical space-time background. We argue that this question must have an answer, regardless of what the underlying microscopic theory of space-time may turn out to be, and further, that the most natural answer is to introduce an absolute simultaneity, associated with a preferred foliation of classical space-time (flat and curved).

The introduction of an absolute simultaneity, to accommodate the non-locality of quantum theory over macroscopic distances, was suggested in particular by Popper (1982), Bohm and Hiley (1984), and Bell (1986, 1987). This proposal is often regarded as unsatisfactory, because quantum non-locality cannot in fact be used for practical signaling at a distance, making the preferred rest frame undetectable in practice. As Bell put it (1986: 50):

> It is as if there is some kind of conspiracy, that something is going on behind the scenes which is not allowed to appear on the scenes. And I agree that that's extremely uncomfortable.
>
> (Bell 1986)

However, Bell missed the point that, in (deterministic) hidden-variables theories, the inability to use quantum non-locality for remote signaling is not a fundamental constraint. It is, rather, a peculiarity of a special "quantum equilibrium" distribution of hidden variables. For more general "non-equilibrium" distributions, practical non-local signaling is indeed possible (Valentini 1991a, 1991b, 1992, 1996, 2001, 2002a, 2002b, 2002c). From this perspective, our inability to detect the preferred rest frame is *not* an uncomfortable conspiracy seemingly built into the laws of physics; it is simply an accident of our living in a state of quantum equilibrium, whose statistical noise masks the underlying non-locality.

In our view, if one wishes to appraise the structure of space-time at a more fundamental level, then this should be done taking into account the wider, explicitly non-local physics of quantum non-equilibrium, rather than merely in terms of the statistical predictions of quantum theory, which are not fundamental but merely contingent on a special distribution of hidden variables.

We shall argue that non-equilibrium instantaneous signaling defines an absolute simultaneity, within the approximately classical space-time defined by macroscopic rods and clocks, and that fundamental local Lorentz invariance should be abandoned. It will also be suggested that the widespread excessive reluctance to consider abandoning (local) Minkowski space-time has its origin in the unfortunate "operational" approach to relativity taken by Einstein in 1905.

Status of Lorentz invariance in contemporary physics

Locally speaking, the relativity of simultaneity is usually regarded as a consequence of (local) Lorentz invariance. Before considering quantum non-locality, then, let us first briefly review the current status of Lorentz invariance in other areas of physics.

In high-energy physics, the status of Lorentz invariance is certainly open to question. The divergences of quantum field theory can be most straightforwardly eliminated by introducing a short-distance cutoff, which breaks Lorentz invariance. This suggests that it would be an advantage if Lorentz invariance were not fundamental. It is also sometimes argued that exact Lorentz invariance is experimentally inaccessible because the boost parameter (for the non-compact Lorentz group) has an infinite range which can never be probed uniformly (Jacobson and Mattingly 2001).

However, like renormalizability, Lorentz invariance did play a key historical role in the development of the standard model of particle physics. Yet, the fact that only renormalizable terms appear in the Lagrangian of the standard model is now generally regarded as merely an accident of the low-energy limit, where non-renormalizable terms are screened off in the infrared (Weinberg 1995: 519). Clearly, the mere fact that a property played a crucial historical role in constructing our current theories is not a conclusive argument for that property to be fundamental.

Possibly, in high-energy physics, Lorentz invariance will eventually acquire a similar status to that of renormalizability, as a mere low-energy symmetry (Nielsen and Ninomiya 1978; Chadha and Nielsen 1983; Allen 1997; Moffat 2003). In any case, non-Lorentz-invariant extensions of the standard model have been considered in detail (Colladay and Kostelecký 1998; Coleman and Glashow 1999), where terms in the Lagrangian breaking Lorentz symmetry might come from deeper physics beyond the standard model. A number of experiments searching for such effects have been performed, while further experiments are underway or being planned (for reviews, see Kostelecký 2002).

Further questioning of Lorentz invariance comes from quantum gravity, in which the possibility of a minimum length at the Planck scale suggests that Lorentz invariance might emerge only as an approximation on larger scales (Kostelecký 2002). Indeed, it has been suggested that peculiarities in cosmic-ray data, together with other astrophysical anomalies, might be a sign of a breakdown of standard special-relativistic kinematics, possibly due to quantum gravity effects (Amelino-Camelia 2002). In addition, models of classical gravitation with a "dynamical preferred frame" have been considered (Jacobson and Mattingly 2001).

One could certainly question the above motivations for considering a breakdown of local Lorentz invariance. Still, it is clear that Lorentz invariance is far from being a dogma in the context of high-energy physics or quantum gravity. Rather, it is often regarded as one important symmetry among others, whose status (approximate or fundamental) is a matter for experiment. And in comparing experiments with theory, it is helpful to have models incorporating violations of Lorentz invariance (as well as models incorporating violations of other important symmetries such as CPT invariance; Mavromatos 2004).

Quantum non-locality

In contrast, in the context of quantum foundations, attachment to Lorentz invariance tends to be more dogmatic. A number of authors insist that a realistic quantum physics should be "seriously Lorentz invariant," in the sense that Lorentz invariance should be fundamental, and not merely phenomenological or emerging in some limit. This contrast is remarkable, because it is precisely in quantum foundations that there is arguably the strongest motivation of all for abandoning fundamental Lorentz invariance: the experimental detection of quantum non-locality, through the observed violations of Bell's inequalities.

As emphasised by Bell (1987), quantum theory is incompatible with locality, *independently* of any assumption about the existence of hidden variables. Given a pair of spin-1/2 particles in the singlet state, a quantum measurement of z-spin at one wing B allows the experimenter at B to predict in advance the outcome of a quantum measurement of z-spin at the distant wing A (in ideal conditions). As was first argued by Einstein *et al.* (1935), using a somewhat different example, if locality is assumed then changing what is done at B (from a z-spin measurement to some other measurement) cannot affect the outcome at A, and therefore the z-spin outcome at A must be determined in advance regardless of what measurement is performed at B. Having reached the conclusion that the outcomes at A and B are locally determined, one can then run a Bell-type argument, showing that their statistical correlation is incompatible with the predicted (and observed) quantum correlation. If we leave aside the many-worlds interpretation,[1] it follows that locality is incompatible with quantum theory. Note that in this argument, determinism at each wing is not assumed, but *deduced* from the assumption of locality (Bell 1987: 143).

There is then strong evidence (again, if we leave aside the possibility of many worlds) that in the above set-up the physical processes at A and B are not independent, no matter how remote A and B may befrom each other. This raises the question of how such non-locally connected processes may be embedded into the structure of standard relativistic space-time.

It is sometimes suggested that, instead of accepting the existence of superluminal influences, the whole issue could be avoided by assuming that our classical space-time is merely emergent. Now, it may well be true that classical space-time is emergent (for example from a deeper discrete structure). However, this does not affect the issue at all. The EPR-Bell correlations observed in the laboratory take place at macroscopic distances (for example 12 m; Aspect *et al.* 1982), involving photons with quite ordinary energies (for example visible photons of wavelength $\sim$500 nm; Aspect *et al.* 1982). The detection events are recorded as taking place in a region of space and time whose structure may be operationally defined by macroscopic rods and clocks, in the laboratory where the experiment is performed. There is no doubt that the structure of space-time in that laboratory, as defined by macroscopic rods and clocks, is to very high accuracy well-described by standard relativity theory. One may then ask, *in the approximation* where the background space-time is approximately classical, how the events or outcomes recorded at A and B are to be embedded in the background space-time. Whatever the final theory underlying space-time (if there is one) turns out to be, this question must have an answer, and the aim of this article is to provide one.

Non-equilibrium hidden variables

We shall consider the issue from the standpoint of deterministic hidden-variables theories. These provide a mapping from initial hidden parameters λ to final outcomes of quantum measurements. The mapping depends on the macroscopic settings defining the experimental set-up. For entangled quantum states, the mapping is non-local, in the sense that outcomes at one wing depend on experimental settings at the distant wing (in at least one direction; Bell 1964). Thus, the non-locality is clearly present in the underlying dynamics associated with the mapping. Instantaneous signaling is not possible in such theories, however, *provided* the initial hidden parameters λ have a special "quantum equilibrium" distribution $\rho_{QT}(\lambda)$. This distribution is chosen so that the resulting statistics of quantum measurement outcomes agree with quantum theory.

As we shall discuss in the section entitled "Instantaneous signaling in quantum non-equilibrium," once one is given a deterministic hidden-variables theory for individual systems – where mathematically the theory is defined by the mapping from λ to outcomes – then there is no conceptual reason why one should not consider the physics of more general "non-equilibrium" distributions $\rho(\lambda) \neq \rho_{QT}(\lambda)$. For such distributions, non-locality is present not only for individual outcomes, but also at the statistical level: the marginal statistics at one wing of an entangled state do (generically) depend on measurement settings at the distant wing (in at least one direction). In such circumstances, with $\rho(\lambda) \neq \rho_{QT}(\lambda)$, practical non-local signaling would be possible (Valentini 1991a, 1991b, 1992, 1996, 2001, 2002a, 2002b, 2002c).

If one takes deterministic hidden-variables theories seriously, then one is driven to conclude that our inability to send instantaneous signals is merely an accident of our living in a time and place where the parameters λ have the special distribution $\rho_{QT}(\lambda)$, for which statistical noise happens to erase (on average) the effects of non-locality. This state is roughly analogous to a state of global thermal equilibrium in classical physics, in which it would be impossible to convert heat into work (as this requires differences of temperature). In such a world – in a state of thermodynamic "heat death" – the inability to convert heat into work is not a law of physics, but rather a contingent feature of the state of thermal equilibrium. Similarly, in our view, the absence of superluminal signaling in our world is not a law of physics, but rather a contingent feature of the state of quantum equilibrium (Valentini 1991a, 1991b, 1992, 1996, 2001, 2002a, 2002b, 2002c).

Non-equilibrium deviations $\rho(\lambda) \neq \rho_{QT}(\lambda)$ might have existed in the very early universe, with the relaxation $\rho(\lambda) \to \rho_{QT}(\lambda)$ taking place during the violence of the big bang (Valentini 1991a, 1991b, 1992, 1996, 2001, 2002a, 2002b, 2002c; Valentini and Westman 2005). In effect, a hidden-variables analogue of Boltzmann's "heat death" may have actually taken place in our observable universe. However, relic cosmological particles that decoupled at sufficiently early times might still be out of equilibrium today (Valentini 1996, 2001). It has also been suggested that quantum non-equilibrium might be generated in systems that are entangled with degrees of freedom located behind the event horizon of a black hole (Valentini 2004a, 2004b).

In any case it is certainly true that, from a hidden-variables perspective, quantum theory is merely the phenomenological description of the statistics of a special state with $\rho(\lambda) = \rho_{\mathrm{QT}}(\lambda)$. In principle, there exists a wider (and explicitly non-local) physics of non-equilibrium with $\rho(\lambda) \neq \rho_{\mathrm{QT}}(\lambda)$.

Physical structure of space-time

The structure of space-time at the most fundamental level should be defined in terms of the physics at the most fundamental level. In a deterministic hidden-variables theory, emergent properties of the quantum equilibrium state (such as locality) have no fundamental status. The truly fundamental and non-local physics is visible only in non-equilibrium. Therefore, a fundamental appraisal of space-time structure must be in terms of non-equilibrium physics, taking into account instantaneous signaling.

Kinematics and dynamics

This might seem problematic, it still being common among physicists to describe superluminal effects as "acausal". But superluminal signaling violates causality – that is, gives rise to backwards-in-time signals in some frames – *if* one assumes a locally Minkowski structure for space-time. Historically, the Minkowski structure was developed for a local physics. If Nature turns out to be non-local, then one should consider revising that structure.

This may seem an obvious point. Yet, many physicists tend to think of Minkowski space-time as a prior ("God-given") background or stage on which physics takes place (at least locally, ignoring gravitation for the moment). A common view is that laws such as Maxwell's equations possess Lorentz symmetry "because" space-time has a Minkowski structure. It is as if we were first given the stage of space-time, and afterwards we wrote laws on it. But one could *equally* take the view that space-time has a Minkowski structure "because" the known laws all have a Lorentz symmetry.[2] This would certainly be closer to the historical facts: first one discovers certain symmetries in the behavior of matter, then one postulates a space-time structure that incorporates those symmetries.

From this last perspective, one should be open to the possibility that, in the future, new phenomena might break old symmetries, or, that new symmetries might emerge; and in either case, the structure of space-time might have to be revised. In a word, one should bear in mind that new laws of physics might demand a new structure for space-time.

Kinematics and dynamics are two sides of the same coin (Brown 2005). As we discover new dynamical effects, we should be prepared to modify our kinematics (or space-time geometry) if necessary or convenient. In the section entitled "Instantaneous signaling in quantum non-equilibrium," we shall describe an effect whose observation in the future is, in the author's opinion, to be expected from a hidden-variables perspective, and which would, we argue, lead us to modify our current relativistic kinematics.

The rise of relativity theory should have taught us the lesson that the structure of space-time is not *a priori*, but depends on physics – just as more recently, with the rise of quantum computing, we have come to learn that the theory of computation is not *a priori* but depends on physics. Unfortunately, after 1905, the dogma of Newtonian space-time was quickly replaced by the dogma of (local) Minkowski space-time.

The replacement of one rigid view by another was perhaps due in part to Einstein's unfortunate "operational" presentation in his first relativity paper of 1905 (Einstein 1905), which treated macroscopic rods and clocks as if they were fundamental entities. This led to a widespread misunderstanding, according to which the resulting kinematics was somehow logically inevitable, when in fact it was highly contingent on properties of the physical dynamics known at the time.

Einstein and Poincaré in 1905

Einstein himself acknowledged the conceptual mistake in his autobiographical notes of 1949:

> The theory [special relativity]. . . introduces two kinds of physical things, i.e., (1) measuring rods and clocks, (2) all other things, e.g., the electromagnetic field, the material point, etc. This, in a certain sense, is inconsistent; strictly speaking measuring rods and clocks would have to be represented as solutions of the basic equations (objects consisting of moving atomic configurations), not, as it were, as theoretically self-sufficient entities (Einstein 1949b).

In 1905 Einstein had treated rods and clocks as primitive entities, independent of theory ("theoretically self-sufficient"). But in fact, as Einstein later recognized, rods and clocks are phenomenological entities arising out of some underlying theory (perhaps involving particles and/or fields). In reality, we need some body of theory to tell us how to construct reliable rods and clocks and to analyze their behavior. For example, using theory we can calculate the effect of acceleration on a real clock, and so use theory to design more robust clocks. Rods and clocks are not simply "given" to us.

The modern view of special relativity, used in high-energy physics for example, makes no mention of rods and clocks. It concerns particles and fields on Minkowski space-time. The essence of Lorentz invariance is simply that the Lagrangian density appearing in quantum field theory should be a Lorentz scalar (resulting in a Lorentz-covariant *S*-matrix). Nor do classical light waves play any special role: what matters are the symmetries of the fundamental equations, not the speed of propagation of some particular particle or field. After all, the photon might turn out to have a small mass. That we first discovered Lorentz invariance via the classical electromagnetic field is merely a historical accident, and Einstein's 1905 approach – based on macroscopic rods, clocks and classical light waves – is merely a historical (and fundamentally inconsistent) heuristic.

The popularity of Einstein's "operational" approach to special relativity had the effect of introducing a deep and widespread confusion between phenomenological and fundamental entities. This confusion seems to have encouraged an overly-rigid philosophy of space and time, in which Einstein's kinematics came to appear as an inevitable – *a priori*, and theoretically self-sufficient – background to the laws of dynamics.[3] Today, despite the discovery of quantum non-locality, there is still a reluctance in some quarters even to consider changing our view of space-time structure.

It is often claimed that Einstein's 1905 approach should be regarded as not merely a historical curiosity, but as the proper way to understand special relativity. After all, it was this approach that in fact first led us to special relativity. And how else could special relativity have been discovered? But as a matter of historical fact, building on earlier work by Lorentz and others, the formal structure of special relativity – the relativity principle, the universality of the Lorentz group, the relativistic addition of velocities, and even 4-vectors with the associated 4-dimensional invariant interval (later taken up by Minkowski in 1908) – was independently arrived at by Poincaré in his paper "On the Dynamics of the Electron" (1906). This paper was submitted to a mathematical journal in Palermo, in the same summer (of 1905) as Einstein's first relativity paper was submitted to the *Annalen der Physik*; it was published in 1906. A summary of the results was published in 1905, in a short paper of the same title (Poincaré 1905b).[4]

The importance of Poincaré's "Palermo" paper has been underestimated, even by some historians. Certainly, most physicists are not even aware of its existence. (An incomplete translation appears in Kilmister (1970); a modernized presentation of most of the paper is given in Schwartz (1971, 1972). For detailed analyses of the paper, see Miller (1973) and Zahar (1989). More recent discussions of Poincaré's extensive contributions to special relativity have been given by Darrigol (1995, 1996) and Granek (2000).)

Among physicists, Pauli was exceptional in being careful to credit Poincaré's Palermo paper properly throughout his celebrated treatise on relativity (Pauli 1958; first published in 1921). For example, with reference to the Palermo paper, Pauli notes that:

> The formal gaps left by Lorentz's work were filled by Poincaré. He stated the relativity principle to be generally and rigorously valid. Since he... assumed Maxwell's equations to hold for the vacuum, this amounted to the requirement that all laws of nature must be covariant with respect to the 'Lorentz transformation'. The terms 'Lorentz transformation' and 'Lorentz group' occurred for the first time in this paper by Poincaré.[5]. ... Poincaré further corrected Lorentz's formulae for the transformations of charge density and current and so derived the complete covariance of the field equations of electron theory.
>
> (Pauli 1958: 3)

Pauli correctly credits Poincaré, not only for postulating the Lorentz group as a universal symmetry group, but also for the first use of 4-vectors and of the associated 4-dimensional invariant interval. Pauli writes:

> As a precursor of Minkowski one should mention Poincaré... He already introduced on occasion the imaginary coordinate $u = ict$ and combined, and interpreted as point coordinates in R_4, those quantities which we now call vector components. Furthermore, the invariant interval plays a rôle in his considerations.
>
> (Pauli 1958: 21)

How had Poincaré done it? The answer is, along the lines that most workers in high-energy physics would probably take today. (Poincaré was concerned with the detailed structure and dynamics of the electron, the "elementary particle physics" of the time.) He first notes the experimental failure to detect the absolute motion of the Earth, and proposes that this is "a general law of Nature," which he calls the "Relativity Postulate" (Poincaré 1906: 129). Further, following and perfecting the extensive work of Lorentz, Poincaré notes that Maxwell's equations have the Lorentz group as an exact symmetry group, and *postulates* that this is a universal symmetry applicable to all forces (including gravitation). Poincaré recognizes that this postulate suffices to explain the observed invariance of phenomena under a boost. Citing Lorentz, Poincaré writes:

> If one can impart a common boost to the whole system without any of the apparent phenomena being modified, this is because the equations of an electromagnetic medium are not changed by certain transformations, which we shall call *Lorentz transformations*; two systems, one at rest, the other in motion, thus become exact images of each other. ... According to him [Lorentz], all forces, whatever their origin, are affected by the Lorentz transformation (and therefore by a boost) in the same manner as electromagnetic forces.
>
> (Poincaré 1906: 130; translation by the author)

Poincaré then deduces the detailed structure of the Lorentz group, including the relativistic addition of velocities, noting that the group leaves invariant the quadratic form $x^2 + y^2 + z^2 - t^2$. There follows an extensive discussion of relativistic electron dynamics. In the final section of the paper, Poincaré formulates a Lorentz-covariant generalization of Newtonian gravitation, with gravitational interactions propagating at the speed of light.[6] This last theory is formulated by finding Lorentz-invariant functions of the velocities and relative positions of the masses (as well as of time). To find these, Poincaré uses the fact that the Lorentz group may be regarded as the group of rotations in a 4-dimensional space with coordinates x, y, z, it. As Poincaré put it:

> We see that the Lorentz transformation is nothing but a rotation of this space around the origin.
>
> (Poincaré 1906: 168; translation by the author)

Independently of Einstein and Minkowski, then, in 1905 Poincaré arrived at the formal, mathematical structure of Minkowski space-time and the Lorentz group.

One may argue over the extent to which Poincaré understood the new kinematics defined by his formalism. According to Darrigol (1995: 35; 1996: 280), Poincaré did understand that the Lorentz-transformed coordinates were to be identified with the actual readings of boosted rods and clocks, since he regarded Lorentz invariance as a physical (not just a mathematical) symmetry, whereby "apparent phenomena" in a moving system follow the same laws as phenomena in a system at rest. Similarly, according to Janssen and Stachel (2004): "Unlike Lorentz, Poincaré realized that the auxiliary quantities are the measured quantities for the moving observer." In fact, as early as 1900, Poincaré understood that if experimenters moving with speed v were to assume that the speed of light is c in every direction, then (to lowest order in v/c) they would synchronize clocks separated by a distance x such that the settings differ by $-vx/c^2$ (see the section entitled "Flat space-time"). At least to lowest order in v/c, Poincaré had already understood in 1900 that the Lorentz-transformed time corresponded to the actual readings of moving clocks.[7]

Any suggestion that Poincaré viewed the Lorentz transformation as a purely mathematical change of variables seems untenable. After all, Poincaré asserted that Lorentz invariance alone sufficed to explain the invariance of apparent phenomena under a boost, so the transformed quantities in question must indeed have been regarded as those measured by a moving observer. (In contrast, for Lorentz, his "theorem of corresponding states" – which was mathematically almost the same as Lorentz invariance – had to be supplemented by further physical assumptions to explain the failure to detect ether drift (Janssen and Stachel 2004).) Further, in his Palermo paper, Poincaré derives real physical corrections to Newton's law of gravity, from the requirement that the law of motion for gravitating bodies should be covariant with respect to rotations in what we would now call Minkowski space (with coordinates x, y, z, it). For Poincaré, this symmetry clearly had real, observable physical consequences.

One may also ask if Poincaré (like Lorentz) took the view that there was a true rest frame. According to Darrigol (1995: 40), for example, Poincaré did indeed maintain this view (which Darrigol sees as the only essential difference between Poincaré and Einstein in 1905). On this point it should be remembered that (as we shall discuss in the section entitled "Discussion and conclusion") for Poincaré, the geometry of space-time is not a fact about the world but merely a convenient convention, so that if one finds it convenient one may indeed think in terms of absolute space and time.[8] In any case, this interpretation of Poincaré's made no empirical difference. Further, we argue that in the light of quantum non-locality it may well be the better interpretation after all.

It does seem fair to say – despite (limited) anticipations by Fitzgerald, Lorentz, and Larmor[9] – that a clear and complete statement of universal time dilation and length contraction is first found in Einstein's paper of 1905. Poincaré's Palermo paper discusses length contraction for spherical electrons, but does not explicitly mention time dilation, despite extensive use of the Lorentz-transformed time variable. As Pauli observed, regarding time dilation:

> While this consequence of the Lorentz transformation was already implicitly contained in Lorentz's and Poincaré's results, it received its first clear statement only by Einstein.
>
> (Pauli 1958: 13)

It was claimed by Pais (1982: 164, 167–68) that even after 1905 Poincaré did not understand special relativity, because, judging from the text of his lectures at Göttingen in 1909 (Poincaré 1910), he did not understand that length contraction was a consequence of Einstein's two postulates (the relativity principle and the light postulate), but instead insisted on including length contraction as a third postulate. In the author's opinion, this issue is confused because Einstein's 1905 approach actually contains an implicit third postulate: that under a boost from one rest frame to another, unit rods are transformed into unit rods, and similarly for unit clock ticks. Einstein himself admitted this, in a footnote to a review he published in 1910, where he writes:

> It should be noted that we will always implicitly assume that the fact of a measuring rod or a clock being set in motion or brought to rest does not change the length of the rod or the rate of the clock.
>
> (Einstein 1993: 130)

To the author's knowledge, the only other place in the historical literature where Einstein's implicit third postulate is mentioned is in Born's relativity text (1962).[10] In fact, Born discusses this postulate in some detail, and regards it as of crucial importance. He writes:

> ... it is assumed as self-evident that a measuring rod which is brought into one system of reference S and then into another S' under exactly the same physical conditions would represent the same length in each... A fixed rod that is at rest in the system S and is of length 1 cm. will, of course, also have the length 1 cm. when it is at rest in the system S'... Exactly the same would be postulated for the clocks... We might call this tacit assumption of Einstein's theory the "principle of the physical identity of the units of measure"... This is the feature of Einstein's theory by which it rises above the standpoint of a mere convention and asserts definite properties of real bodies.
>
> (Born 1962: 251–52)

It might be thought that the third postulate could be dispensed with, by using the relativity principle to deduce that any specific process for constructing rods and clocks must give the same results in all inertial frames. Certainly, using the light postulate as well, one could then deduce that the Lorentz transformation relates the readings of *different* rods and clocks tha have been constructed (by a similar process) in different inertial frames. However, one would still have deduced nothing about what happens when the *same* rod or clock is boosted (or accelerated) from one inertial frame to another. (As an example one might, in principle, envisage a theory satisfying the relativity principle and the light postulate, but with the additional property that once a rod or clock has been constructed in a given inertial frame it is destroyed by any subsequent arbitrarily small acceleration.)

Thus, despite widespread opinion to the contrary, length contraction and time dilation under a boost do *not* follow from Einstein's two postulates alone. A further postulate is required, to relate the readings of rods and clocks boosted from one inertial frame to another.[11]

In view of the crucial importance of the third assumption implicitly used by Einstein, it must be regarded as regrettable that Einstein did not mention it explicitly in his first relativity paper. In the author's opinion, it is quite possible that Poincaré was aware of this lacuna, explaining why in his lectures of 1909 (Poincaré 1910) – where he sketches an axiomatic basis for the "new mechanics," in terms of simple physical postulates independent of the details of Maxwell's equations, much as Einstein did in 1905 – he added the third postulate of length contraction, which was not as elegant as the third postulate implicitly used by Einstein, but effective nonetheless.[12]

In any case, such detailed questions of priority, or of who understood exactly what and when, while historically interesting, are not strictly relevant here. What really matters, for our purpose, is that the approach taken in Poincaré's Palermo paper – in which the Lorentz group is first discovered through Maxwell's equations and then postulated to be a universal (physical) symmetry group – quite plainly *could have been* the historical route to special relativity. Regardless of the extent to which Poincaré did or did not understand it at the time, the fact is that the kinematics of Minkowski space-time was contained in the formal structure put forward in Poincaré's paper.

Minkowski, in his famous lecture on "Space and Time" delivered in 1908 (Minkowski 1952), appears to express a preference for this sort of approach, which actually goes back to 1887 when Voigt (1887) derived the Lorentz transformation, up to an overall constant factor, as a symmetry of the (scalar) wave equation.[13] According to Minkowski:

> Now the impulse and true motive for assuming the group G_c [that is, the Poincaré group, which leaves invariant the 4-dimensional interval] came from the fact that the differential equation for the propagation of light in empty space possesses that group G_c. An application of this fact in its essentials has already been given by W. Voigt, Göttinger Nachrichten, 1887, p. 41.[14]
>
> (Minkowski 1952: 81)

It is sometimes argued that Einstein's operational approach has the advantage of being independent of the details of specific equations such as Maxwell's. This may be so, but Einstein's approach also has the disadvantage of giving a special status to classical light waves, and of being conceptually inconsistent with regard to the nature of rods and clocks. As for Poincaré's approach, as a scientific methodology there is nothing wrong with discovering a symmetry in certain equations and then postulating that the symmetry is universal (regardless of whether those equations turn out to be fundamental or not). This is, after all, common practice in high-energy physics today. Clearly, Einstein's operational approach was not necessary, and special relativity could have been (and arguably essentially was) discovered without appeal to a fundamentally inconsistent operationalism.

In the author's opinion, if Poincaré's approach had in fact been the generally accepted historical route to special relativity, then physicists today might be more keenly aware that space-time geometry is not "prior to" dynamics but is rather a reflection of symmetries of the currently known dynamics.[15] From this standpoint, as physics progresses, the structure of space-time is as subject to possible revision as are the laws of dynamics themselves.

Instantaneous signaling in quantum non-equilibrium

In this section we show how, in deterministic hidden-variables theories, a non-standard distribution of hidden variables (generically) gives rise to instantaneous signaling at the statistical level. We first discuss this for general theories (Valentini 2002a, 2002b), then for the specific example of the pilot-wave theory of de Broglie and Bohm (Valentini 1991b, 2002c).

General (deterministic) hidden-variables theories

For a 2-state system, consider quantum observables of the form $\hat{\sigma} = \mathbf{m} \cdot \hat{\boldsymbol{\sigma}}$, where $\mathbf{m}$ is a unit vector specifying a point on the Bloch sphere and $\hat{\sigma}$ is the Pauli spin operator. The values $\sigma = \pm 1$ are obtained upon performing a quantum measurement of $\hat{\sigma}$. Over an ensemble with density operator $\hat{\rho}$, the quantum expectation value of $\hat{\sigma}$ is given by the Born rule as $\langle \hat{\sigma} \rangle = \mathrm{Tr}(\hat{\rho} \mathbf{m} \cdot \hat{\boldsymbol{\sigma}}) = \mathbf{m} \cdot \mathbf{P}$, where $\mathbf{P} = \langle \hat{\boldsymbol{\sigma}} \rangle$ (with norm $0 \leq P \leq 1$) is the mean polarization. The quantum probabilities $p^{\pm}_{\mathrm{QT}}(\mathbf{m})$ for outcomes $\sigma = \pm 1$ are then fixed as

$$p^{\pm}_{\mathrm{QT}}(\mathbf{m}) = \frac{1}{2}(1 \pm \mathbf{m} \cdot \mathbf{P}) \tag{1}$$

In a (deterministic) hidden-variables theory, for every run of the experiment with measurement axis $\mathbf{m}$, there are hidden parameters collectively denoted λ that determine the outcome $\sigma = \pm 1$ according to some mapping $\sigma = \sigma(\mathbf{m}, \lambda)$. Over an ensemble of experiments, the observed distribution of outcomes is explained by some assumed distribution $\rho_{\mathrm{QT}}(\lambda)$ of parameters λ, where $\rho_{\mathrm{QT}}(\lambda)$ is such that expectations

$$\langle \sigma(\mathbf{m}, \lambda) \rangle_{\mathrm{QT}} = \int d\lambda \; \rho_{\mathrm{QT}}(\lambda) \sigma(\mathbf{m}, \lambda)$$

agree with the quantum prediction $\langle \mathbf{m} \cdot \hat{\boldsymbol{\sigma}} \rangle$. The values of λ are usually defined at some initial time, say at the time of preparation of the quantum state. The outcomes $\sigma = \sigma(\mathbf{m}, \lambda)$ are defined at the time of measurement.

Now, there is a clear conceptual distinction between the initial values λ and the mapping $\sigma = \sigma(\mathbf{m}, \lambda)$ to final outcomes σ. In particular, the former amount to what are usually called "initial conditions," while the latter would usually be called a "dynamical law" that maps initial conditions to final states. Therefore, once such a theory has been constructed, one may contemplate arbitrary initial conditions – over an ensemble, distributions $\rho(\lambda) \neq \rho_{\mathrm{QT}}(\lambda)$ – while retaining the mapping $\sigma = \sigma(\mathbf{m}, \lambda)$. Generically, such "non-quantum" or "non-equilibrium" distributions will yield expectation values

$$\langle \sigma(\mathbf{m}, \lambda) \rangle = \int d\lambda \; \rho(\lambda) \sigma(\mathbf{m}, \lambda)$$

that disagree with quantum theory, and the statistics of outcomes will generally violate the standard quantum-theoretical constraints. Note the key conceptual point: we have the same deterministic mapping $\sigma = \sigma(\mathbf{m}, \lambda)$ for each system, regardless of the (equilibrium or non-equilibrium) distribution for the ensemble.

Many of the supposedly fundamental constraints of quantum theory, such as statistical locality, are (from a hidden-variables perspective) merely contingent features of the special distribution $\rho_{QT}(\lambda)$. As noted in the section entitled "Non-equilibrium hidden variables," there is an analogy here with the contingent constraints that arise in classical physics in a state of global thermal equilibrium: the inability to convert heat into work is not fundamental, but a contingency due to all systems having the same temperature.

Consider a pair of widely separated 2-state systems with spatial locations A and B. Quantum measurements of $\hat{\sigma}_A \equiv \mathbf{m}_A \cdot \hat{\boldsymbol{\sigma}}_A$, $\hat{\sigma}_B \equiv \mathbf{m}_B \cdot \hat{\boldsymbol{\sigma}}_B$ can yield outcomes σ_A, $\sigma_B = \pm 1$. For the singlet state

$$|\Psi\rangle = (|{+\mathbf{n}}, -\mathbf{n}\rangle - |{-\mathbf{n}}, +\mathbf{n}\rangle)/\sqrt{2}$$

(for any axis $\mathbf{n}$) quantum theory predicts that outcomes σ_A, $\sigma_B = \pm 1$ occur in the ratio 1 : 1 at each wing, with a correlation

$$\langle \Psi | \hat{\sigma}_A \hat{\sigma}_B | \Psi \rangle = -\mathbf{m}_A \cdot \mathbf{m}_B \tag{2}$$

Nevertheless, the distant settings have no effect on the expectation values ($\langle \hat{\sigma}_{A,B} \rangle = 0$) or on the probabilities ($p^{\pm}_{QT}(\mathbf{m}_{A,B}) = 1/2$) at each wing, making non-local signaling impossible.

However, from a hidden-variables perspective, Bell's theorem (1964) tells us that to reproduce this correlation a hidden-variables theory must take the non-local form

$$\sigma_A = \sigma_A(\mathbf{m}_A, \mathbf{m}_B, \lambda), \qquad \sigma_B = \sigma_B(\mathbf{m}_A, \mathbf{m}_B, \lambda) \tag{3}$$

in which the individual outcomes σ_A, σ_B do depend on the distant measurement settings. Only with such non-local dependence can the theory reproduce the quantum correlation

$$\langle \sigma_A \sigma_B \rangle_{QT} \equiv \int d\lambda \; \rho_{QT}(\lambda) \sigma_A(\mathbf{m}_A, \mathbf{m}_B, \lambda) \sigma_B(\mathbf{m}_A, \mathbf{m}_B, \lambda) = -\mathbf{m}_A \cdot \mathbf{m}_B \tag{4}$$

for some ensemble distribution $\rho_{QT}(\lambda)$. More precisely, at least one of σ_A, σ_B must depend on the distant setting, and without loss of generality we shall assume that σ_A has a non-local dependence on $\mathbf{m}_B$.

Now, for an arbitrary ensemble with $\rho(\lambda) \neq \rho_{QT}(\lambda)$, in general

$$\langle \sigma_A \sigma_B \rangle \equiv \int d\lambda \; \rho(\lambda) \sigma_A(\mathbf{m}_A, \mathbf{m}_B, \lambda) \sigma_B(\mathbf{m}_A, \mathbf{m}_B, \lambda) \neq -\mathbf{m}_A \cdot \mathbf{m}_B \tag{5}$$

and the outcomes σ_A, $\sigma_B = \pm 1$ at each wing will occur in a ratio generally differing from 1 : 1. Further, under a change in the measurement setting at one wing, the outcome statistics at the distant wing will generally change, amounting to a non-local signal at the statistical level. The key point here is that, assuming a non-local dependence of σ_A on $\mathbf{m}_B$, the "transition sets"

$$T_A(-,+) \equiv \{\lambda | \sigma_A(\mathbf{m}_A, \mathbf{m}_B, \lambda) = -1,\ \sigma_A(\mathbf{m}_A, \mathbf{m}'_B, \lambda) = +1\}$$
$$T_A(+,-) \equiv \{\lambda | \sigma_A(\mathbf{m}_A, \mathbf{m}_B, \lambda) = +1,\ \sigma_A(\mathbf{m}_A, \mathbf{m}'_B, \lambda) = -1\}$$

cannot be empty for arbitrary settings $\mathbf{m}_A$, $\mathbf{m}_B$, $\mathbf{m}'_B$. Some outcomes at A must change under a shift $\mathbf{m}_B \rightarrow \mathbf{m}'_B$ at B. In quantum equilibrium, the ratio of outcomes $\sigma_A = \pm 1$ is $1:1$ for all settings, therefore we must have "detailed balancing"

$$\mu_{\mathrm{QT}}[T_A(-,+)] = \mu_{\mathrm{QT}}[T_A(+,-)]$$

with respect to the equilibrium measure $d\mu_{\mathrm{QT}} \equiv \rho_{\mathrm{QT}}(\lambda)d\lambda$. In other words, in quantum equilibrium, the fraction of the ensemble making the transition $\sigma_A = -1 \rightarrow \sigma_A = +1$ under $\mathbf{m}_B \rightarrow \mathbf{m}'_B$ must equal the fraction making the reverse transition $\sigma_A = +1 \rightarrow \sigma_A = -1$. (This is analogous to the principle of detailed balance in statistical mechanics: thermal equilibrium is maintained if the mean transition rate from state i to state j is equal to the mean transition rate from j to i.) Since $T_A(-,+)$ and $T_A(+,-)$ are fixed by deterministic equations, they are independent of the ensemble distribution of λ. Thus, for a hypothetical non-equilibrium ensemble $\rho(\lambda) \neq \rho_{\mathrm{QT}}(\lambda)$, in general

$$\mu[\mathrm{T_A}(-,+)] \neq \mu[\mathrm{T_A}(+,-)]$$

where $d\mu \equiv \rho(\lambda)\mathrm{d}\lambda$. In other words, the fraction of the non-equilibrium ensemble making the transition $\sigma_A = -1 \rightarrow \sigma_A = +1$ will not in general balance the fraction making the reverse transition; the ratio of outcomes at A will in general change under $\mathbf{m}_B \rightarrow \mathbf{m}'_B$ and there will be instantaneous signals at the statistical level from B to A (Valentini 2002a, 2002b).

In any deterministic hidden-variables theory, then, hypothetical non-equilibrium distributions $\rho(\lambda) \neq \rho_{\mathrm{QT}}(\lambda)$ generally make it possible to use non-locality for instantaneous signaling (just as, in ordinary statistical physics, differences of temperature make it possible to convert heat into work) (Valentini 2002a, 2002b).

The example of pilot-wave theory

Non-equilibrium signaling at a distance was first noted (Valentini 1991b, 2002c) in the hidden-variables theory of de Broglie and Bohm (de Broglie 1928; Bohm 1952a, 1952b). In this "pilot-wave theory" (as it was originally called by de Broglie), a system with wave function $\Psi(X, t)$ satisfying the Schrödinger equation

$$i\frac{\partial \Psi}{\partial t} = \hat{H}\Psi \tag{6}$$

has an actual configuration $X(t)$ whose motion is given by the first-order differential equation

$$\dot{X}(t) = \frac{J(X,t)}{|\Psi(X,t)|^2} \tag{7}$$

where $J = J[\Psi] = J(X,t)$ (which in quantum theory is called the "probability current") satisfies the continuity equation

$$\frac{\partial |\Psi|^2}{\partial t} + \nabla_X \cdot J = 0 \tag{8}$$

(which follows from (6)). In pilot-wave theory, Ψ is regarded as an objective physical field guiding the system.

For example, for a system of N particles with 3-vector positions $\mathbf{x}_i(t)$ and masses m_i $(i = 1, 2, \ldots, N)$ the wave function $\Psi(X,t)$ on $3N$-dimensional configuration space $(X \equiv (\mathbf{x}_1, \mathbf{x}_2, \ldots, \mathbf{x}_N))$ is a complex field obeying the Schrödinger equation

$$i\frac{\partial \Psi}{\partial t} = \sum_{i=1}^{N} -\frac{1}{2m_i}\nabla_i^2\Psi + V\Psi \tag{9}$$

and the particle velocities are given by

$$\frac{d\mathbf{x}_i}{dt} = \frac{1}{m_i}\mathrm{Im}\left(\frac{\nabla_i \Psi}{\Psi}\right) = \frac{\nabla_i S}{m_i} \tag{10}$$

where $\Psi = |\Psi|e^{iS}$ and we take $\hbar = 1$.

Equations (6) and (7) determine the motion $X(t)$ of an *individual* system, given the initial configuration $X(0)$ and wave function $\Psi(X,0)$ at $t = 0$. If we are given an arbitrary initial distribution $P(X,0)$, for an ensemble of systems with the same wave function $\Psi(X,0)$, then the evolution of $P(X,t)$ is necessarily given by the continuity equation

$$\frac{\partial P}{\partial t} + \nabla_X \cdot (P\dot{X}) = 0 \tag{11}$$

This same equation is satisfied by $|\Psi|^2$, as follows from (8). Thus, if $P(X,0) = |\Psi(X,0)|^2$ at some initial time, then $P(X,t) = |\Psi(X,t)|^2$ at all times t. As shown by Bohm (1952a, 1952b), one then recovers the statistical predictions of quantum theory.

In pilot-wave theory, the outcome obtained in a given experiment is determined by $X(0)$ and $\Psi(X, 0)$, so that one may identify λ with the pair $X(0)$, $\Psi(X, 0)$. For an ensemble of experiments with the same $\Psi(X, 0)$, in effect λ is just $X(0)$, and the distribution $\rho_{\mathrm{QT}}(\lambda)$ is given by $P_{\mathrm{QT}}(X, t) = |\Psi(X, t)|^2$. As in the general discussion above, we may retain the same deterministic dynamics for individual systems, and consider a non-standard distribution of initial conditions. Here, this means we retain the dynamical equations (6), (7) and consider an arbitrary initial ensemble with $P(X, 0) \neq |\Psi(X, 0)|^2$. The evolution of $P(X, t)$ will be determined by (11).

In appropriate circumstances, (11) leads to relaxation $P \to |\Psi|^2$ on a coarse-grained level (Valentini 1991a, 1992, 2001; Valentini and Westman 2005), much as the corresponding classical evolution on phase space leads to thermal relaxation. However, for as long as the ensemble is in non-equilibrium, the statistics of outcomes of quantum measurements will disagree with quantum theory.

As required by Bell's theorem, pilot-wave theory is fundamentally non-local. For two particles whose wave function $\Psi(\mathbf{x}_A, \mathbf{x}_B, t)$ is entangled, $\dot{\mathbf{x}}_A(t) = \nabla_A S(\mathbf{x}_A, \mathbf{x}_B, t)/m_A$ depends instantaneously on $\mathbf{x}_B$, and ordinary operations on particle B – such as switching on a local potential – have an instantaneous effect on the motion of particle A. But for a quantum equilibrium ensemble $P(\mathbf{x}_A, \mathbf{x}_B, t) = |\Psi(\mathbf{x}_A, \mathbf{x}_B, t)|^2$, such operations on particle B have no statistical effect on particle A: the individual non-local effects are masked by quantum noise.

As in the general case discussed above, non-locality is (generally speaking) hidden by statistical noise only in quantum equilibrium. For an ensemble of entangled particles with initial distribution $P(\mathbf{x}_A, \mathbf{x}_B, 0) \neq |\Psi(\mathbf{x}_A, \mathbf{x}_B, 0)|^2$, a local change in the Hamiltonian of particle B generally induces an instantaneous change in the marginal distribution $p_A(\mathbf{x}_A, t) \equiv \int d^3\mathbf{x}_B \, P(\mathbf{x}_A, \mathbf{x}_B, t)$ of particle A. For example, in one dimension, a sudden change $\hat{H}_B \to \hat{H}'_B$ in the Hamiltonian of particle B induces a change $\Delta p_A \equiv p_A(x_A, t) - p_A(x_A, 0)$ of the form (for small t) (Valentini 1991b)

$$\Delta p_A = -\frac{t^2}{4m} \frac{\partial}{\partial x_A} \left(a(x_A) \int dx_B \, b(x_B) \frac{P(x_A, x_B, 0) - |\Psi(x_A, x_B, 0)|^2}{|\Psi(x_A, x_B, 0)|^2} \right) \tag{12}$$

(Here $m_A = m_B = m$, the factor $a(x_A)$ depends on $\Psi(x_A, x_B, 0)$, while $b(x_B)$ also depends on $\hat{H}'_B$ and vanishes if $\hat{H}'_B = \hat{H}_B$.) In general, the signal is non-zero if $P_0 \neq |\Psi_0|^2$ (that is, if $\rho(\lambda) \neq \rho_{\mathrm{QT}}(\lambda)$).

Elsewhere (Valentini 2002c), using the example of pilot-wave theory, we have described how non-equilibrium particles might be detected in practice, by the statistical analysis of random samples (taken, for example, from a parent population of relic particles left over from the early universe). Once such particles have been identified, they may be used as a resource for superluminal signaling; further, they may be used to perform "subquantum measurements" on ordinary, equilibrium systems (Valentini 2002c).

Absolute simultaneity in flat and curved space-time

We have seen that, even at ordinary laboratory distances and energies, quantum non-equilibrium would unleash instantaneous signals between entangled systems. This raises the question of how these signals could mesh with the surrounding approximately classical space-time. As we emphasized in the Introduction, this question must have an answer, *irrespective* of the underlying microscopic theory of space-time.

If experimenters at space-time events A and B had access to non-equilibrium systems entangled between A and B, then they would be able to signal back and forth to each other instantaneously. In an arbitrarily short time (as measured at each wing), a long conversation could in principle take place, during which (for example) the experimenters agree to set their clocks to read time $t = 0$. They could signal to each other to confirm that they have done so. In such conditions, A and B must be regarded as simultaneous events, and the agreed-upon time variable t would define an absolute simultaneity. Thus, using non-equilibrium matter, experimenters at remote locations could set their clocks to read the same instantaneous time t.

There are, however, some differences depending on whether gravitation is absent or present. Let us discuss these in turn.

Flat space-time

In the absence of gravitation (where the kinematics is usually represented by flat Minkowski space–time), remote experimenters may use entangled non-equilibrium systems to set their clocks to read the same time t. However, they must be careful to bear in mind that clocks in motion drift out of synchronization with clocks at rest. For if a clock undergoes a spatial displacement $d\mathbf{x}$ in a time dt , then the "proper" time $d\tau$ ticked by the clock is given by

$$d\tau^2 = dt^2 - d\mathbf{x}^2$$

Thus, a clock moving through space with speed $v = |d\mathbf{x}/dt|$ is slowed by the factor $1/\sqrt{1 - v^2/c^2}$. Here, this "time dilation" may be regarded as a dynamical effect of motion on the rate of evolution of physical systems, as originally anticipated by Larmor and Lorentz. (An instructive account of this viewpoint was given by Bell (1987: 67–80).)

One must then distinguish between *simultaneity* and *synchronicity*. The first refers to events that exist "in unison," in a sense that could be verified by non-local communication. The second refers merely to the coincidence of readings of certain (usually classical, macroscopic) systems called "clocks," where for dynamical reasons the rate of evolution of such systems depends on how fast they are moving through space. Simultaneity is not equivalent to synchronicity.

For example, if two clocks are initially close together and synchronized in a standard inertial frame with time function t, and if one clock is accelerated and eventually returns close to its partner, then finally the two clocks will be out of step, as the accelerated clock will have been slowed down. If t coincides with our absolute time, the final clock readings will correspond to simultaneous events, yet, the readings will not be synchronous.

Note that this dynamical effect of motion occurs at the classical macroscopic level, as well as at the statistical level for ensembles of microscopic quantum systems, but it is not necessarily relevant to the deeper level of hidden variables. (For example, atomic decay rates are affected by time dilation, but such rates apply to quantum ensemble averages and not to individual systems.) Therefore, there is no reason why this dynamical effect should be built into the fundamental kinematics (as it usually is).

The objection might be raised that superluminal signals in a given frame would "violate causality," since in other frames the signals could travel backwards in time, leading to paradoxes. But as we discussed in the section entitled "Physical structure of space-time," this argument assumes that the structure of space-time is fundamentally Minkowskian. There is no reason to assume this. At the non-local hidden-variable level, there may well be a preferred slicing of space-time, with a time function t that defines a fundamental causal sequence (Popper 1982; Bohm and Hiley 1984; Bell 1986, 1987).

Clearly, in a given preferred frame with standard Lorentzian coordinates t, x, y, z, instantaneous signaling between distant experimenters would not in itself be problematic. But what about the Lorentz transformation? One might be disturbed by the idea that an experimenter moving along (for example) the x-axis could "see" such signals propagating "backwards in time". However, a real experimenter does not simply "see" the global time of his Lorentz frame. Rather, the experimenter has a collection of clocks distributed over space, which have to be *set* according to some chosen procedure. The time associated with an event occurring at some point in space is just the reading of the clock in the neighborhood of that event. If an event B is for some physical reason regarded as "causing" a spatially distant event A (for example a message is sent from B to A), and if the reading of a clock at B is larger than the reading of a clock at A, then before declaring this paradoxical one ought to ask how the clocks at A and B were set in the first place.

If the moving experimenter chooses Einstein's so-called "synchronisation," using light pulses whose speed is taken to be isotropic, then at (preferred) time t the moving clock located at x, y, z will read a time

$$t' = \frac{t - vx/c^2}{\sqrt{1 - v^2/c^2}} \tag{13}$$

From our perspective, the interpretation of this formula is very simple. The moving clocks distributed along $x > 0$ have been initially set (for example at $t = 0$) to read progressively earlier times, with a lag proportional to x; while the moving clocks along $x < 0$ have been similarly set to read later times. These settings have been chosen precisely so as to make a light pulse (with speed c in the preferred frame[16]) *appear* to have a speed c, along both $+x$ and $-x$, in the moving frame. This is the origin of the term $-vx/c^2$ (to lowest order in v/c). If one includes the effect of motion, which as we have said slows clocks down, one also obtains the factor $1/\sqrt{1 - v^2/c^2}$.

If the moving experimenter adopts the Einstein convention for the synchronization of clocks, then the settings (13) have the following peculiarity: an instantaneous signal propagating along $+x$ in the preferred frame appears to be going "backwards in time" as judged by the moving clocks with settings t'. That is, if the signal starts at x_B and propagates to $x_A > x_B$, then if $v > 0$ the readings t'_A, t'_B of the moving clocks at the events A, B have the property $t'_B > t'_A$. But there is nothing mysterious or paradoxical here: for the moving clocks were initially set with a time-lag proportional to x, and the result $t'_B > t'_A$ is a direct and immediate reflection of this initial set-up. Indeed, this phenomenon is exactly the same as the familiar "jet lag" which occurs when an experimenter moves rapidly from one time zone to another on the Earth's surface. Clocks distributed over the Earth's surface have been set according to a convention related to the locally observed position of the Sun in the sky, and it is in no way surprising or problematic that a jet passenger may in a formal sense "travel backwards in time".

Note that, from this point of view, time dilation is a real physical effect of motion which may be unambiguously verified by experiment (for example by taking one clock on an accelerated round trip and comparing it with an unaccelerated clock before and after). Whereas, the so-called relativity of simultaneity is merely the result of a convention about the way clocks are synchronized in different frames.

It is worth remarking that, as already mentioned in the section entitled "Einstein and Poincaré in 1905," the origin of the term $-vx/c^2$ in (13) was clearly understood by Poincaré well before 1905. In a paper published in 1900 (Poincaré 1900), concerned mainly with action and reaction in electrodynamics, Poincaré (who works to lowest order in v/c) writes:[17]

> I assume that observers situated at different points set their watches with the aid of light signals; that they try to correct these signals by the

transmission time, but that ignoring their translatory motion and therefore believing that the signals are transmitted with equal speed in both directions, they content themselves with crossing the observations, sending a signal from A to B, then another from B to A. The local time t' is the time shown by watches set in this way.

If then $V = 1/\sqrt{K_0}$ is the speed of light, and v the speed of translation of the Earth which I assume parallel to the positive x-axis, one will have:

$$t' = t - \frac{vx}{V^2}$$

(Poincaré 1900: 483; translation by the author)

Poincaré understood that moving experimenters who assume that the speed of light is still c in all directions would adjust their clocks at different points in space with settings that differ by the term $-vx/c^2$ (to lowest order in v/c).

If instead distant clocks are synchronized by non-local means, then the speed of light will be measured to be isotropic only in the preferred rest frame. In quantum equilibrium, of course, such non-local signaling is impossible and the true rest frame cannot be detected.

Note that, in the specific hidden-variables theory given by pilot-wave dynamics, even leaving non-locality aside, the natural kinematics of the theory is arguably that of Aristotelian space–time $E \times E^3$, with a preferred state of rest (Valentini 1997). This is essentially because the dynamics is first order in time, so that rest is the only reasonable definition of "natural" or "unforced" motion. Pilot-wave theory then has a remarkable internal logic: both the structure of the dynamics, and the operational possibility of non-local signaling out of equilibrium, independently point to the existence of a natural preferred state of rest.

Curved space-time

In the presence of gravitation, the above discussion may be extended to any classical background space-time possessing at least one global time function t. This is hardly a restrictive requirement. For it is widely assumed that, classically, any physical space-time must be globally hyperbolic[18] – that is, must possess a Cauchy surface (a space-like slice on which initial data determine the entire space-time) – and it is a theorem that any globally hyperbolic space-time has topology $\mathbb{R} \times \Sigma$ (where Σ is a Cauchy surface) (Hawking and Ellis 1973).

Consider, then, a curved space-time that can be foliated (in general non-uniquely) by space-like hypersurfaces Σ labeled by a global time function t. The classical space-time metric may then be written in the form

$$d\tau^2 = {}^{(4)}g_{\mu\nu}dx^\mu dx^\nu = N^2dt^2 - g_{ij}dx^i dx^j$$

where we have set the shift vector $N^i = 0$, so that lines $x^i =$ const. are normal to Σ. (This may always be done, as long as the lines $x^i =$ const. do not run into singularities.) The lapse function $N(x^i, t)$ measures the proper time lapse normal to Σ per unit of coordinate time t.

It may now be assumed that non-locality acts instantaneously with respect to one of these foliations, denoted $\Sigma(t)$. There is then a true slicing, and space-time is really the time evolution of the (absolute) 3-geometry $\mathcal{G}(t)$ of $\Sigma(t)$, with metric $g_{ij}(x^k, t)$ (Valentini 1992, 1996).

On this view, a small rod at time t has proper length

$$dl = (g_{ij}dx^i dx^j)^{1/2}$$

while a clock at rest in 3-space ticks a proper time

$$d\tau = N(x^i, t)dt$$

If a clock moves a spatial distance dl in a time dt it will tick a proper time

$$d\tau^2 = N^2 dt^2 - dl^2$$

Some remarks are in order.

First, there is an asymmetry here between space and time. It is assumed that ordinary rods faithfully realize the true distance element dl of space. Whereas, we assume that ordinary clocks do not faithfully register true time t; rather, their rate of ticking is affected by the local lapse field $N(x^i, t)$.

Second, note the difference from the case where gravity is absent. There we saw that moving clocks are slowed down. The same effect occurs here, but in addition, the rate of ticking of clocks is affected by their spatial location. There is a field $N(x^i, t)$ on 3-space which has a dynamical effect on the rate of clocks even when they are at rest.

Third, this interpretation does not necessarily involve the introduction of an independent field N on 3-space. For this field could be determined by the geometry of 3-space; N could, for example, be a simple fixed function of the 3-metric g_{ij} such as

$$N = g^{-1/2} \qquad (g \equiv \det g_{ij})$$

(as in unimodular gravitation with $\det^{(4)} g_{\mu\nu} = 1$ (Unruh 1989)). Presumably, N will be merely an effective field, emerging from some more fundamental theory (possibly a quantum or subquantum theory of gravity). In this way, there could be an underlying dynamical origin for the phenomenological distortion of clock rates by the field N.

As in the flat case, one must be careful to distinguish between simultaneity and synchronicity. Clocks located at different spatial points x^i on the same hypersurface (with label t) record simultaneous events, but the field N causes even stationary clocks to tick at different rates and lose their synchrony. Thus, for example, let clocks at events A_1, B_1 at time t_1 (on the preferred space-like hypersurface Σ_1) move along time-like lines to events A_2, B_2 at time $t_2 > t_1$ (on the preferred space-like hypersurface Σ_2). Assume for simplicity that the clocks remain at rest in space. Then each clock will tick a proper time

$$\Delta\tau = \int_{t_1}^{t_2} N(x^i, t)dt$$

where the integral is taken along the respective path. The lapse function N will generally differ along the two paths $A_1 - A_2$, $B_1 - B_2$. Thus, clocks synchronized at the simultaneous events A_1 and B_1 (using non-local signals) will no longer be synchronized at A_2 and B_2, even though A_2 and B_2 are also simultaneous.

From a conventional perspective, this view will certainly seem eccentric, and indeed it would be in the absence of any evidence for non-locality. But if one takes seriously Bell's deduction that non-local influences do occur in Nature, and if one further accepts that our current inability to control these events is merely a contingency of a particular distribution of hidden variables, and bearing in mind that these effects occur at ordinary energies and macroscopic distances, then the above view provides a consistent phenomenological means of embedding such non-locally connected quantum events within the surrounding approximately classical space-time.

Again, one need not view the above construction as fundamental. A microscopic theory of space-time may well provide a very different picture at the fundamental level. But if one accepts the existence of non-locality, then it seems natural that the above construction should emerge in some approximation.

In quantum equilibrium, of course, non-locality and the true slicing cannot be detected, as in the case of flat space-time. Possibly, the observed cosmological rest frame is a relic of early non-locality – arising from quantum non-equilibrium in the early universe – and coincides with true rest (Valentini 1991b, 1992, 1996).

Discussion and conclusion

We have presented a means of embedding quantum non-locality within a background classical space-time (flat or curved), by introducing an absolute simultaneity associated with a preferred foliation by space-like hypersurfaces (where the preferred foliation defines a preferred local state of rest). It should be noted that this is unlikely to be the *only* way of constructing such an embedding. For as emphasized by Poincaré, the choice of geometry to be used in physics is really dictated by convenience. There is no question of proving that the most convenient choice is the only one possible, because one may always adopt a different geometry by adding appropriate compensating factors to the dynamics.

In his book *Science and Hypothesis* (Poincaré 1902), Poincaré illustrated this point in terms of an analogy with measuring rods affected by thermal expansion. Consider, for example, metal rods on a heated flat metal plate.[19] If the temperature of the plate is non-uniform, and if all the rods have the same expansion coefficient, then (assuming the rods reach thermal equilibrium instantly) measurements within the surface using these rods will simulate the geometry of a curved 2-surface – that is, a non-Euclidean geometry. Creatures living on such a surface could believe it to be curved, as long as all their rods were affected by temperature in the same way. Equally, they could believe their surface to be flat, with all rods being universally distorted (expanded or contracted) by means of some agency acting upon them. There would be no way of telling the difference. However, the creatures may well come to think that, because the required distortions are the same for all rods, it is more convenient to ascribe the distortions to the geometry of space itself; that is, if the apparent geometry of the 2-surface is the same no matter which rods are used, then one may as well define the apparent geometry to be the actual geometry of space. As Poincaré put it:

> Experiment... tells us not what is the truest, but what is the most convenient geometry.
>
> (Poincaré 1902)

The situation is no different in present-day physics. For example, instead of interpreting general relativity in terms of a curved space-time with metric $g_{\mu\nu}$, it is possible to interpret it in terms of a Minkowski space-time, with flat metric $\boldsymbol{\eta}_{\mu\nu}$, containing a field $h_{\mu\nu} = g_{\mu\nu} - \boldsymbol{\eta}_{\mu\nu}$ which distorts rods and clocks so as to give the appearance of curved space-time (Weinberg 1972). It cannot be proved that space-time is really curved; but, because the effects of the field $h_{\mu\nu}$ are universal – the same for all rods and clocks – it is more convenient to regard those effects as purely kinematical, that is, as part of the geometry with metric $g_{\mu\nu}$.

Similarly, classical special relativity may equally be interpreted in terms of a preferred (yet unobservable) rest frame, where motion with respect to the preferred frame has the dynamical effect of slowing clocks and contracting rods. As emphasized for example by Bell (1987), this is an equivalent formulation of the same physics. One may find it objectionable to have an underlying preferred frame which can never be detected (classically), but nevertheless this formulation of special-relativistic physics is consistent. It often happens that the same physics can be formulated in equivalent, empirically indistinguishable ways. Instead of insisting that non-standard formulations are "wrong," it might be wiser to bear in mind that they might prove useful in some situations, and that in the future, as new physics is discovered, they might even turn out to be closer to the truth. The preferred-frame interpretation of special relativity certainly comes into its own in the face of quantum non-locality.

These examples illustrate a general point. The division between kinematics and dynamics cannot be determined uniquely. There is a "shifty split" between the two. Yet, it is convenient to define the kinematics (or space-time geometry) so that it contains or summarizes universal physical effects which are independent of (for example) the mass and composition of bodies. For this reason, universal symmetries such as Lorentz invariance are usually regarded as part of the kinematics, so that space-time is defined as locally Minkowskian. However, with the discovery of new effects such as quantum non-locality, the most convenient choice of space-time geometry may have to be revised, as we have argued here.

From this "Poincaréan" point of view, it seems misguided to try to argue that a certain kinematics – with or without an absolute simultaneity – *must* be adopted. One can only propose a certain kinematics and argue that it provides the simplest and most natural description of the phenomena.[20]

We claim, then, that the above construction, with an absolute simultaneity (associated with a preferred foliation and a preferred local state of rest), is the natural one given the known facts; and, we suggest that it should emerge from a more fundamental theory in the limit of an approximately classical space-time background.

Alternatively, one might try to develop a theory of non-local interactions on Minkowski space-time. In itself, the mere fact of superluminal interaction is not necessarily incompatible with fundamental Lorentz invariance. For example, the interactions might be instantaneous in the centre-of-mass frame (a manifestly Lorentz-invariant statement). But then one must somehow make sense of backwards-in-time signals in other frames. This last question becomes particularly poignant if one is willing to consider quantum non-equilibrium and the associated practical signaling at a distance. Some workers, however, maintain that backwards-in-time effects should be allowed, arguing that these provide a loophole through which non-locality may be avoided (Price 1996).[21] Attempts have been made to formulate a de Broglie-Bohm-type theory of particle trajectories with fundamental Lorentz invariance, but it would appear that the dynamics (and the quantum equilibrium distribution) must be defined on a preferred space-like slice, that is, in a preferred rest frame (Hardy 1992; Berndl and Goldstein 1994; Berndl *et al.* 1996). (See, however, Dewdney and Horton (2002) for an attempt to avoid this problem.) A similar result has been shown for any preferred local quantum observable (not necessarily particle positions) (Myrvold 2002). In evaluating the advantages and disadvantages of all these approaches, in our view, it ought to be remembered that space-time structure is not a metaphysical *a priori* background onto which dynamics is to be grafted at all costs; rather, it is as subject to possible revision as dynamics itself.

It may well be that the issue of non-locality vis à vis relativistic space-time will be settled only upon making further progress in physics. From our perspective, for as long as we are confined to a state of statistical equilibrium that hides the underlying non-locality from direct view, it seems probable that the argument will continue to be unresolved. On the other hand, if quantum non-equilibrium were to be discovered and used in practice for instantaneous signaling over remote distances, then in such circumstances it seems likely that physicists would see the convenience of adopting a global definition of absolute simultaneity.

Acknowledgement. I am grateful to Harvey Brown for allowing me to see the manuscript of his book (Brown 2005) prior to publication, and for detailed comments and correspondence regarding the second section of this paper entitled "Physical structure of space-time."

Notes

1 The Bell inequalities do not apply in the many-worlds theory, because their derivation assumes that a quantum measurement has only one outcome.

2 Arguably, these are two different ways of saying the same thing. The kinematical structure of space-time cannot be disentangled from the dynamics taking place within it (Brown 2005).

3 As we shall discuss elsewhere, the confusion between phenomenology and fundamentals also led to inconsistencies in quantum theory, in the form of the "measurement" or "reality" problem.

4 According to its original title page, Poincaré's long paper "On the Dynamics of the Electron" (1906) was accepted for publication by the *Rendiconti del Circolo Matematico di Palermo* on 23 July 1905, printed on 14 December, and officially published in 1906. The short summary with the same title was (according to Pais 1982) communicated to the Académie des Sciences in Paris on 5 June 1905; it was published in 1905 in the *Comptes Rendus de l'Académie des Sciences de Paris* (Poincaré 1905b). Einstein's first relativity paper was received by the *Annalen der Physik* on 30 June 1905 and published in 1905 (Einstein 1905).

5 This sentence appears as a footnote in the original text.

6 Poincaré's short summary (1905b) refers to "gravitational waves" propagating between gravitating bodies. For a detailed discussion of Poincaré's 1905 theory of gravity, see Zahar (1989: 192–200).

7 Brown (2005), however, questions whether in 1905 Poincaré fully understood the physical significance of the transformed coordinates to higher orders in v/c. On the other hand, Darrigol (1995: 37–40) shows that, in lectures delivered at the Sorbonne in 1906–7, Poincaré (apparently independently of Einstein) generalized his 1900 discussion of clock synchronization (taking into account length contraction) to obtain the full Lorentz-transformed time to all orders in v/c.

8 We are inclined to agree with Zahar (1989: 150): ". . . . that Poincaré did discover special relativity, that his philosophy of science provided him with heuristic guidelines, but that certain ambiguities within that same philosophy prevented both his contemporaries and many historians from appreciating the true value of his contribution."

9 A limited form of time dilation was anticipated by Larmor (in a paper of 1897, and in his book of 1900), and by Lorentz (in a paper of 1899). See Brown (2005: Section 4.5), and Janssen and Stachel (2004).

10 The recent book by Brown (2005: Section 2.4) calls this assumption the "boostability" of rods and clocks, and regards it more as a "stipulation" (or convenient convention) than an assumption.
11 In fact, a still further assumption of spatial isotropy is also needed – see Brown (2005: Section 5.4.3).
12 As noted by Darrigol (1995: 39), Poincaré had also used length contraction as a hypothesis in his Sorbonne lectures of 1906–7. Again, in the author's view, Poincaré may well have understood that some such extra hypothesis was needed to relate measurements in different frames.
13 Voigt's paper is briefly discussed by Pais (1982: 121–22), and in great detail by Ernst and Hsu (2001), who also provide an English translation of it.
14 This sentence appears as a footnote in the original text.
15 Even if Poincaré himself, for philosophical reasons of his own, seemed to prefer retaining the old notions of space and time in the background.
16 The speed c in the preferred frame will of course be independent of the motion of the source, as expected of a wave phenomenon.
17 For a detailed analysis of this paper by Poincaré, as well as for a reconstruction of Poincaré's argument in the cited passage, see the paper by Darrigol (1995).
18 See, for example, Penrose (1979).
19 Poincaré's example actually involved a 3-sphere within which the temperature varies as a certain function of radius.
20 This is, in fact, arguably true not only regarding space-time geometry, but also regarding physical laws in general, since it is always possible to write alternative formulations of the same physics.
21 The derivation of Bell's inequality assumes that the initial parameters λ are unaffected by the future settings of the equipment.

References

Allen, R. E. (1997) 'A Statistical Superfield and its Observable Consequences', *Int. J. Mod. Phys. A* 12, 2385–2412.

Amelino-Camelia, G. (2002) 'Space-Time Quantum Solves Three Experimental Paradoxes', *Phys. Lett. B* 528, 181–87.

Aspect, A., J. Dalibard and G. Roger (1982) 'Experimental Test of Bell's Inequalities Using Time-Varying Analyzers', *Phys. Rev. Lett.* 49, 1804–7.

Bell, J. S. (1964) 'On the Einstein-Podolsky-Rosen Paradox', *Physics* 1, 195–200.

Bell, J. S. (1986) 'interview', in P. C. W. Davies and J. R. Brown (eds.), *The Ghost in the Atom*, Cambridge: Cambridge University Press.

Bell, J. S. (1987) *Speakable and Unspeakable in Quantum Mechanics*, Cambridge: Cambridge University Press.

Berndl, K., D. Dürr, S. Goldstein and N. Zanghì (1996) 'Nonlocality, Lorentz Invariance, and Bohmian Quantum Theory', *Phys. Rev. A* 53, 2062–73.

Berndl, K. and S. Goldstein (1994) 'Comment on "Quantum mechanics, local realistic theories, and Lorentz-invariant realistic theories"', *Phys. Rev. Lett.* 72, 780.

Bohm, D. (1952a) 'A Suggested Interpretation of the Quantum Theory in Terms of "Hidden" Variables. I', *Phys. Rev.* 85, 166–79.

Bohm, D. (1952b) 'A Suggested Interpretation of the Quantum Theory in Terms of "Hidden" Variables. II', *Phys. Rev.* 85, 180–93.

Bohm, D. and B. J. Hiley (1984) 'Measurement Understood Through the Quantum Potential Approach', *Found. Phys.* 14, 255–74.

Born, M. (1962) *Einstein's Theory of Relativity*, New York: Dover.

Brown, H. R. (2005) *Physical Relativity: Space-Time Structure from a Dynamical Perspective*, Oxford: Oxford University Press.

Chadha, S. and H. B. Nielsen (1983) 'Lorentz Invariance as a Low Energy Phenomenon', *Nucl. Phys. B* 217, 125–44.

Coleman, S. and S. L. Glashow (1999) 'High-Energy Tests of Lorentz Invariance', *Phys. Rev. D* 59, 116008.

Colladay, D. and V. A. Kostelecký (1998) 'Lorentz-Violating Extension of the Standard Model', *Phys. Rev. D* 58, 116002.

Darrigol, O. (1995) 'Henri Poincaré's Criticism of *fin de siècle* Electrodynamics', *Stud. Hist. Phil. Mod. Phys.* 26, 1–44.

Darrigol, O. (1996) 'The Electrodynamic Origins of Relativity Theory', *Hist. Stud. Phys. Biol. Sci.* 26, 241–312.

de Broglie, L. (1928) 'La Nouvelle Dynamique des Quanta', in J. Bordet (ed.), *Électrons et Photons: Rapports et Discussions du Cinquième Conseil de Physique*, Paris: Gauthier-Villars, pp. 105–32. [English translation: G. Bacciagaluppi and A. Valentini (forthcoming), *Quantum Theory at the Crossroads: Reconsidering the 1927 Solvay Conference*, Cambridge: Cambridge University Press.]

Dewdney, C. and G. Horton (2002) 'Relativistically Invariant Extension of the de Broglie – Bohm Theory of Quantum Mechanics', *J. Phys. A* 35, 10117–27.

Einstein, A. (1905) ' Zur Elektrodynamik Bewegter Körper', *Ann. der Phys.* 17, 891–921.

Einstein, A. (1949a) 'Remarks Concerning the Essays Brought Together in this Cooperative Volume', in P. A. Schilpp (ed.), *Albert Einstein: Philosopher-Scientist*, Illinois: Open Court, p. 678.

Einstein, A. (1949b) 'Autobiographical Notes', in P. A. Schilpp (ed.), *Albert Einstein: Philosopher-Scientist*, Illinois: Open Court, pp. 21–23.

Einstein, A. (1993) *The Collected Papers of Albert Einstein*, volume 3, Princeton: Princeton University Press. [Original paper: A. Einstein (1910) 'Le Principe de Relativité et ses Conséquences dans la Physique Moderne', *Archives des Sciences Physiques et Naturelles* 29, 5–28.]

Ernst, A. and J.-P. Hsu (2001), in J.-P. Hsu and Y.-Z. Zhang (eds.), *Lorentz and Poincaré Invariance: 100 Years of Relativity*, Singapore: World Scientific.

Einstein, A., B. Podolsky and N. Rosen (1935) 'Can Quantum-Mechanical Description of Physical Reality be Considered Complete? ', *Phys. Rev.* 47, 777–80.

Granek, G. (2000) 'Poincaré's contributions to relativistic dynamics', *Stud. Hist. Phil. Mod. Phys.* 31, 15–48.

Hardy, L. (1992) 'Quantum Mechanics, Local Realistic Theories, and Lorentz-Invariant Realistic Theories', *Phys. Rev. Lett.* 68, 2981–84.

Hawking, S. W. and G. F. R. Ellis (1973) *The Large Scale Structure of Space-Time*, Cambridge: Cambridge University Press.

Jacobson, T. and D. Mattingly (2001) 'Gravity with a Dynamical Preferred Frame', *Phys. Rev. D* 64, 024028.

Janssen, M. and J. Stachel (2004), 'The Optics and Electrodynamics of Moving Bodies', preprint 265, Max Planck Institute for the History of Science.

Kilmister, C. W. (1970) *Special Theory of Relativity*, New York: Pergamon.

Kostelecký, V. A. (2002) (ed.) *Proceedings of the Second Meeting on CPT and Lorentz Symmetry*, Singapore: World Scientific.

Mavromatos, N. E. (2004) 'CPT Violation and Decoherence in Quantum Gravity', gr-qc/0407005.

Miller, A. I. (1973) 'A Study of Henri Poincaré's "Sur la Dynamique de l'Électron"', *Arch. Hist. Exact Sci.* 10, 207–328 (1973). [Reprinted (1986), in Miller, A. I. *Frontiers of Physics: 1900–911*, Boston: Birkhäuser, pp. 29–150.]

Minkowski, H. (1952) 'Space and Time', in *The Principle of Relativity*, New York: Dover, pp. 75–91.

Moffat, J. W. (2003) 'Spontaneous Violation of Lorentz Invariance and Ultra-High Energy Cosmic Rays', *Int. J. Mod. Phys. D* 12, 1279–87.

Myrvold, W. (2002) 'Modal Interpretations and Relativity', *Foundations of Physics* 32, 1773–84.

Nielsen, H. B. and M. Ninomiya (1978) 'β-Function in a Non-Covariant Yang-Mills Theory', *Nucl. Phys. B* 141, 153–77.

Pais, A. (1982) *Subtle is the Lord: the Science and the Life of Albert Einstein*, Oxford: Oxford University Press.

Pauli, W. (1958) *Theory of Relativity*, London: Pergamon Press.

Penrose, R. (1979) 'Singularities and Time-Asymmetry', in S. W. Hawking and W. Israel (eds.), *General Relativity: an Einstein Centenary Survey*, Cambridge: Cambridge University Press, pp. 581–638.

Poincaré, H. (1900) 'La théorie de Lorentz et le Principe de Réaction', *Archives néerlandaises des Sciences Exactes et Naturelles*, 5, 252–78. [Reprinted (1954), in *Oeuvres de Henri Poincaré*, Paris: Gauthier-Villars, pp. 464–88.]

Poincaré, H. (1902) *La Science et l'Hypothèse*, Paris: Flammarion. [English translation (2001): *The Value of Science: Essential Writings of Henri Poincaré*, New York: Modern Library.]

Poincaré, H. (1905a) *La Valeur de la Science*, Paris: Flammarion. [English translation (2001): *The Value of Science: Essential Writings of Henri Poincaré*, New York: Modern Library.]

Poincaré, H. (1905b) 'Sur la Dynamique de l'Électron', *C. R. Ac. Sci. Paris* 140, 1504–8. [Reprinted (1954), in *Oeuvres de Henri Poincaré*, Paris: Gauthier-Villars, pp. 489–93.]

Poincaré, H. (1906) 'Sur la Dynamique de l'Électron', *Rend. Circ. Matem. Palermo* 21, 129–76. [Reprinted (1954), in *Oeuvres de Henri Poincaré*, Paris: Gauthier-Villars, pp. 494–550.]

Poincaré, H. (1910) 'La Mécanique Nouvelle', in *Sechs Vorträge über Ausgewählte Gegenstände aus der Reinen Mathematik und Mathematischen Physik*, Leipzig: Teubner, pp. 51–58.

Popper, K. R. (1982) *Quantum Theory and the Schism in Physics*, London: Unwin Hyman.

Price, H. (1996) *Time's Arrow and Archimedes' Point*, Oxford: Oxford University Press.

Schwartz, H. M. (1971) 'Poincaré's Rendiconti Paper on Relativity. Part I', *Am. J. Phys.* 39, 1287–94.

Schwartz, H. M. (1972) 'Poincaré's Rendiconti Paper on Relativity. Part II', *Am. J. Phys.* 40, 862–72.

Unruh, W. G. (1989) 'Unimodular Theory of Canonical Quantum Gravity', *Phys. Rev. D* 40, 1048–52.

Valentini, A. (1991a) 'Signal-Locality, Uncertainty, and the Subquantum *H*-Theorem. I', *Phys. Lett. A* 156, 5–11.

Valentini, A. (1991b) 'Signal-Locality, Uncertainty, and the Subquantum *H*-Theorem. II', *Phys. Lett. A* 158, 1–8.

Valentini, A. (1992) *On the Pilot-Wave Theory of Classical, Quantum and Subquantum Physics*, PhD thesis, International School for Advanced Studies, Trieste, Italy. [http://www.sissa.it/ap/PhD/Theses/Valentini.pdf]
Valentini, A. (1996) 'Pilot-Wave Theory of Fields, Gravitation and Cosmology', in J. T. Cushing, A. Fine, and S. Goldstein (eds.), *Bohmian Mechanics and Quantum Theory: an Appraisal*, Dordrecht: Kluwer, pp. 45–66.
Valentini, A. (1997) 'On Galilean and Lorentz Invariance in Pilot-Wave Dynamics', *Phys. Lett. A* 228, 215–22.
Valentini, A. (2001) 'Hidden Variables, Statistical Mechanics and the Early Universe', in J. Bricmont *et al.* (eds.), *Chance in Physics: Foundations and Perspectives*, Berlin: Springer, pp. 165–81.
Valentini, A. (2002a) 'Signal-Locality in Hidden-Variables Theories', *Phys. Lett. A* 297, 273–78.
Valentini, A. (2002b) 'Signal-Locality and Subquantum Information in Deterministic Hidden-Variables Theories', in T. Placek and J. Butterfield (eds.), *Non-locality and Modality*, Dordrecht: Kluwer, pp. 81–103.
Valentini, A. (2002c) 'Subquantum Information and Computation', *Pramana – J. Phys.* 59, 269–77.
Valentini, A. (2004a) 'Black Holes, Information Loss, and Hidden Variables', hep-th/0407032.
Valentini, A. (2004b) 'Extreme Test of Quantum Theory with Black Holes', astro-ph/0412503.
Valentini, A. and H. Westman (2005) 'Dynamical Origin of Quantum Probabilities', *Proc. Roy. Soc. A* 461, 253–72.
Voigt, W. (1887) 'Uber das Dopplersche Prinzip,' *Nachr. Ges. Wiss. Göttingen* 41. [English translation (2001), in Ernst and Hsu, pp. 10–19.]
Weinberg, S. (1972) *Gravitation and Cosmology*, New York: Wiley.
Weinberg, S. (1995) *The Quantum Theory of Fields*, vol. I, Foundations, Cambridge: Cambridge University Press.
Zahar, E. (1989) *Einstein's Revolution: a Study in Heuristic*, La Salle: Open Court.

5 Non-local correlations in quantum theory: how the trick might be done

Tim Maudlin

It has long been remarked that "The Theory of Relativity" is a poor name for the Theory of Relativity. The usual justification for the name looks backward to pre-relativistic space-time structure: the absolute temporal and spatial structures of (Neo-)Newtonian space-time (namely, simultaneity, lapse of time between events, and spatial dimension of objects) all become "relative to the observer" in Einstein's theory. But this masks the radical nature of the shift to special and general relativity. In those theories, simultaneity, lapse of time between events, and spatial dimensions of objects rather become *physically non-existent*. It is entirely incidental to the physics that in special relativity the choice of an "observer" (which is nothing more than the choice of a time-like vector at a point in space-time) allows for the definition of a unique global reference system (a Lorentz frame) which in turn supplies a global time function and a spatial metric on the simultaneity slices of that function. In a generic general relativistic space-time, choice of an "observer" does not pick out any such global reference system, and indeed, no such system may exist. That is, it is evident in general relativity that "simultaneity" is not a physically-interesting-but-observer-dependent notion; it is rather absent altogether from the physics. And the same holds in special relativity, even though one can, as a purely formal exercise, associate a global time function with an observer-at-an instant. The moral of the theories of relativity is not that classical spatio-temporal notions are rendered merely relative, but that they are expunged from physics altogether.

What replaces the classical space-time structure is also well known: the relativistic metric. Once this is taken on board, it is easy to arrive at the proper definition of a relativistic theory: it is not one that is Lorentz invariant (a notion that has no content at all in general relativity) or "generally covariant" (a purely formal mathematical requirement that has no physical content), but rather a theory that postulates only the relativistic metric as space-time structure. The question is whether one can formulate the basic laws of the theory using only the Minkowski metric (special relativity) or the Lorentzian metric supplied by a solution to the Einstein Field Equations (general relativity).

Einstein himself wanted something more. He believed that physics had to employ a "principle of contiguity":

> The following idea characterizes the relative independence of objects far apart in space (A and B): external influence on A has no direct influence on B; this is known as the "principle of contiguity," which is used consistently in the field theory. If this axiom were to be completely abolished, the idea of the existence of (quasi-) enclosed systems, and thereby the postulation of laws which can be empirically checked in the accepted sense, would become impossible.
>
> (Born 1971: 171)

Einstein here rejects all action-at-a-distance: if one event is to have a physical influence on another, and they are situated in different regions of space-time, then there must a continuous chain of causally mediating events that connect them. And, most famously, no such chain can "exceed the speed of light," that is, the chain must everywhere lie on or within the light cones. Since the light-cone structure is determined by the relativistic metric, this constraint can be given a properly relativistic formulation.

The classical results concerning determinism in relativity were derived within Einstein's framework. If Einstein's constraints hold, then any physical condition that can influence an event must lie in the event's past light cone, and can only have an influence by means of a continuous chain that reaches the event. The complete physical specification on a surface that cuts the past light cone must suffice, in a deterministic theory, to determine the event at the apex. Data on a Cauchy surface, a surface that intersects every inextendible time-like curve exactly once, would determine the physical state of the entire space-time.

In a non-deterministic setting, the situation is a bit more subtle. Complete physical data on a surface that cuts the past light cone of an event will not, of course, determine what happens at the apex, but will suffice to assign probabilities for various possibilities. But those probabilistic predictions could be improved by conditionalizing on events within the past light cone that lie between the surface and the apex, since those events can carry information about how various chance occurrences came out. Furthermore, even conditionalizing on events at space-like separation from the apex can improve one's predictions, as illustrated in Figure 5.1.

Suppose we have a complete specification of the physical state on the surface σ, and we wish to make a prediction about what will happen at event P. On σ there is a muon headed directly toward P, but there is also some non-zero chance that the muon will decay into an electron and two neutrinos before reaching P, as is shown on the right. Then our prediction about what will happen at P can be improved by getting information about the state at point Q, since that will tell us whether or not the muon decayed there. Furthermore, even information about the physical state at R, space-like

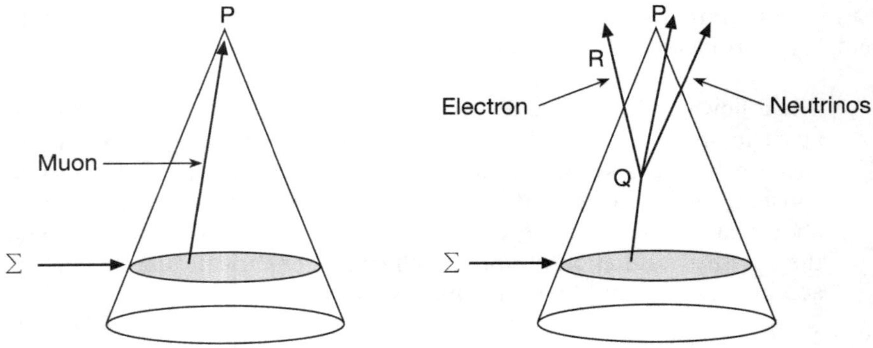

Figure 5.1 Getting information at space-like separation.

separated from P, can improve our prediction, since an electron at R informs us that the muon decayed at Q. Even though what happens at R has no physical influence on P, it can carry information in virtue of a classical common-cause structure. None of this would violate Einstein's principles.

There is, however, a clean way to formulate Einstein's demand in the context of indeterministic theories. Suppose that the event at P is indeterministic: what happens at P is not determined by its causal antecedents. Then we can conditionalize on the physical state of the *entire past light cone* of P, and the laws will still only deliver probabilities for what happens at P. And if Einstein's constraints hold, *that prediction cannot be altered by conditionalizing on any event at space-like separation from P.* For no such event can have an immediate influence on P (by the principle of contiguity) nor can it have an influence mediated by a subluminal physical process. Nor can the event carry information about the causal antecedents of P that goes beyond the information already taken into account: the entire past light cone of P.

So it is possible to have both deterministic and indeterministic theories that satisfy Einstein's constraint. Maxwellian electrodynamics, for example, does, and furthermore the basic dynamics of that theory can be formulated in a way that makes no use of the notion of "simultaneity" at all. Initial data can be specified on any Cauchy surface, whether or not it is a "simultaneity slice" according to some observer, and the laws will then determine a unique global solution. Only the relativistic metric is required to formulate the dynamical laws. We can easily imagine a stochastic theory with the same structure: initial data on any Cauchy surface would yield a Markov process on any foliation into Cauchy surfaces that includes the initial slice. The notion of simultaneity, or a preferred foliation of the space-time, would play no role in the statement of the laws, and so would have no physical significance. One would not have to decide which space-like separated events are "simultaneous" with a given event exactly because those events would be *irrelevant* for physical purposes once one had conditionalized on the past light cone. We have not proved a theorem, but it is easy to see how satisfaction of Einstein's constraint can help render the notion of distant simultaneity otiose, since it renders the physical conditions at space-like separated events physically irrelevant to one another.

But it is precisely Einstein's constraint that Bell used in proving his famous inequality, and it is this condition that is ruled out by violations of that inequality (see, in particular, "The Theory of Local Beables" in Bell 1987: 52–62.) It cannot be that full information about the past light cone of an event is maximal information for physical predictions about what will happen at that event. By looking outside the past light cone one can improve one's bets about what happens at the apex. Bell proved that *no empirically adequate theory can avoid this result*, even though Einstein's constraint rules it out.

What options are available for theories that violate Bell's inequality for events at space-like separation? One could hold on to the principle of continuity but relax the restriction on superluminal velocities: this would yield a theory with tachyons that carry information from one wing of a Bell-correlation experiment to the other. This route has not been pursued by any actual theory. It would, in particular, be difficult to implement this strategy without allowing for straightforward superluminal energy transmission, a phenomenon not predicted by quantum mechanics.

Rather, the theories we actually have violate continuity: events at one location can influence events at space-like separation without the help of a continuous chain of intermediaries. In all of these theories, the connection between events is mediated by the wave function, rather than by spatially localized physical conditions or particles that propagate faster than light. What the wave function secures – in one way or another – is correlations among the *local beables* of the theory (to use Bell's phase), that is, among the physical entities that have unproblematic locations in space-time. If a theory were to lack such local beables – if there were nothing, according to the theory, that can be assigned to a particular region in space-time – then it becomes exceedingly difficult to understand exactly how space-time structure is supposed to play a physical role in the theory at all, and also exceedingly difficult to understand why we should give much weight to the theory of relativity, given that we take whatever knowledge we have of space-time structure to be mediated by interactions with localized objects in space-time. Were we to become convinced that no such object exists, we would have to revisit our grounds for putting stock in the theory of relativity in the first place.

So for the purposes of this paper, we will adhere to two conditions for the physical theories we consider: first, each theory must have some local beables, and second, the physics must predict violations of Bell's inequality for some possible experimental situations involving experiments performed on the localized objects at space-like separation. Bell's result proves that this can occur only if the physics is irreducibly non-local, in the sense that the physics governing a beable at a particular space-time point cannot be exhausted by considering only the physical state in the past light cone of that point. Even once we have conditionalized on the past light cone, there must be further predictive improvements to be made by conditionalizing on events at space-like separation.

The central physical question, then, is *which* events at space-like separation must be taken into account and *how* they must be taken into account.

Until very recently, the only available fully articulated theories took a particular line on the "which?" question, a line that put the theories in tension with relativity. These theories proceed in the most straightforward way to adapt non-local, non-relativistic theories to the relativistic space-time, but in doing so, they are forced to *add* space-time structure, or the equivalent, to the physical picture. So if one interprets relativity as demanding that the Lorentzian metric be all the space-time structure there is, these theories are not fully relativistic. The theories need not *reject* the physical significance of the Lorentzian metric, but they do need to appeal to further space-time structure to formulate their laws. We will examine two examples of such theories first, and then turn to a recently discovered alternative theory, which is completely relativistic. Our goal will be to assess, as far as we can, the advantages and demerits of these theories.

If we begin with a non-relativistic theory that makes essential use of absolute simultaneity, the most obvious (or perhaps crude and flat-footed) way to adapt the theory to a relativistic space-time is to add a foliation to the space-time, a foliation that divides the space-time into a stack of space-like hyperplanes. One then employs these hyperplanes in place of absolute simultaneity in the original theory. If no attempt is made to produce a further account of the foliation, and it is accepted as intrinsic space-time structure, then such a theory will clearly fail, in the sense described above, to be relativistic. But the way it fails is worthy of note: no positive part of the relativistic account of space-time is being *rejected*: rather, in addition to the Lorentzian metric, a new structure is being *added*. Because of this, there is a straightforward sense in which no successful relativistic account of any physical phenomenon need be lost or revised: if something can be accounted for without the foliation, then one need not mention it. So there is no danger that existing adequate relativistic accounts of phenomena will somehow be lost: in this sense, the content of relativity is not being rejected at all.

Furthermore, the addition of such a foliation to a relativistic space-time need not be, in any deep sense, a vindication of the classical space-time theory. It is tempting to call the new foliation "absolute simultaneity," but there is little positive justification for using that term. The foliation is being added to account for certain subtle physical phenomena involving distant correlations, phenomena completely unknown in classical physics and even more unknown to those who first introduced the notion of simultaneity. If one thinks four-dimensionally, then absolute simultaneity induces a foliation of the space-time, but it does not follow that any such foliation deserves the name "absolute simultaneity". In particular, the foliation introduced here evidently plays no role at all explaining the synchronization of clocks, or anything of that sort. So our immediate aim is not to reintroduce absolute simultaneity to physics, but just to introduce a structure needed to do a particular kind of physics. It would at least take further argumentation to establish that this new structure should be interpreted as absolute simultaneity.

If one begins with non-relativistic Bohmian mechanics one notices that distant correlations in that theory are explained directly by the dynamics because the dynamics governs *the total configuration of particle positions by a global law* rather than governing the positions of individual particles by a local law. That means that where a particular particle goes may depend on where an arbitrarily distant particle is and, most strikingly, on what is being done to the distant particle. In the non-relativistic theory, determining where the distant particle *is* requires the use of absolute simultaneity: we mean where the distant particle is *at the very same moment for which the velocity of the local particle is to be determined.*

This feature of the theory is best illuminated by an example discussed by David Albert (1992: 155–60). In Bohmian mechanics, when one does a spin measurement on a particle that is not in an eigenstate of spin (so quantum mechanics makes only probabilistic predictions), the outcome of the measurement will depend, first, on exactly how the spin-measuring device is constructed and, second, on the exact initial *location* of the particle. Suppose we have a specifically constructed device designed to measure the x-spin of an electron. An electron is fed into the device, and, depending on how the device is constructed, the trajectory of the particle as it leaves the device will be interpreted as an indication of its x-spin. Suppose, in particular, that particles exiting with trajectories headed toward the ceiling are called "x-spin up" and those exiting with trajectories headed towards the floor are called "x-spin down". Since Bohmian mechanics is a deterministic theory, and since the physics of the device is fixed, the outcome must depend on the exact physical state of the incoming particle. And indeed, for a single unentangled particle, the outcome depends solely on the *position* of the particle in space. Given the usual physical symmetry of the measuring apparatus, it is easy to show that particles that exit the device headed toward the ceiling (and so found to have x-spin up) entered the device in the upper part of the region allowed by the wave function, and particles that exit the device headed toward the floor entered the device in the lower region.

To be concrete, suppose we prepare a beam of particles in the state y-spin up, and then subject the particles in the beam to an x-spin measurement. Quantum mechanics predicts that half the particles in the beam will exit the device going up and half going down. Bohmian mechanics further implies, for an apparatus like a Stern-Gerlach device, that the particles that exit going up were all initially located in the upper region of their wave function, and the half that go down were originally located in the lower half.

But the outcome of an x-spin measurement cannot always be determined, in such a straightforward way, from the initial location of the particle being measured. As Albert points out, the situation becomes much more interesting if we make x-spin measurements on pairs of particles that are entangled. If, for example, we create a pair of electrons in a singlet state and then measure the x-spin of both, quantum mechanics predicts that the outcomes will always be anti-correlated: one electron will exhibit x-spin up and the other x-spin down. But whether a particular electron exhibits a particular result *cannot* be determined simply by the initial location of that particle: if it could, then there would be a completely *local* account of the spin measurements, and they could not violate Bell's inequality (which they do). Suppose, for example, each of the pair of electrons is initially in the upper spatial region consistent with the wave function. Then it cannot be that each electron will exit its device headed toward the ceiling if it enters the device in the upper region: that would violate the perfect anti-correlation. So what determines which electron will go up and which go down?

As Albert shows, the exact outcome of the experiment depends on *which electron goes through its device first*. If the right-hand electron is measured first, it will be found to have x-spin up and the left-hand electron x-spin down, but if the left-hand electron is measured first, one will get the opposite outcome. And this holds *no matter how far apart* the two electrons are, and it holds without the action of any intermediary particles or fields traveling between the two sides of the experiment. So the behavior of the right-hand electron at some moment depends on what has happened (arbitrarily far away) to the left-hand electron. The dynamical non-locality of Bohm's theory is thereby manifest.

Since the exact outcome of the experiment depends on which x-spin measurement is made first, the notion of "first" and "second" has an ineliminable physical role in Bohm's theory. In the non-relativistic theory, which measurement comes first and which second is determined by absolute simultaneity. And if one is to transfer the Bohmian dynamics to a space-time with a Lorentzian structure, one needs there to be something fit to play the same dynamical role. Since no such structure is determined by the Lorentzian metric, the simplest thing to do is to add the required structure: to add a foliation relative to which the relevant sense of "first" and "second" (or "before" and "after") is defined. The foliation would then be invoked in the statement of the fundamental dynamical law governing the particles.

The connection between the foliation and the trajectories is, as Albert's example shows, very tight. Suppose that the foliation were "hidden" from us, in the sense that we could not directly observe it, but we did have access to the precise particle trajectories. Then by a suitable series of experiments we could determine the exact structure of the foliation experimentally. All we would need to do is to perform a large number of x-spin measurements on singlet pairs, varying the locations in space-time of the x-spin measurements on each side. Every time the particles both begin in the upper part of the region allowed by the wave function, we could tell that the measurement which yields an "up" outcome must occur before (relative to the foliation) the measurement that yields a "down" outcome. With enough such experiments, we could locate the foliation to any required degree of accuracy.

Similarly, if we could create pairs of electrons in the singlet state in a way guaranteed to produce only particles located in the upper region of the wave function, we could empirically determine the foliation by means of x-spin measurements. In this sense, the foliation in Bohm's theory would be *revealed by the local beables*: given as data only the space-time *without* the foliation and the exact disposition of the local beables of the theory (particle locations), we could recover the location of the foliation to any degree of accuracy (given enough varied experiments). The only reason, in this theory, that the foliation would not be determinable by experiment is because *the exact disposition of the local beables is not determinable by experiment*. And that, according to Bohm's theory, is a consequence of the *physics that governs interactions between the observer and the environment*, i.e. it is an analytic consequence of Bohmian mechanics itself, as applied to measurement situations. To sum up, the only reason we can't "see" the foliation is because we can't "see" the local beables with arbitrary accuracy (without disturbing the wave function), and the reason in turn for this is given by the structure of the basic dynamical laws that govern all physical interactions.

When John Bell came to consider how to do relativistic quantum theory in "Beables for Quantum Field Theory" (1987: 171–80) he constructed a version of Bohmian Quantum Field Theory that uses fermion number density (at a point in space-time) as the local beable, and also employs a foliation of the space-time. At the end of that paper, Bell remarks:

> As with relativity before Einstein, there is a preferred reference frame in the formulation of the theory ... but it is experimentally indistinguishable. It seems an eccentric way to make a world.
>
> (Ibid. 180)

We have seen how this situation comes about: non-local interactions are required by the violations of Bell's inequality, and the simplest way to dynamically implement the non-locality is via a foliation. Dynamical laws are written in the most obvious fashion using that foliation, and then it turns out *as amatter of physical analysis* that the foliation is not empirically accessible. It is time to directly confront Bell's evaluation: is this an "eccentric" way to make a world?

There is a long tradition in philosophy that denigrates the postulation of empirically undeterminable facts. In the days of the logical empiricists, such a postulation would be considered literally meaningless, but we are logical positivists no more. In Quine's phrase, the Bohmian theories we have been considering confront experience "as a corporate body," and the foliation is a member of the corporation. The postulation of the foliation, and the role it plays, is intelligible, and might possibly be physically correct. So the objection to empirically inaccessible matters of fact must take the form of a *suspicion* rather than a *semantic prohibition.* The question is: what exactly is suspicious or "eccentric" about an empirically inaccessible structure.

I think that there are, in fact, two quite different legitimate sources of suspicion. The first is that the postulated entity is *otiose*, the second is that the theory is somehow *conspiratorial*. Let's consider these in turn.

Newton's theory of Absolute Space postulated certain physical facts – most famously degrees and directions of absolute motion – that according to Newtonian mechanics were empirically inaccessible. If Absolute Space exists, every laboratory has a certain degree of motion through Absolute Space, but that motion would be completely indetectible: experiments carried out in any inertially moving laboratories would come out exactly the same. This suggests – but does not prove – that the absolute velocities are playing *no substantive physical role* in the theory. And indeed, this suspicion turns out to be correct: Newtonian dynamics does not need the full resources of Newtonian Absolute Space to be formulated. Neo-Newtonian (a.k.a. Galilean) space-time has enough structure to support the dynamics without the need for absolute velocities. The extra structure in Newtonian space-time can be *subtracted without physical consequence.* Since that structure is a free rider in the theory, it appears to be good methodological advice to do the subtraction, and commit only to the Neo-Newtonian structure.

But we have already seen that the foliation in the Bohmian theory is not, in this sense, otiose. The basic ontological posits of the theory are the local beables (e.g. particles), the wave function, the space-time *cum* relativistic metric, and the foliation. But one *cannot* just subtract the foliation from this package and leave the rest unchanged. Indeed, as we have seen, from knowledge of the first three alone, one could recover the last. The foliation is not like a temporary scaffold that can be used to construct the theory and then renounced: marks of it remain in the particle trajectories themselves. So the first ground of suspicion does not apply in this case. The foliation is *physically essential* to the theory.

The second ground of suspicion is harder to state in the clean way. It raises its head whenever terms like "conspiratorial" or "concocted" (or perhaps "eccentric") are used to describe a theory. The underlying notion is something like this: if one employs a certain structure in formulating a physical theory, then *ceteris paribus* one should expect that the structure will be accessible through physical interaction. Otherwise, the existence of the structure is being somehow "covered up" by some complex mechanism or adjustment of free parameters: the rest of the theory must be fixed, or delicately balanced, to "hide" the structure from view. In a natural, unconcocted theory one can expect the essential physical structures to *show themselves* to the determined observer.

This principle has some *prima facie* plausibility, but its general plausibility clearly cannot outweigh the consideration of particular cases. And in the case of the Bohmian theory we have been discussing, or of Bell's version of Bohmian Quantum Field Theory, the charge is simply false. When trying to formulate a Bohmian theory in a relativistic domain, there is an evident need for some structure that will give rise to non-locality, and the postulation of a foliation is the simplest, most natural way to be able to write down non-local dynamical laws. *And once the foliation is postulated, no particular effort or adjustment of parameters is made with the purpose of hiding the foliation.* Rather, one writes down the simplest dynamical equations that look like versions of the non-relativistic equations, and it then *turns out* that foliation will not be empirically accessible. Once the equations are in place, all the rest is just analysis.

The same can be said, in non-relativistic Bohmian mechanics, about the empirical accessibility of the particle locations. In that theory, one can't determine exactly where particles are without disturbing the (effective) wave function of the particle: in that sense, the exact particle locations are "hidden".[1] As we have seen, if exact particle locations were not "hidden" in this way, we could empirically determine the foliation. This also might strike one as conspiratorial: the theory postulates exact particle locations, but you can't find out what they are without disturbing or changing the (effective) wave function of the system. But again, the theory was not *constructed* or *adjusted* to give this result, no *mechanisms* were introduced to "cover up" the trajectories. Rather, the theory is mathematically the *simplest* theory one can write down if one postulates that there are particles and that their trajectories are determined by the wave function (see Dürr *et al.* 1992). In this theory, "absolute uncertainty," the empirical inaccessibility of physical structures by means of physical interaction, is a consequence of the physics, but the physics itself is not *motivated* or *designed* to produce it. A glance at the fundamental equations of Bohm's theory demonstrates that the theory is not "concocted" or "artificial" or "jerry-rigged". It is, in fact, remarkably simple.

It is perhaps surprising that such simple physical postulates could have as a consequence that certain physical structures or facts are "hidden from view" ... but then, what exactly did one expect? Gaining information about the world requires the availability of certain sorts of physical interactions. Why should one think, *a priori*, that physics must take just the right form to make all physical structure accessible to an observer? One might even think that such physical *transparency* of the world to investigation would require some sort of "conspiracy".

Indeed, it is hard to state the intuition behind the second ground for suspicion without resorting to theological metaphors. Wouldn't it be odd, or uncharitable, for God to write the laws of physics in such a way that it would be impossible for His creatures to determine the physical facts? If there must be a "conspiracy" to render a physical structure empirically inaccessible, must not God be the conspirator? And wouldn't He revise His plans once He realized what a fix human physicists were likely to be in? Even Bell writes of "an eccentric way to make a world" as if someone, or something, that could qualify as eccentric *did* make the world.

Personally, I am an atheist. I don't think that anyone, or anything, made the world, or knew or cared about the consequences of physics for empirical accessibility. I don't think that any guiding hand would make sure that we should be able to pin down all physical facts in our laboratories. I have faith only that the laws of physics, suitably formulated, should be simple – and this faith is not shaken when I contemplate the basic equations of Bohmian mechanics.

In any case, if one fetishes empirically accessibility too much, then quantum theory is already a lost cause. For on *any* interpretation of quantum mechanics, the *wave function* of a particle is *also* empirically inaccessible. Suppose, for example, we are presented with an electron. There is absolutely no experiment we can perform on it that will reveal its exact quantum state – whether, for example, it is in a spin eigenstate in some direction. If we knew how it was prepared we might be able to say, but direct physical interaction with the particle will not inform us. So if you won't accept any empirical inaccessibility you won't accept wave functions, and it is hard to do quantum theory without them.

Let's turn next to the collapse theory of Ghirardi, Rimini and Weber (GRW). While in Bohmian mechanics, non-locality is achieved by the way the wave function choreographs particle behavior, so that what one particle does may depend on how a distant particle is treated, in the GRW theory non-locality is achieved through the collapse of the wave function itself. It is this which, in the non-relativistic theory, is instantaneous and insensitive to the spatial separation between particles. Interacting with a particle at one end of the universe can cause a collapse that alters the physical state of an entangled particle at the other end. So one question is how, in a relativistic space-time, such collapses are to be implemented.

But even before that question arises, there is another to address. Discussion of how the physics affects the disposition of things in space-time requires that there *be* things disposed in space-time, that there be some local beables. In Bohmian theories, the local beables are postulated at the very beginning: they are the particles, or the local field values, or the fermion number density at a point in space-time. The GRW theory, in contrast, was originally developed solely as an account of the dynamics of the wave function, and the wave function does not exist in space-time. The wave function is defined on a much higher dimensional space than space-time, on the configuration space of the system. In Bohmian mechanics, this is not so surprising: there is a well-defined configuration space because there are well-defined *configurations*, determined by the disposition of the particles in space at any moment. If one only has the wave function in the ontology, then it appears that there are no local beables at all, and it also becomes something of a curiosity that one is using a configuration space without there being any configurations. In any case, without local objects existing in space-time, the project of understanding the space-time structure (what it is, how we could know about it, why it matters) becomes somewhat acute.

In the past several decades there have been two different approaches toward supplying the GRW theory with local beables, one mentioned by Bell and the other pioneered by GianCarlo Ghirardi. Bell's idea was to identify the local beables with certain point-like events in space-time, which have come to be called the *flashes*. Ghirardi, on the other hand, has sought to associate a continuous matter density (density of "stuff") in space-time with the wave function. We will take up Ghirardi's approach first, and a novel variant of Bell's later.

The wave function, as we have seen, is defined on a much higher dimensional space than space-time. But given the wave function, there is a way – many ways, in fact – to "project down" the wave function to a density in space-time. Ghirardi makes one choice of such a projection, with the sorts of consequences one would expect (see Bassi and Ghiardi 2003). For a single particle wave function, the mass density at a point is essentially the square-amplitude of the wave function at that point. So in a two-slit experiment, when one fires a "single electron" at the pair of slits, the mass density associated with the electron spreads out in space, and half of it goes through each slit. If a single particle is in an equal superposition of "heading toward the ceiling" and "heading toward the floor," then half of the mass density will be moving upward and half downward.

Suppose we now do an experiment designed to "locate" the particle – we put up a detector that will correlate the target particle's location with that of a large number of other particles (the "pointer"). Then according to the GRW dynamics, the whole system will very rapidly suffer a GRW collapse, and one term or the other of the superposition will be reduced to almost nothing. Correspondingly, the mass density of the electron will go from being equally distributed in the two regions of space-time to being almost entirely concentrated in one region, with the other region being almost entirely emptied. So if we attend only to the mass density of the electron, in the non-relativistic GRW theory the mass density of the electron will behave as depicted in Figure 5.2.

As is evident from the picture, the shifts in mass density in this GRW theory immediately reveal the surfaces of simultaneity in the space-time: if the simultaneity slices were "hidden" in themselves, the disposition of the local beables would reveal them. The instantaneous jumps of the wave function induce simultaneity-revealing shifts in the mass density. And the jumps really must alter the wave function everywhere at the same moment (i.e. along some space-like hypersurface) if Bell's inequality is to be violated for events at space-like separation.

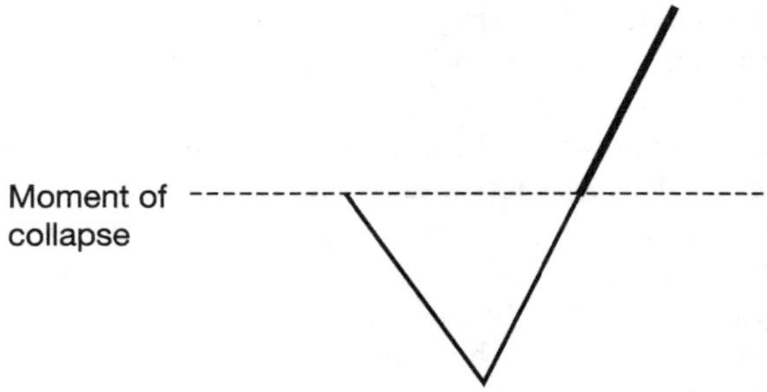

Figure 5.2 Distribution of mass density in a GRW collapse.

Fitting this form of the GRW theory into a relativistic space-time poses the same problem that Bohm's theory faced, and admits of the same solution: introduction of a foliation in terms of which the collapses can be specified. Just as in the Bohmian theory, the foliation would be evident if we could know the exact disposition of the local beable – the mass density. But again, just as in the Bohmian theory, the exact disposition of the local beable is *not* empirically accessible. No experiment will ever show the mass distribution indicated above – no experiment will ever reveal half of the mass density on one side of the experiment and half on the other, because the very physical interaction required to couple to the mass density and amplify it to macroscopic proportions will induce a GRW collapse. Even if half the mass density is here and half there, when we look we will either find it all here or all there – each location 50% of the time. In particular, no experiment will reveal the exact location in space-time where the mass density suddenly disappears or suddenly increases. If we could, then we could map out the foliation: perform an experiment on one side and see where, on the other side, the sudden jump occurs. This empirical inaccessibility is already built into the non-relativistic GRW theory, and the empirical inaccessibility of the foliation by means of this sort of experiment is already implied.

The general question of the empirical accessibility of the foliation in this GRW theory is, however, a bit more subtle. In principle, GRW theory makes slightly different empirical predictions than standard quantum mechanics. These deviations are, so far as anyone now knows, not within our experimental grasp: otherwise either GRW or standard quantum theory would have been decisively refuted. But if one could detect the deviation of GRW from standard quantum theory, and if one used a foliation for the collapses, then the foliation would become empirically accessible.

This issue has been discussed more thoroughly by David Albert (2000), but a simple example can illustrate the point. Here is a way that GRW theory differs, in principle, from standard quantum theory. Suppose we perform a two-slit experiment by putting an electron in a superposition of being in two locations and then recombining the beams to show an interference pattern. In standard quantum theory, it is irrelevant how long the superposition persists before recombination: the interference pattern will always be the same. But in the GRW theory, the longer the superposition persists, the more likely it becomes that the system will suffer a spontaneous collapse to one location or the other. And if such a collapse occurs, then there will no longer be any interference when the sides are recombined.

So here is a way to empirically distinguish GRW from standard quantum mechanics: put a beam of electrons into a superposition of two different locations, wait a while, then recombine the beams and look for interference. If there is still interference, GRW is refuted, if there isn't, standard quantum theory is. The only practical problem is that, for single particles, the wait has to be many millions of years. (If we could put many-particle objects, like viruses, into such superpositions, then the wait could be correspondingly reduced. So this is, in a sense, merely a technical obstacle.)

The sort of experiment just sketched also shows that in principle the foliation in the relativistic version of GRW would be, to some extent, observable. For we can turn the method just described around to produce a sort of clock: if one were to create a collection of particles in a spatial superposition and then later recombine them, the degree to which the interference pattern degrades can serve as a measure of how long (in the frame of the foliation) the superposition was maintained. This "foliation clock" could be compared to regular clocks in inertially moving laboratories. There ought to be one laboratory in which the rate of collapse is slowest, and this will correspond to a laboratory at rest in the foliation frame: all other laboratories will, due to relativistic effects, record a faster collapse rate. So in the version of GRW that employs a foliation, not only is the foliation immediately evident at the level of the local beable, it is even empirically accessible in principle. Accessing it, however, would be far beyond present technical capacities.

How does relativistic GRW with a foliation score on the plausibility meter? The foliation is not empirically accessible For All Practical Purposes (FAPP), so we would not be able, as far as we can tell, to determine what it is. The first ground of suspicion, though, is completely eliminated: the theory evidently cannot be purged of the foliation leaving everything else the same. The second ground, however, is a more delicate matter. The foliation is not *easily* empirically accessible, and that is because the predictive deviations from standard quantum theory are not easily accessible. And this, in turn, is somewhat a matter of design. The GRW theory has two new parameters – the average time between collapses and the degree of localization of a collapse – and those parameters have been deliberately set to render the predictive deviations of the theory from standard quantum mechanics small. The reason for that is empirical: deviations from the predictions of quantum mechanics have not yet been observed, so if these parameters had been set to make the differences between the theories large, GRW would have been empirically refuted. A theory certainly cannot be faulted for setting fundamental parameters at values that correspond to observations – that is evidently what must be done in any case. What seems more uneasy is that there are rival theories that employ no such parameters at all, yet still give predictions in accord with all present data. The fact that GRW has to fine-tune to match the predictive success of these theories is a ground for uneasiness.

But the uneasiness, if any, was already there in the non-relativistic version of the theory. It is hard to argue that the addition of the foliation in the relativistic version has made matters *worse*. The addition of the foliation, to repeat, is not unmotivated: one needs a way to account for the non-local aspects of the theory. It turns out that the simplest way to do so involves postulation of a structure that is not easily empirically accessible. But the structure really is necessary for the theory, and the difficulty for empirical accessibility was not designed – it is a natural consequence of this way of formulating the theory in a space-time with a Lorentzian metric. Perhaps one can do better than this theory, and even manage without a foliation at all, but it will not be by a simple adjustment or reinterpretation of this ontology.

So far, we have not attempted to demonstrate that the addition of a foliation is the *only* way to account for non-local correlations in a relativistic domain, just that it is the most straightforward way to adapt a non-relativistic theory that assigns a real physical role to distant simultaneity (unlike a theory such as Maxwellian electrodynamics, in which simultaneity plays no such role). Both Bohmian mechanics and non-relativistic GRW do so, and so can be fairly easily altered to fit a space-time with a Lorentzian metric and a foliation. As we noted at the outset, there is a sense in which the resulting theories reject relativity, interpreted as the claim that the Lorentzian metric is all the space-time structure there is. Ironically, though, it is the very empirical inaccessibility of the foliation (FAPP) that explains why the Lorentz metric was considered a complete account of space-time: until the discovery of Bell's inequality, the need for non-local correlations, and the physics to produce them, was not recognized and the foliation itself, if it exists, would not make itself known by direct empirical means. Still, one rightly wonders whether there is really a necessity to introduce a foliation to account for violations of Bell's inequality at space-like separation. Could one make do with only the Lorentzian metric after all?

In *Quantum Non-Locality and Relativity* (1994) I noted that one can, in principle, construct theories with superluminal action using only relativistic space-time resources. The trick is to only use space-like separated loci that can be defined using the metric alone. In Minkowski space-time, this amounts to using certain hyperbolae of revolution, and in a generic general relativistic space-time there would be analogous structures (cf. Maudlin 1994: 104–8). Recently, the first (to my knowledge) fully relativistic theory capable of accounting for violations of Bell's inequality at space-like separation has been constructed by Roderick Tumulka in "A Relativistic

Version of the Ghirardi-Rimini-Weber Model" (2006). This is a development of the first magnitude, since it allows us to consider, in one concrete case, the sorts of adjustments to the physics required to produce a completely relativistic theory.

There are some important restrictions on Tumulka's theory, which we will mention later, but it is best to begin with an exact ontological accounting. Tumulka begins with the GRW model, but makes a different choice than Ghirardi regarding the local beables of the theory. Ghirardi, recall, uses the wave function to define a continuous mass distribution in space-time, with cats being made up of full, cat-shaped regions. Following a suggestion of Bell's, Tumulka uses a different beable. The GRW wave function collapses are associated with particular points of space-time: the center of the Gaussian by which the wave function is multiplied. Bell proposed using these discrete space-time points as the locations of the local beables of the theory, and the resulting version of the theory has recently come to be known as "flashy GRW". In this theory, the spatio-temporal manifold is populated by a collection of such point-like events and, in Bell's words, "[a] piece of matter is then a galaxy of such events" (Bell 1987: 205). One advantage of using the flashy ontology is that the locations of the events provide markers from which relativistic intervals may be defined (using only the Lorentz metric), and those intervals may in turn be used in specifying the dynamics of the theory.

The flashes are divided into families, with each family corresponding, intuitively, to a single particle, which "appears," discontinuously, from time to time. The role of the wave function is to provide a *conditional probability measure* over the flashes: given the wave function, together with a set of flash locations (one from each family), one first wants to calculate the probability that the *next* flash in a given family will be at some specified space-time location. Then, given that new flash, one wants to calculate a new wave function, so the process can be repeated.[2]

The technical apparatus Tumulka uses is a bit unfamiliar – for example, he uses a multi-time wave function defined over N copies of space-time (for N families of flashes), and he models the "collapse of the wavefunction" as a change from one multi-time wave function defined over the *whole* multi-time space to a different wave function defined over that whole space – but these refinements need not detain us. What we can already see is how Tumulka's choice of local beable aids in rendering the theory completely relativistic.

What, in the GRW theory with a foliation, is the foliation used for? An immediate clue to this is provided by the non-relativistic theory: what, in that theory, is absolute time used for? One role it plays is in defining the probability measure for the GRW collapses. Those collapses have a *fixed rate per unit time* of occurring. So to calculate the chance that the next collapse will be centered at a particular point in space-time, one must have a measure of the time that has elapsed since the last collapse. Hence, as we saw above, these GRW collapses can be utilized as a sort of clock, to measure elapsed time in the frame of the foliation.

When we turn to Tumulka's theory, we find that the punctate ontology permits definition of a surrogate for the foliation theory's universal clock. Given a point in a relativistic space-time, there are well-defined regions of constant invariant relativistic intervals from the point. The loci of zero intervals are, of course, the light cones, and regions of fixed time-like intervals form hyperbolae that foliate the interiors of the light cones. Since the "particles" are not allowed to reappear at space-like intervals from their previous location, these hyperbolae can serve the role of the universal time function for this particle: the likelihood of the next flash being at a certain location is a function of the time-like interval between the last flash and that location. Such a "flash-centered clock" can be defined using only the Lorentzian metric.

Furthermore, given a locus of constant time-like intervals from a given flash, the Lorentzian metric on space-time will induce a pure spatial metric on that locus (which will be a space-like hypersurface). This pure spatial metric can in turn be used to define the Gaussian used to specify the GRW collapse of the wave function, conditional on the next flash occurring at a specified location on the hypersurface. So the basic tools used in the original GRW theory can be adapted to this milieu using only the Lorentzian metric and the location of the last flash. In the multi-time formalism, the spatio-temporal structures needed to calculate both the conditional probability of the next flash and the new (post-collapse) wave function can be constructed using only the Lorentzian metric and the location of the last flash.

Since only the relativistic metric has been used in the construction, we have a completely relativistic theory.[3] There is no worry that a special foliation can be recovered from examining the complete space-time ontology (the distribution of flashes in space-time) since no foliation was introduced in the construction. What we are left with is just the set of flashes and a rule (which depends on the initial wave function) for calculating the probability for the next flash in a family to occur at a particular location given a particular set of other flashes (which can be chosen as one likes) that have occurred in that and all the other families. And the "initial" wave function can be given relative to any Cauchy surface through the multi-time configuration space: the later, "collapsed," function is generated by collapsing successively on the flashes that are given.[4]

To take a concrete example, suppose we have a "pair of particles in the singlet state" that are each going to be subjected to an x-spin measurement. We know that the results of the measurements will be anti-correlated (one particle will "go up" and the other will "go down"), and in the GRW theory there is no predetermining cause for which goes up and which goes down. The set-up in this formalism would be, first, the initial singlet wave function for the pair, and also an initial pair of flashes, one for each particle. We take the initial pair to occur before either "particle" has "reached" its detector. The scare quotes are to remind us that, unlike Bohmian particles, these "particles" do not have continuously evolving positions: they appear from time to time, and there is no fact about where they are in the interim. Given this initial data, we can then calculate the probability that the next flash for either particle will be in the upper spatial region, indicating an "x-spin up" result: that probability (for each particle) will be 50%. And we can also ask for conditional probability that the next flash on the right will be in the upper region *given that the next flash on the left is in the upper region*: that probability will be 0. It is important to keep in mind that the theory yields conditional probabilities: the likelihood that the next flash in a family will be in a particular location given a specified set of flashes in the other families. The probabilities are not conditional on the wave function alone, they are conditional on the wave function and a set of flashes. Change the set and, as above, the probability changes. (The wave function will also change by "collapse" when we conditionalize on the newly specified flash.)

Note that the probability structure defined above is *completely symmetric between the right and left sides*. Conditional only on the pre-measurement flashes, each "particle" has a 50% chance of displaying x-spin up and 50% x-spin down. Conditional on one particle having displayed a given result, the other particle will be assured of displaying the opposite result. Unlike the example with Bohmian mechanics discussed above, the actual outcome cannot be used to determine which measurement "really came first": the notion of temporal or causal priority of one measurement over the other never appears in the theory. And unlike the GRW theory with a continuous mass distribution, a collapse in one location never causes a sudden change in mass distribution, or any change in the local beable, in another location. When we conditionalize on a flash we get new *probabilities* for other flashes, but unlike the mass distribution case, this does not pick out any foliation. It can't since, as we noted, no foliation was used in the theory at all.

What exactly is the ontological status of the multi-time wave function in this theory? This is a subtle question, but some remarks can be made. First, the role of the wave function is to provide the conditional probabilities for flashes given other flashes. There are certainly physical degrees of freedom reflected in the wave function: if the pair of "particles" discussed above were in a triplet state rather than a singlet state at the outset, then the probabilities for the subsequent flashes would have been different. Or, in a more mundane case, consider a single "free particle" that flashes at a particular location. Where is the next flash likely to be? That depends on the particle's "velocity" or "momentum" (where is it going?), but that information is not contained in the flash: it is contained in the wave function. So the set of physically possible wave functions corresponds to a set of physically possible conditional probability measures, and the role of the wave function in the theory is to allow the conditional probability measure to be calculated. Not every mathematical detail of how that calculation is accomplished need correspond to a physical fact, nor must the mathematical form employed in representing the wave function directly correspond to a physical degree of freedom. As Bell remarked, perhaps too cavalierly, in a similar context:

> One of the apparent non-localities of quantum mechanics is the instantaneous, all over space, 'collapse of the wave function' on 'measurement'. But this does not bother us if we do not grant beable status to the wave function. We can regard it simply as a convenient but inessential mathematical device for formulating correlations between experimental procedures and experimental results, i.e., between one set of beables and another. Then its odd behavior is as acceptable as the funny behaviour of the scalar potential of Maxwell's theory in Coulomb gauge.
>
> (Bell 1987: 53)

Now one can't simply discount the wave function out of the ontology altogether: different wave functions correspond to different probabilities and hence to different physical situations. But the essential object is the conditional probability functions over the local beables, and there may be various mathematical schemes for representing those functions that all have the same physical content. In this sense, we should treat the particular mathematical details in this presentation with caution.

Note, however, that even this degree of insouciance regarding the wave function is only possible because there are the local beables, the flashes: something other than the wave function over which a probability distribution is to be defined. If one were to insist that there are no such local beables, that the wave function is *all there is*, then its "funny" behavior could certainly not be discounted in this way.

What are the essentials of Tumulka's theory that allow it to live within the narrow means of the Lorentz metric? It seems essential that the local beable be point-like: the point is a locus that privileges no reference frame. The point also has the virtue of determining, in conjunction with the Lorentz metric, a set of hyperplanes that can be used in place of an imposed foliation. It seems essential that the dynamics be stochastic: we have seen how the determinism of Bohmian mechanics forces a distinction between the two sides of the singlet experiment: one measurement or the other must "come first". And it seems essential that the local beable be *intermittent*, or flashy, so that experiments done on one side need not immediately register on the local beables on the other side, as they do in Ghirardi's theory. Perhaps some of these features can be eliminated, but it is not obvious to me how this could be done.

There are some technical shortcomings of Tumulka's theory at this point. The multi-time formalism can only be employed when there is no interaction Hamiltonian between the various families, so a lot of existent physics cannot be done in this form. The multi-time wave function is defined for a fixed collection of "particles," so particle creation and annihilation phenomena are not covered. And there seems to be an initiation problem: the calculation of probabilities depends on specifying the "most recent" flash for *each* "particle," but presumably right after the Big Bang most of the "particles" *had had no flashes at all*. It is not clear to me how to get the whole calculation of probabilities off the ground in this circumstance.

But these shortcomings are vastly overshadowed by Tumulka's accomplishment: he has figured out how to construct a theory that displays quantum non-locality without invoking anything but the Lorentz metric. For years it has been known that no one had proven that the trick *can't* be done, and many people suspected that somehow or other it could be done, but one could not be sure. Now one can, since Tumulka has shown at least one way to do it.

We have in possession, then, three theories with different local beables, different dynamical laws, and different postulated space-time structures, all capable of reproducing the predicted violations of Bell's inequality for events at space-like separation. To illustrate how diverse these theories are, it might help to consider a simple system: a single hydrogen atom in its ground state. Let's consider, first, the electron in the atom, and focus exclusively on the space-time ontology. If we could see only the local contents of space-time, but could see them in perfect detail, how would the electron look according to each theory?

In Bohm's theory, the electron is a point-like particle, corresponding in that way to the old "planetary" model of the atom. But unlike the planets orbiting the Sun, the electron in the ground state is at rest relative to the nucleus. If we examined many such atoms in the ground state, we would find the electrons resting in different places, at different distances from the nucleus (which is also a collection of point-like quarks). The distribution of these many electrons would have a density proportional to the squared amplitude of the wave function: the fuzzy picture of the electron smeared out in space would rather represent the distribution of many definitely located particles.

In Ghiradri's version of GRW, the single electron *would* be smeared out: its mass would be distributed about the nucleus in the familiar shape of the *s* orbital. But again, there would be no motion: the mass density would be static since the square amplitude of the wave function is. The nucleus would be a very compact mass distribution at the center of the atom.

In Tumulka's version of GRW, we would have to wait a very long time before noticing anything at all in space-time. About once every 10^8 years, there would be a single point-like event – a flash – and if we had the patience to wait for a large collection of such flashes, they too would eventually trace out the *s* orbital.[5] Similarly, the nucleus would only make its presence felt in space-time sporadically. Evidently, vastly most of the "particles" that make up your body would not leave a mark in space-time throughout your lifetime. Still, there are enough particles in your body that a person-shaped "galaxy," or cloud, of punctate events would constantly reveal your location to the attentive observer.

This is another way in which Ghirardi's and Tumulka's versions of GRW differ: if we imagine a cat (very briefly) in a superposition of two macroscopically different states, in Ghirardi's version a single collapse on any particle in the cat will shift *the whole mass density*, the mass of every single particle, to the favored side of the superposition. In Tumulka's theory, the single collapse will produce only a single flash in space-time, but the *probabilities* for flashes of all the other particles will change conditional on that flash.

None of these accounts of the local physical contents of space-time is familiar. Some may even seem more repellant to common sense than others, but it is unclear by what rights common sense could assert any authority here. Each theory does provide an objective, mind-independent physical world in space-time, a world whose macroscopic outline corresponds to the world we think we live in. If one has an unreflective preference for one picture or another, that preference cannot be grounded in the gross features of experience.

Similarly, the three theories posit different structures to space-time itself, or different degrees of empirical access to space-time. All of the theories postulate the Lorentzian metric of (special or general) relativity, and so all have available the explanatory resources of that metric. The light-cone structure, for example, is there to play a role in accounting for optical phenomena and "the constancy of the speed of light". The time-like intervals between events can play the same role in explaining "time dilation" effects, and so on. Insofar as the theories differ, it is by what they add, if anything, to the space-time structure. And adding is *adding*: it cannot result in the elimination of explanatory resources.

What the Bohmian theory and Ghirardi's version of GRW add is a foliation, a slicing of the space-time into space-like hypersurfaces. The slicing has the appeal of familiarity: it corresponds in structure to Newtonian Absolute time. Whether such a foliation should be taken to play a central role in a philosophical account of time is a question I have not addressed: our focus has been instead solely on the role the foliation plays in the physics. As we have seen, the foliation is an ineliminable part of these theories: the foliation would be determinable, in appropriate circumstances, from the exact distribution of the local beables in space-time. But in each of these theories, that exact distribution is not empirically accessible: the very physics puts limits on how much we can determine about the distribution by physical interaction. These limits are consequences of the fundamental dynamics of the theories.

In the Bohmian case, the limit is absolute: the foliation would remain empirically indescriminable to the experimenter no matter what. In Ghirardi's version of GRW, the limit is rather practical: the foliation could, in principle, be empirically determined (at least to some degree), but the sorts of experiments needed to see the foliation are (as far as is now known) not practically feasible. In the eyes of the logical positivists, this difference between the theories would be all the difference in the world: it would be the difference between speaking sense and nonsense. But logical positivism expired for good reason. Both Bohmian mechanics and Ghirardi's GRW with a foliation are clearly legitimate theories: ways the world might be. Neither has, as we have seen, excess physical structure that can be purged from the theory without loss to the remaining contents of the theory. Neither theory is artificial or *ad hoc*. And it is hard to see, between the two, why one should prefer the theory in which the foliation is *in principle* empirically accessible to one in which it is not, given that *in practice* we have not observed it and have no immediate prospect of doing so.

Tumulka's theory, of course, keeps the Lorentzian metric neat: it need postulate no additional foliation. In the eyes of many, this alone would constitute good reason to prefer it. But again, the exact methodological principle involved here is not easy to state. Surely it cannot be a naïve faith that our *empirical engagement* with the world has revealed that the Lorentzian metric – and only the Lorentzian metric – is all there is to space-time structure. After all, according to Tumulka's theory our empirical engagement with the world has not, at least until now, led us even vaguely close to the *precise spatio-temporal structure of electrons and quarks*: their flashy, intermittent existence was not heretofore imagined, much less directly revealed by experiment. And if experiment cannot directly reveal the spatio-temporal distribution of *matter*, why think that it can directly (or even indirectly) reveal the spatio-temporal structure of space-time? If we can't tell by observation whether electrons inhabit space-time as point-like particles, or distributed mass fields, or intermittent point events, why put so much faith in relativity as the last word in intrinsic space-time structure?

One can, of course, appeal of Occam's razor: Tumulka gets along with less space-time structure. But less is not always more: physicists did not flock to action-at-a-distance theories over field theories simply because the former had strictly less ontology. Weighing the advantages and disadvantages of a theory, its plausibilities and implausibilities, its elegance and simplicity, its naturalness, is a subtle business. And it is certainly a global business. One might assert, correctly, that *ceteris paribus* one would prefer a theory without a foliation to one with one, but the *cetera* are, in real theories, never *paria*. Until we have a theory that can live without a foliation on the table, we can't begin to ask whether the price of avoiding the foliation is worth paying. At last, with Tumulka's theory, we can begin to have a concrete discussion of the issue.

My guess is that many people will not much like Tumulka's ontology, but for the very bad reason that they have gotten used to employing quantum theory without postulating any clear ontology at all, neither an exact account of the local beables, or an account of how to do without any local beables (a space-time with no contents?), or an account of the status of the wave function. It is only when one gets serious about detailing the physical ontology that one confronts the hard choices we have just been examining.

No doubt, there are unpleasant aspects of Bohm or GRW with a foliation, and also some unpleasant aspects of Tumulka's flashy ontology, but these are real, precise, clear theories. If theorists can do better, they are invited to do so.

It has been a constant complaint against Bohmian mechanics, from its inception, that it "has no Relativistic version". The reason that the theory is hard to reconcile with relativity is clear: it is because of the way the non-locality of theory is implemented. In fact, the easiest way to extend that implementation to a space-time with a Lorentzian metric is to add a foliation, as we have seen. There may be some other way, but no one has discovered it yet.

But the non-locality of Bohm's theory is not a peculiar feature of that theory: it is instead a feature of the world. Any theory adequate to explain the phenomena will have to have some non-locality. As we have seen, Ghirardi's version of GRW, which takes a rather different approach to the non-locality than Bohmian mechanics, also is most easily adapted to a Lorentzian space-time by adding a foliation. One can create a fully relativistic version of GRW, as Tumulka has shown, but only by a very different choice of local beable than Ghirardi made. So these are real, physical problems, with implications for the physical ontology, not merely matters of rewriting a theory in one way or another.

If one decides, in the end, that avoiding a foliation in space-time outweighs all other considerations in theory choice, so be it. But this is a decision that must be made with eyes open, understanding the consequences for the rest of the physical ontology. This means, in particular, consequences for the local beables, for the spatio-temporal distribution of matter. For myself, I can't see any reason to accord avoidance of a foliation such a high priority, but, like everyone else, I have not had any live options to consider until Tumulka's theory. We can only now begin a serious discussion of what must be paid for the privilege of maintaining relativity as Einstein proposed it.[6]

Notes

1 One must be very careful about the sense in which particle locations or trajectories are empirically inaccessible in Bohm's theory. The gross particle locations are, far from hidden, quite manifest. You can tell where the particles in a cat are, and hence whether it is alive or dead, just by looking, and thereby correlating the positions of particles in your brain to those in the cat. But in doing so, you must necessarily alter the *effective wave function* of the cat. What you can't do is find out about the particle positions in the system and leave the effective wave function of the system unaltered. And that is what would have to be done to discover the foliation using the method outlined.

2 In fact, what Tumulka does is to calculate the joint probability for all of the families to have their next flash at specified locations, and then updates the wave function after the whole new set, but there seems to be no obstacle in principle to doing things flash by flash.

3 Assuming, of course, we have a *coherent* theory! One has to check, for example, that the probability one gets for a new flash given a set of other flashes does not

depend on the order in which the other flashes are conditionalized on. Tumulka shows this.

4 Since the space on which the multi-time wave function is defined is N copies of a regular four-dimensional space-time, the notion of a Cauchy surface must be similarly modified: it is composed of N regular Cauchy surfaces, one for each space-time. These need not be the same for each space-time.

5 This ignores the increase in energy due to the collapse: we assume the electron remains in the ground state.

6 I am very grateful to Shelly Goldstein for an extremely detailed vetting of this paper, and to comments from members of the Mirror Lake Institute.

References

Albert, David (1992) *Quantum Mechanics and Experience*, Cambridge, Mass.: Harvard University Press.

Albert, David (2000) 'Special Relativity as an Open Question', in H.-P. Breuer and F. Petruccione (eds.), *Relativistic Quantum Measurement and Decoherence*: *Lectures of aworkshop*, held at the Istituto Italiano per gli Studi Filosofici, Naples, April 9–10, 1999, Berlin: Springer, pp. 1–13.

Bassi, Angelo and GianCarlo Ghirardi (2003) 'Dynamical reduction Models'. quant-ph/0302164

Bell, J. S. (1987) *Speakable and Unspeakable in Quantum Mechanics*, Cambridge: Cambridge University Press.

Born, Max (1971) *The Born-Einstein Letters* (trans. I. Born), New York: Walker,

Dürr, Detlef, Sheldon Goldstein and Nino Zanghí (1992) 'Quantum Equilibrium and the Origin of Absolute Uncertainty', *Journal of Statistical Physics* 67, 843–907.

Maudlin, Tim (1994) *Quantum Non-Locality and Relativity*, Oxford: Blackwell.

Tumulka, Roderich (2006) A Relativistic Version of the Ghirardi-Rimini-Weber Model. quant-ph/0406094 (forthcoming in *Journal of Statistical Physics*).

6 The zero acceleration discontinuity and absolute simultaneity

Franco Selleri

Summary

In the transformation of time between inertial reference frames the coefficient of the space variable, e_1, describes the possible choices of clock synchronization with its different values. In the first part of the paper we review the proof of a zero acceleration discontinuity for rotating platforms: the velocity of light relative to inertial reference frames agrees with the zero acceleration limit of the velocity of light relative to rotating platforms only by assuming $e_1 = 0$, that is by replacing the Lorentz by the "inertial" transformations based on absolute simultaneity. The second part of the paper contains the proof that the same discontinuity exists also for linearly accelerated reference frames. Absolute simultaneity again provides the cure.

Introduction

According to Poincaré (1898: 1); Reichenbach (1958); Mansouri and Sexl (1977: 497, 515, 809); and Jammer (1979: 202–36) the clock synchronization in inertial systems is conventional and the choice based on the invariance of the one way velocity of light made in the theory of special relativity (TSR) is only based on simplicity. Einstein explicitly agreed in considering this part of his TSR conventional (Einstein 1952: 37–65). In 1995 (Selleri 1995: 25; 1996a: 641) I introduced a suitable parameter e_1 (the coefficient of the space variable in the transformation of time) to describe, with its different values, the various clock synchronization procedures in inertial reference frames. The TSR is obtained for a particular $e_1 \neq 0$.

The theory was applied to the rotating platform and to the Sagnac effect (Sagnac 1913: 708, 1410; 1914: 177) with the result that only the choice $e_1 = 0$ could give a satisfactory explanation (Selleri 2004b: 57–77). This means that the space variable disappears from the transformation of time so that the "equivalent" transformations (reviewed below, in the second section) become the "inertial" transformations based on absolute simultaneity. In fact, if time transforms like this: $t = t_0 R$ (R is the usual velocity dependent square root factor) two events in different points happening at the same t_0 obviously take place also at the same t.

A consequence of this research was the discovery of a relativistic discontinuity between inertial systems and slowly accelerated systems (Selleri 1997: 73). The discontinuity goes away only if $e_1 = 0$. Its existence in the TSR is the root of the difficulties met by Langevin (1921: 831; 1937: 304); Post (1967: 475); Anandan (1981: 338); and Landau and Lifschitz (1996: 82–87) in explaining the physics on a rotating platform. These difficulties gave rise to confusion, to the point that Hasselbach and Nicklaus, describing their own experiment (1993: 143), listed about twenty different explanations of the Sagnac effect and commented: "This great variety (if not disparity) in the derivation of the Sagnac phase shift constitutes one of the several controversies ... that have been surrounding the Sagnac effect since the earliest days of studying interferences in rotating frames of reference."

There are several other good reasons to adopt $e_1 = 0$, the so called "absolute synchronization" of Mansouri and Sexl:

(a) The relativistic "explanation" of aberration works poorly as experimentally it is evident that the phenomenon does not depend on the *relative* velocity star-Earth, as explicitly assumed by Einstein in his 1905 article. In a theory with $e_1 = 0$, instead, aberration becomes dependent on the Earth velocity in space and everything is explained (Puccini and Selleri 2002: 283).
(b) The phenomenon of differential retardation between separating and reuniting clocks ("clock paradox") was never given a satisfactory description in relativity (despite many opposite claims), while the theory with $e_1 = 0$ provides a complete explanation in terms of absolute motion (Builder 1957: 246; Selleri 2004a).
(c) The growing evidence for the existence in nature of superluminal signals can easily be accommodated in the theory with $e_1 = 0$, while it is incompatible with relativity due to the presence of the famous causal paradox in which events belonging to the future of an observer can actively modify the past of the same observer (Selleri 2002: 63).
(d) Practically all paradoxes of the special theory of relativity disappear in a theory based on the inertial transformations.

The outcome of all this is that a satisfactory theory of the physics of space and time has to be based on absolute simultaneity. The present paper is a review of the reasons leading to $e_1 = 0$, limited to the zero acceleration discontinuity. The second part of the paper extends the discontinuity argument to linear accelerations.

The inertial transformations imply absolute simultaneity: two events taking place in different points of an inertial frame S_0 at the same time are judged to be simultaneous also in any other inertial frame S, this property being consequence of the absence of space variables in the transformation of time. Of course the existence of absolute simultaneity does not imply that time is absolute: on the contrary, the velocity dependent factor in the transformation of time gives rise to clock retardation phenomena similar to those of the TSR. In another sense if $e_1 = 0$ clock retardation is, however, also absolute: a clock at rest in S is seen from S_0 to run slower, but a clock at rest in S_0 is seen from S to run *faster.* Both observers agree that motion relative to S_0 slows down the pace of clocks, and the phenomenon loses the relativistic flavor it has in the TSR. The difference with respect to the TSR is, however, more apparent than real: a meaningful comparison of rates implies that a clock at rest in S_0 must be compared with clocks at rest in different points of S, and the result depends on the synchronization adopted for the latter clocks.

Absolute length contraction is also obtained from the inertial transformations. A rod at rest on the x-axis of S is seen in S_0 to have end points x_{02} and x_{01} at a common time t_0 such that it appears to be contracted. The reasoning can be inverted by considering the rod at rest in S and observed from S_0, and using the inverse inertial transformations. One gets then an elongated rod with the conclusion (on which both observers agree) that motion relative to S_0 leads to contraction. This is obviously an absolute effect, but again the discrepancy with the TSR is conventional: the length of a moving rod can only be obtained by marking the simultaneous positions of its end points, and therefore depends on the definition of simultaneity of distant events.

The paper is organized as follows. The second section entitled "The equivalent transformations" contains a short review of the transformations ("equivalent") containing the parameter e_1. The third section entitled "Rotating disks" gives an elementary derivation of the ratio of light velocities relative to a rotating disk for propagations concordant and discordant with the disk rotation. The ratio turns out to be different from unity and dependent only on the rotational velocity (not on acceleration). The zero acceleration limit leads to inertial systems (see section entitled "Absolute simultaneity in inertial systems"), where a ratio of velocities of light in opposite directions should thus be different from unity, in contradiction with the assumptions of the TSR. Only a theory accepting absolute simultaneity ($e_1 = 0$) can avoid the discontinuity. The zero acceleration discontinuity argument is extended in the following section entitled "General proof of absolute simultaneity" to a more general set of transformations. In this way absolute simultaneity is proven to be a necessary consequence of the continuity of physical quantities in passing from slowly rotating disks to inertial systems, quite independently of the assumptions from which the equivalent transformations are deduced. In the section entitled "The linear acceleration discontinuity" it is shown that a relativistic discontinuity between slowly accelerated and inertial systems exists also in the case of linear motions and that absolute simultaneity ($e_1 = 0$) is again the only way to save the continuity of the velocity of light. The section entitled "Export of relativistic simultaneity fails" deals with the concrete production of a new inertial system, made with a network of clocks initially at rest in a given system and accelerating equally until at rest in the new one. As the physical actions on the clocks are the same, one could naively expect to be able to "export" to the new system whatever synchronization existed in the old one. It is not so: relativistic simultaneity cannot be exported while it is easy to do so for absolute simultaneity (explained in section entitled "Export of absolute simultaneity succeeds"). The possibility to resynchronize clocks in the new system does not help as the time marked by irreversible clocks cannot be modified (see section entitled "Irreversible processes cannot be synchronized"). The final section entitled "Resynchronization of clocks" reviews recent results about the necessity of a weaker formulation of the relativity principle.

The equivalent transformations

In this section previous results are reviewed which provide a generalization of the Lorentz transformations. Given the inertial frames S_0 and S one can set up Cartesian coordinates and make the following assumptions:

(i) Space is homogeneous and isotropic and time homogeneous, at least if judged by observers at rest in S_0;
(ii) In the isotropic system S_0 the velocity of light is " c " in all directions, so that clocks can be synchronized in S_0 and one-way velocities relative to S_0 can be measured;
(iii) The origin of S, observed from S_0, is seen to move with velocity $v < c$ parallel to the $+x_0$ axis, that is according to the equation $x_0 = vt_0$;
(iv) The axes of S and S_0 coincide for $t = t_0 = 0$;
(v) The two-way velocity of light is the same in all directions and in all inertial systems;
(vi) Clock retardation takes place with the usual velocity dependent factor when clocks move with respect to S_0.

The isotropic reference system S_0 turns out to have a privileged status in all theories satisfying the assumptions (i)–(vi), with the exception of the TSR.

These conditions were shown [6] to imply the following transformations of the space and time variables from S_0 to S:

$$\begin{cases} x = \dfrac{x_0 - vt_0}{R} \\ y = y_0; z = z_0 \\ t = Rt_0 + e_1(x_0 - vt_0) \end{cases} \tag{2.1}$$

where

$$R = \sqrt{1 - v^2/c^2} \tag{2.2}$$

The transformations (2.1) contain only a free parameter, e_1, the coefficient of x in the transformation of time, which remains undetermined after the assumptions (i)–(vi) are made, and which can be fixed by choosing the clock synchronization in S. Different choices of e_1 imply different theories of space and time which are empirically equivalent to a very large extent. In fact, Michelson-type experiments (Selleri 1995: 25; 1996a: 641), Doppler effect and aberration (Hasselbach and Nicklaus 1993: 143), occultations of Jupiter satellites and radar ranging of planets (Selleri 2002b: 413–28) and elementary particle kinematics (Selleri 1996b: 43) were shown to be insensitive to the choice of e_1.

Using (ii) and eq. (2.1), the one-way velocity of light relative to the moving system S for light propagating at an angle θ from the velocity $\vec{v}$ of S relative to S_0 can be shown to be (Selleri 1995: 25; 1996a: 641):

$$c_1(\theta) = \frac{c}{1 + \Gamma cos\theta} \tag{2.3}$$

with

$$\Gamma = \frac{v}{c} + e_1 Rc \tag{2.4}$$

while, of course, the two-way velocity of light relative to S is c in all directions.

Notice that for all theories with $\Gamma \neq 0$ the velocity of light in S is necessarily anisotropic, so that the assumed isotropic frame S_0 is unique. In this sense S_0 has a privileged status. This seems important, as one would be led to conclude that if a theory describes correctly the physical reality, a particular inertial system has to exist in which simultaneity and time are not conventional but truly physical. This would be expected to be the system in which the Lorentz ether is at rest, of course. It has been shown, however, that the equivalent transformations admit a relativistic property according to which the "privileged" system can be chosen arbitrarily: this point, anticipated by Mansouri and Sexl (1977: 497, 515, 809) and more recently by other authors (Ghosal *et al.* 2004: 457; Rizzi *et al.* 2005: 1835) is reviewed in the last section.

The TSR is a particular case, obtained for

$$e_1 = -\frac{v}{c^2 R} \tag{2.5}$$

giving $\Gamma = 0$ and $c_1(\theta) = c$ and reducing (2.1) to their Lorentz form.

Rotating disks

In this section we review earlier results (Ghosal *et al.* 2004: 457; Rizzi *et al.* 2005: 1835) showing that the comparison between the relativistic descriptions of rotating disks and inertial reference frames gives rise to a fundamental difficulty. In the next section it will be shown that this difficulty can be overcome only by substituting the Lorentz transformations between inertial frames with the "inertial" ones.

It is well known that no perfectly inertial frame exists in practice because of Earth rotation and orbital motion, and of Galactic rotation. All knowledge about inertial systems has been obtained in frames having small but non-zero acceleration a. For this reason the mathematical limit $a \to 0$ taken in the theoretical schemes should be smooth and no discontinuities should arise. From such a point of view the existing relativistic theory has been shown to be inconsistent.

Consider the isotropic inertial reference system S_0 and a circular disk with radius r and center constantly at rest in S_0 rotating around its axis with constant angular velocity ω and peripheral velocity $v = \omega r$. On its rim consider a single clock C_Σ (marking the time t) and assume it to be set as follows: when a clock of the laboratory momentarily very near C_Σ shows time $t_0 = 0$ then also C_Σ is set at time $t = 0$. If the disk does not rotate, C_Σ constantly shows the same time as the laboratory clocks. When it rotates, however, motion modifies the pace of C_Σ and the relationship between the times t and t_0 is taken to have the general form

$$t_0 = t\,F(v, \ldots) \tag{3.1}$$

where F is a function of velocity v and eventually acceleration a and higher derivatives of position (not shown). Eq. (3.1) is a consequence of the isotropy of S_0. Its validity can be shown in three steps:

1. In the inertial frame S_0 all directions are physically equivalent. If a clock is moving in a straight line with a given speed, the rate of advancement of its hands cannot depend on the orientation of the line.
2. Similar is the case of the clock C_Σ at rest on the rim of a disk rotating with constant angular velocity. If space is isotropic the speed of its hands cannot depend on the clock's instantaneous velocity.
3. This conclusion was confirmed by the 1977 CERN measurements of the anomalous magnetic moment of the muon (Bailey *et al.* 1977: 301). The decay of muons, observed in different parts of a storage ring, was found to be everywhere the same.

We are, of course, far from ignorant about the function F. There are strong experimental indications (Bailey *et al.* 1977: 301) that the dependence on a is totally absent and that:

$$F(v, \ldots) = \frac{1}{\sqrt{1 - v^2/c^2}} \tag{3.2}$$

Important as it is, eq. (3.2) is irrelevant for our present needs, because the results obtained below hold for all possible factors F.

On the rim of the disk there is also a light source Σ placed in a fixed position very near the clock C_Σ. Two light flashes leave Σ at the time t_1 of C_Σ and are forced to move on a circular path by "sliding" on the internal surface of a cylindrical mirror placed at rest on the disk, all around it and very near its border. The light flashes propagate in the vacuum. The motion of the mirror cannot modify the velocity of light, because a mirror is a virtual source. Thus, relative to the laboratory, the light flashes propagate with the usual velocity c.

The laboratory description of light propagation is the following: two light flashes leave Σ at time t_{01}. The first one propagates in the sense discordant from the disk rotation and comes back to Σ at time t_{02} after circling around the disk. The second flash propagates in the sense concordant with the disk rotation, and comes back to Σ at time t_{03}. These laboratory times, all of events taking place on the disk very near C_Σ, are related to the C_Σ times via (3.1):

$$t_{0i} = t_i F(v, \ldots) \qquad (i = 1, 2, 3) \tag{3.3}$$

The circumference length is L_0 and L measured in the laboratory S_0 and on the disk, respectively. Light propagating in the direction opposite to the disk rotation, must cover a distance smaller than L_0 by $x = \omega r(t_{02} - t_{01})$, the shift of Σ during the time $t_{02} - t_{01}$ taken by light to reach Σ. Therefore

$$L_0 - x = c(t_{02} - t_{01}); \qquad x = \omega r(t_{02} - t_{01}) \tag{3.4}$$

From these equations it follows:

$$t_{02} - t_{01} = \frac{L_0}{c(1 + \beta)} \tag{3.5}$$

with $\beta = \omega r/c$. Light propagating in the disk rotational direction, must instead cover a distance larger than L_0 by $y = \omega r(t_{03} - t_{01})$, the shift of Σ during the time $t_{03} - t_{01}$ taken by light to reach Σ. Therefore

$$L_0 + y = c(t_{03} - t_{01}); \qquad y = \omega r(t_{03} - t_{01}) \tag{3.6}$$

One now gets

$$t_{03} - t_{01} = \frac{L_0}{c(1-\beta)} \tag{3.7}$$

The difference between (3.7) and (3.5) gives the time delay separating the two flashes back in Σ:

$$t_{03} - t_{02} = \frac{2L_0\beta}{c(1-\beta^2)} \tag{3.8}$$

We show next that these results fix to some extent the velocity of light relative to the disk. In fact eq. (3.3) applied to (3.5) and (3.7) gives

$$(t_2 - t_1)F = \frac{L_0}{c(1+\beta)}; \qquad (t_3 - t_1)F = \frac{L_0}{c(1-\beta)} \tag{3.9}$$

If $\tilde{c}(0)$ and $\tilde{c}(\pi)$ are the light velocities, relative to the disk, for the flash concordant and discordant with disk rotation, respectively, we have

$$\frac{1}{\tilde{c}(\pi)} = \frac{t_2 - t_1}{L} = \frac{L_0/L}{Fc(1+\beta)}; \qquad \frac{1}{\tilde{c}(0)} = \frac{t_3 - t_1}{L} = \frac{L_0/L}{Fc(1-\beta)} \tag{3.10}$$

From (3.10) it follows:

$$\frac{\tilde{c}(\pi)}{\tilde{c}(0)} = \frac{1+\beta}{1-\beta} \tag{3.11}$$

Notice that the function F has disappeared from the ratio.

Clearly, eq. (3.11) gives not only the ratio of the two global light velocities for full trips around the disk, but *the ratio of the instantaneous velocities as well.* In fact the isotropy of the inertial frame S_0 ensures, by symmetry, the instantaneous velocities of light to be the same in all points of the rim of the disk with centre at rest in S_0. With reference to Figure 6.1 we can then write for the instantaneous velocities:

$$\tilde{c}_{\phi_1}(0) = \tilde{c}_{\phi_2}(0); \qquad \tilde{c}_{\phi_1}(\pi) = \tilde{c}_{\phi_2}(\pi) \tag{3.12}$$

where ϕ_1 and ϕ_2 are *arbitrary* values of the angle ϕ.

Therefore the light instantaneous velocities relative to the disk will also coincide with the average velocities $\tilde{c}(0)$ and $\tilde{c}(\pi)$, and eq. (3.11) will apply also to the ratio of the instantaneous velocities:

$$\frac{\tilde{c}_{\phi}(\pi)}{\tilde{c}_{\phi}(0)} = \frac{1+\beta}{1-\beta} \tag{3.13}$$

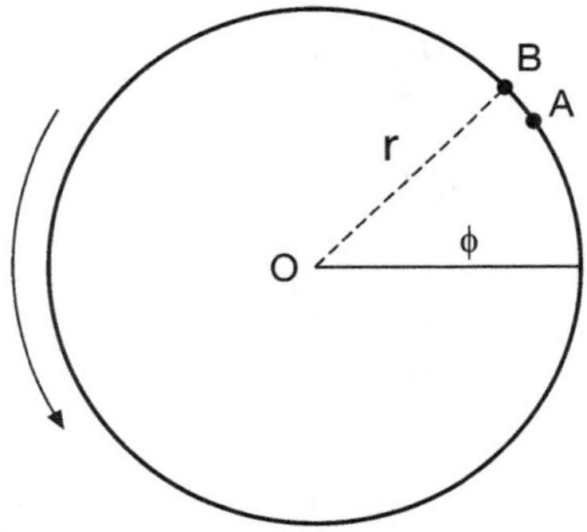

Figure 6.1 The velocity of light relative to the rotating disk between two nearby points A and B.

The consequences of (3.13) are discussed in the next section, where it will be shown that the velocity of light in any inertial system cannot have the constant value predicted by the TSR.

Absolute simultaneity in inertial systems

Eq. (3.13) holds equally for disks having different radii, but the same peripheral velocity v. Let a set of circular disks be given with centers at rest in S_0. Let their radii be $r_1, r_2, \ldots r_i, \ldots$, with $r_1 < r_2 < \ldots < r_i < \ldots$, and suppose they are spinning with constant angular velocities $\omega_1, \omega_2, \ldots \omega_i, \ldots$ such that

$$\omega_1 r_1 = \omega_2 r_2 = \ldots = \omega_i r_i = \ldots = v \tag{4.1}$$

Obviously, then, (3.13) applies to all such disks with the same β ($\beta = v/c$). The centripetal accelerations

$$v^2/r_1, v^2/r_2, \ldots v^2/r_i, \ldots \tag{4.2}$$

decrease regularly with increasing r_i. Therefore, a small part AB of the rim of a disk, having peripheral velocity v and a large radius, for a short time is completely equivalent to a small part of a "co-moving" inertial reference frame (endowed with the same velocity). For all practical purposes the segment AB will belong to that inertial reference frame. But the velocities of light in the two directions AB and BA satisfy (3.13). It follows that the one-way velocity of light relative to the co-moving inertial frame cannot be c and must instead satisfy

$$\frac{c_1(\pi)}{c_1(0)} = \frac{1+\beta}{1-\beta} \tag{4.3}$$

For a better understanding of the reasons why the TSR does not work consider that eq. (3.13) implies that the light velocities parallel and anti-parallel to the disk rotation are different. For the TSR this conclusion is unacceptable, as many disks having the same peripheral velocity, locally approximate an inertial system better and better with increasing radius. The logical situation is shown in Figure 6.2.

Thus the TSR predicts for ρ a discontinuity at zero acceleration (Selleri 2004b: 57–77). We can say that all experiments are performed in the real physical world [where $a \neq 0$, $\rho = (1+\beta)/(1-\beta)$], but the theory has gone out of the world ($a = 0, \rho = 1$)!

This discontinuity is the origin of the problems met with clock synchronization on or near the Earth's surface. This is not a surprise: after all, the Earth is also some kind of rotating disk!

Notice that a velocity of light satisfying eq. (4.3) is required for all inertial frames but one, the isotropic system S_0. In fact, for any small region *AB* of any such system one can imagine a large rotating disk with the center at rest in S_0 and the rim locally co-moving with *AB* and the result (4.3) can be applied. Therefore the velocity of light depends on direction in all inertial systems with the exception of S_0.

General proof of absolute simultaneity

In the present section we will show that the absolute simultaneity condition $e_1 = 0$ can by deduced in the very broad context of the "general transformations" (eqs. (5.1) and (5.2) below). In this way absolute simultaneity will be proven to be a necessary consequence of the continuity of physical quantities in passing from slowly accelerated to inertial systems, independently

Figure 6.2 The ratio $\rho = \tilde{\mathbf{c}}(\pi)/\tilde{\mathbf{c}}(0)$ plotted as a function of acceleration for rotating disks of constant peripheral velocity and decreasing radius (increasing acceleration).

of the particular assumptions leading to the equivalent transformations (Selleri 2004b: 57–77).

Given the inertial frames S_0 and S one can set up Cartesian coordinates and make only the assumptions (i)–(iv) of the first section. The general transformations from S_0 to are then necessarily

$$\begin{cases} x = f_1(x_0 - vt_0) \\ y = g_2 y_0; \qquad z = g_2 z_0 \\ t = e_1 x_0 + e_4 t_0 \end{cases} \tag{5.1}$$

where f_1, g_2, e_4 and e_1 are v dependent parameters. The transformations inverse of (5.1) can easily be shown to be

$$\begin{cases} x_0 = \dfrac{(e_4/f_1)x + vt}{e_4 + e_1 v} \\ y_0 = \dfrac{1}{g_2} y; \qquad z_0 = \dfrac{1}{g_2} z \\ t_0 = \dfrac{t - (e_1/f_1)x}{e_4 + e_1 v} \end{cases} \tag{5.2}$$

A spherical wave front born at time $t_0 = 0$ in the origin of the isotropic reference frame S_0 satisfies the condition

$$x_0^2 + y_0^2 + z_0^2 = c^2 t_0^2 \tag{5.3}$$

The one-way velocity of light relative to the moving system S, $c_1(\theta)$, can be found by introducing the transformations (5.2) into eq. (5.3), which then becomes the following second degree equation in t

$$c^2(1 - \beta^2)t^2 - 2\frac{c^2 e_1 + v e_4}{f_1} xt + \frac{c^2 e_1^2 - e_4^2}{f_1^2} x^2 - (e_4 + v e_1)^2 \frac{y^2 + z^2}{g_2^2} = 0 \tag{5.4}$$

where $\beta = v/c$. From eq. (5.4) it follows

$$t = \frac{c e_1 + \beta e_4}{c(1 - \beta^2) f_1} x + \frac{e_4 + c\beta e_1}{c f_1} \left[\frac{x^2}{(1 - \beta^2)^2} + \frac{f_1^2}{g_2^2} \frac{y^2 + z^2}{1 - \beta^2} \right]^{1/2} \tag{5.5}$$

where the solution with the negative sign in front of the square root has been discarded because it gives negative propagation times. By introducing polar coordinates in which the *x-axis* of the previously introduced Cartesian coordinates is taken as a polar axis, we have

$$x = r\cos\theta; \qquad y = r\sin\theta\sin\phi; \qquad z = r\sin\theta\cos\phi \tag{5.6}$$

and eq. (5.5) gives

$$\frac{t}{r} = \frac{1}{f_1(1-\beta^2)}\left\{\left(e_1 + \frac{e_4}{c}\beta\right)\cos\theta + \left(\frac{e_4}{c} + e_1\beta\right)\left[\cos^2\theta + \gamma^2\sin^2\theta\right]^{1/2}\right\} \tag{5.7}$$

where

$$\gamma = \frac{f_1}{g_2}\sqrt{1-\beta^2} \tag{5.8}$$

Clearly θ is the angle, in S, between the light propagation direction and the absolute velocity $\vec{v}$ of S. Introducing the one-way velocity of light $c_1(\theta) = r/t$, we have

$$\frac{1}{c_1(\theta)} = \frac{1}{f_1(1-\beta^2)}\left\{\left(e_1 + \frac{e_4}{c}\beta\right)\cos\theta + \left(\frac{e_4}{c} + e_1\beta\right)\left[\cos^2\theta + \gamma^2\sin^2\theta\right]^{1/2}\right\} \tag{5.9}$$

Particular cases of eq. (5.9) are

$$\frac{1}{c_1(0)} = \frac{1}{f_1(1-\beta)}\left(\frac{e_4}{c} + e_1\right); \qquad \frac{1}{c_1(\pi)} = \frac{1}{f_1(1+\beta)}\left(\frac{e_4}{c} - e_1\right) \tag{5.10}$$

whence

$$\frac{c_1(\pi)}{c_1(0)} = \frac{\left(\frac{e_4}{c} + e_1\right)(1+\beta)}{\left(\frac{e_4}{c} - e_1\right)(1-\beta)} \tag{5.11}$$

As we know, the continuity condition of the physical quantities in passing from slowly accelerated systems to inertial systems leads to eq. (5.3), which we repeat:

$$\frac{c_1(\pi)}{c_1(0)} = \frac{1+\beta}{1-\beta} \tag{5.12}$$

By comparing (5.11) and (5.12) it necessarily follows

$$e_1 = 0 \tag{5.13}$$

We can conclude that the most general transformations of the space and time variables between inertial systems allowed by the continuity condition are

$$\begin{cases} x = f_1(x_0 - vt_0) \\ y = g_2 y_0; \qquad z = g_2 z_0 \\ t = e_4 t_0 \end{cases} \tag{5.14}$$

This implies the necessary existence of absolute simultaneity. In fact, according to the fourth eq. (5.14), two point-like events with coordinates x_{10} and x_{20} ($x_{01} \neq x_{02}$), taking place at the same time t_0, are judged to happen at the same time also in S. Once more, absolute simultaneity is seen to be unavoidable. With $e_1 = 0$ the velocity of light becomes

$$\frac{1}{c_1(\theta)} = \frac{e_4}{cf_1(1-\beta^2)}\left\{\left[\cos^2\theta + \gamma^2\sin^2\theta\right]^{1/2} + \beta\cos\theta\right\} \tag{5.15}$$

showing that the $\beta\cos\theta$ term in the one-way velocity of light is a fixed ingredient in all theories of inertial systems satisfying the continuity condition with the accelerated ones. Such a term is present in Galilean physics. In fact, the Galilean transformations satisfy (5.1) with

$$f_1^G = g_2^G = e_4^G = 1; \qquad e_1^G = 0 \tag{5.16}$$

Using (5.16) the one-way velocity of light of the Galilean theory follows from (5.15) and turns out to be

$$\frac{1}{c_1^G(\theta)} = \frac{1}{c(1-\beta^2)}\left\{\left[1 - \beta^2\sin^2\theta\right]^{1/2} + \beta\cos\theta\right\} \tag{5.17}$$

As we see, eq. (5.17) contains the term $\beta\cos\theta$, characteristic of all theories treating inertial systems in a way continuous with the accelerated ones. The absence of this term in the TSR, in which $c_1 = c$ is isotropic, gives rise to the discontinuity of Figure 6.2.

We can conclude that the famous synchronization problem is solved by nature itself: it is not true that the synchronization procedure can be chosen freely, as the usually adopted convention leads to an unacceptable discontinuity in the physical theory.

The linear acceleration discontinuity

In the previous sections we showed that the absolute simultaneity condition $e_1 = 0$ can be deduced from the continuity of physical quantities in passing from slowly rotating platforms to inertial systems. We concluded that, in a way, the clock synchronization problem of inertial systems is solved by nature itself: it is not true that the synchronization procedure can be chosen freely, as all conventions but one lead to unacceptable discontinuities in the physical theory. In the present section we will show that exactly the same discontinuity exists in the case of linearly accelerated reference frames and that $e_1 = 0$ is once more the only way to save the continuity of the physical quantities. The only assumptions needed are two well-known consequences of the equivalent transformations, clock retardation and length contraction.

In considering linearly accelerated systems it is necessary to remember that no perfectly rigid body can exist in nature so that the application of a force in a point does not transmit instantaneously to the other points of a system. A good approximation to a system equally accelerating in all its parts can be obtained by considering a set of separated equal objects equipped with similar engines programmed in such a way as to generate the same acceleration in all of them. The set can be as large as one desires, but we will limit our discussion to the simplest case of two objects ("spaceships").

Two identical spaceships ***A*** and ***B*** are initially at rest on the x_0 axis of the isotropic inertial system S_0 at a distance ℓ_0 from one another. Their clocks are synchronized with those of S_0. At time $t_0 = 0$ they start accelerating in the $+x_0$ direction, and they do so equally, having the same velocity $v(t_0)$ at any time t_0 of S_0. With respect to S_0 the positions of ***A*** and ***B*** at time t_0 are given by:

$$\begin{cases} x_{0A}(t_0) = x_{0A}(0) + \int_0^{t_0} dt'_0 v(t'_0) \\ x_{0B}(t_0) = x_{0B}(0) + \int_0^{t_0} dt'_0 v(t'_0) \end{cases} \tag{6.1}$$

so that

$$x_{0B}(t_0) - x_{0A}(t_0) = x_{0B}(0) - x_{0A}(0) = \ell_0 \tag{6.2}$$

As eq. (6.2) shows, the motions described by (6.1) do not modify the distance ℓ_0 separating ***A*** and ***B***, as seen from S_0. In this sense the accelerated reference frame provided by ***A*** and ***B*** can be considered "rigid".

Let a light signal σ leave ***A*** and reach ***B*** along the straight line joining ***A*** and ***B***. The times of the events of emission and detection of σ are:

t_{0A}: S_0 time of departure of σ from $\boldsymbol{A}$;
t_{0B}: S_0 time of arrival of σ at $B(t_{0B} > t_{0A})$;
t_A: time of departure of σ from $\boldsymbol{A}$ as marked by clocks aboard $\boldsymbol{A}$;
t_B: time of arrival of σ at $\boldsymbol{B}$ as marked by clocks aboard $\boldsymbol{B}$.

For simplicity we will later assume that $v(t_0)$ is a linear function of its argument at least in the time interval $t_{0A} \leq t_0 \leq t_{0B}$, namely

$$v(t_0) = v_A + a(t_0 - t_{0A}) \qquad (t_{0A} \leq t_0 \leq t_{0B}) \tag{6.3}$$

where $a \geq 0$ is the acceleration and $v_A = v(t_{0A})$. Of course, owing to the increasing inertial resistance of the kinetic energy, the acceleration has to be kept constant by suitably increasing forces; anyway, we will consider only small accelerations applied during a reasonably short time, so that $v(t_{0B}) < c$.

Defining the usual Lorentz factor for clocks and rods as

$$R(t_0) = \sqrt{1 - v^2(t_0)/c^2} \tag{6.4}$$

we will use the following two basic consequences of the transformations (2.1):

1. Clock retardation. A clock at rest in an accelerated system S_a (e.g. aboard a spaceship), moving with instantaneous velocity $v(t_0)$ with respect to S_0 is retarded by the factor $R(t_0)$. In other words, during the S_0 time interval $(t_0, t_0 + dt_0)$ the clock marks a time increase dt given by

$$dt = dt_0 R(t_0) \tag{6.5}$$

2. Length contraction. A segment at rest in an accelerated system S_a moving with instantaneous velocity $v(t_0)$ with respect to S_0 (the segment is thought to be between the two spaceships on the line joining them) is contracted by the factor $R(t_0)$ according to the observers at rest in S_0. We can show that if during the S_0 time interval $(t_0, t_0 + dt_0)$ a point-like flash of light σ propagates along the $+x_0$ axis over the distances $d\ell_0$ and $d\ell$ (seen from S_0 and S_a, respectively) one has

$$R(t_0)d\ell = d\ell_0 - v(t_0)dt_0 \tag{6.6}$$

Proof of eq. (6.6). Let P and Q be any two points at rest in S_a on the line joining $\boldsymbol{A}$ and $\boldsymbol{B}$, and P_0 and Q_0 two points at rest in S_0 chosen as follows. The point P_0 coincides with P when σ passes near P at time t_0. The point Q_0 coincides with Q when σ passes near Q at time $t_0 + dt_0$. Thus the observer in S_a sees the light signal σ propagating from P to Q, these points defining a segment of length $d\ell$, while the observer in S_0 sees σ propagating from P_0 to Q_0 defining a segment of length $d\ell_0$. At time $t_0 + dt_0$ the points P_0 and P do not coincide anymore, P having shifted by $v(t_0)dt_0$ towards Q_0 (see Figure 6.3). Classically this would suggest $d\ell = d\ell_0 - vdt_0$, but the Lorentz contraction implies that the length measured to be $d\ell$ in S_a is actually seen as $R(t_0)d\ell$ by the observers in S_0. Hence eq. (6.6) is correct.

From eq. (6.5) we get

$$t_A = \int_0^{t_{0A}} dt'_0 R(t'_0); \qquad t_B = \int_0^{t_{0B}} dt'_0 R(t'_0) \tag{6.7}$$

Therefore

$$t_B - t_A = \int_{t_{0A}}^{t_{0B}} dt'_0 R(t'_0) \tag{6.8}$$

By assuming eq. (6.3) we can write

$$t_B - t_A = \int_{t_{0A}}^{t_{0B}} dt'_0 \sqrt{1 - \left[v_A + a(t'_0 - t_{0A})\right]^2 / c^2} \tag{6.9}$$

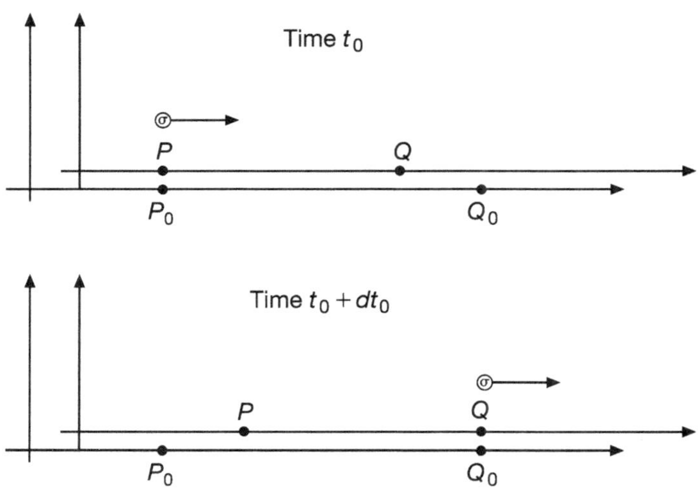

Figure 6.3 The points **P** and **Q** at rest in $\mathbf{S_a}$, while $\mathbf{P_0}$ and $\mathbf{Q_0}$ are at rest in $\mathbf{S_0}$.

This integral can be calculated. A useful change of variable is

$$\beta = \frac{1}{c}\left[v_A + a(t'_0 - t_{0A})\right] \tag{6.10}$$

Defining also

$$\beta_A = \frac{v_A}{c}; \qquad \beta_B = \frac{1}{c}[v_A + a(t_{0B} - t_{0A})] \tag{6.11}$$

we obtain the time difference

$$t_B - t_A = \frac{c}{a}\int_{\beta_A}^{\beta_B} d\beta\sqrt{1-\beta^2} \tag{6.12}$$

Eq. (6.11) shows that in the limit of small accelerations also $\beta_B - \beta_A$ is small. It is easy to see that to the first order in $\beta_B - \beta_A$ eq. (6.12) becomes

$$t_B - t_A \cong \frac{c}{a}(\beta_B - \beta_A)\sqrt{1-\beta_A^2} \tag{6.13}$$

which is finally the same as

$$t_B - t_A \cong (t_{0B} - t_{0A})\sqrt{1 - v_A^2/c^2} \tag{6.14}$$

Next we calculate the distance ℓ separating ***A*** and ***B***. Relative to S_0 the light signal propagates with velocity c, so that $d\ell_0 = cdt_0$. Therefore eq. (6.6) becomes

$$R(t_0)d\ell = [c - v(t_0)]dt_0 \tag{6.15}$$

whence, by integration

$$\ell = \int_{t_{0A}}^{t_{0B}} dt'_0 \frac{c - v(t'_0)}{\sqrt{1 - v^2(t'_0)/c^2}} \tag{6.16}$$

Using eqs. (6.3), (6.10) and (6.11) we get

$$\ell = \frac{c^2}{a}\int_{\beta_A}^{\beta_B} d\beta \frac{1-\beta}{\sqrt{1-\beta^2}} \tag{6.17}$$

In the limit of small acceleration one gets

$$\ell \cong \frac{c^2}{a}(\beta_B - \beta_A)\frac{1-\beta_A}{\sqrt{1-\beta_A^2}} \tag{6.18}$$

which can also be written

$$\ell \cong c(t_{0B} - t_{0A})\frac{1-v_A/c}{\sqrt{1-v_A^2/c^2}} \tag{6.19}$$

The velocity of light relative to S_a, ratio of ℓ given by eq. (6.19) and $t_B - t_A$ given by eq. (6.14), is therefore

$$\tilde{c} = \frac{c}{1+v_A/c} \tag{6.20}$$

This result, rigorous in the limit $a_A \rightarrow 0$, is not the relativistic result $\tilde{c} = c$, but is what one would expect from the inertial transformations, since the straight line connecting ***A*** and ***B*** was taken parallel to their velocity ($\theta = 0$ and $e_1 = 0$ in eq. (2.3)). A similar result, with a negative sign in the denominator, is found for the velocity relative to S of a light pulse traveling from ***B*** to ***A***, again in agreement with the inertial transformations ($\theta = \pi$). These results can perhaps be found surprising, as they are not in agreement

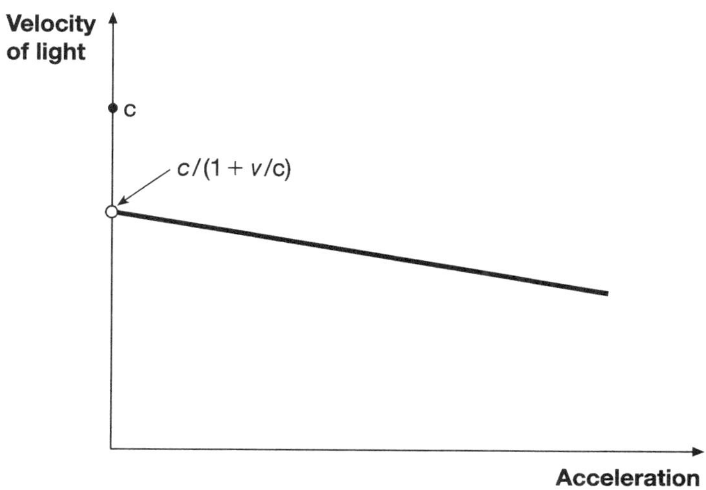

Figure 6.4 The zero acceleration discontinuity for the one-way velocity of light in the forward direction in the case of linear motions.

with the theory of relativity. In fact, the breakdown from the theory has probably been introduced with eq. (6.5) where a linear relation between the times of S_0 and S does not contain the space coordinates. Notice, however, that eq. (6.5) is physically reasonable, besides being a necessary consequence of our assumptions. We can only conclude that the spontaneously established synchronization of clocks during the motion of the accelerated reference frame S_a is incompatible with relativity.

Our spaceships ***A*** and ***B***, initially at rest on the x_0 axis of the isotropic inertial system S_0 at a distance d_0 from one another, were supposed to have clocks aboard synchronized with those of S_0. Let C_A and C_B be two clocks aboard ***A*** and ***B***, respectively. Before the time of departure ($t_0 = 0$) C_A, C_B are used to measure the velocity with which a pulse of light propagates between C_A and C_B. The result is bound to be c, of course. The same experiment is repeated after the acceleration has ended and the spaceships are at rest in the different inertial system S. Now, if the invariance of the velocity of light were a law of nature one should find the same result in S and in S_0, given that the retardation of C_A and C_B due to velocity during the accelerated motion is exactly the same. Instead, as we saw, the velocity of light in S turns out to be given by eq. (6.22). Notice that the equal retardation of C_A and C_B is expressed by the equality of the proper times of C_A and C_B and is therefore an objective property on which all observers must agree. Thus everything goes as if we measured the velocity of light with two clocks, then set backwards their hands by the same amount, then measured again the velocity of light and found a different result. It is a surprise!

The result can only be understood in terms of inequivalence of the inertial systems, in the sense that the second postulate of the theory of relativity does not hold in nature and the velocity of light relative to an inertial system depends on the absolute motion of the latter. This seems to imply the existence of an objectively privileged system.

Export of relativistic simultaneity fails

In the previous section the two spaceships argument was used as a way to present the zero acceleration discontinuity for linear motions. In the present section we will use it again for showing that the Einstein synchronization of clocks cannot be exported from an inertial frame to another. We will assume that at some time the acceleration ends and the two spaceships, in a way, constitute a new inertial frame. As the physical actions on the clocks are the same, one could naively expect that it should be possible to "export" to the new frame whatever synchronization existed in the old one. It is not so: relativistic simultaneity cannot be exported.

As before ***A*** and ***B*** are initially at rest on the x_0 axis of the (privileged) inertial system S_0 at a distance d_0 from one another. Their clocks are synchronous with the clocks at rest in S_0. At time $t_0 = 0$ they start accelerating in the $+x_0$ direction, and they do so equally, having the same velocity $v(t_0)$ at any time t_0 of S_0, until at $t_0 = \bar{t}_0$ they reach a preassigned velocity $v = v(\bar{t}_0)$ parallel to $+x_0$. At time $t_0 = \bar{t}_0$ the acceleration ends. For $t_0 \geq \bar{t}_0$ the spaceships are at rest in a different inertial system S. The positions of ***A*** and ***B*** at time $t_0 \geq \bar{t}_0$ are given by:

$$\begin{cases} x_{0A}(t_0) = x_{0A}(0) + \int\limits_0^{\bar{t}_0} dt'_0 v(t'_0) + (t_0 - \bar{t}_0)v \\ x_{0B}(t_0) = x_{0B}(0) + \int\limits_0^{\bar{t}_0} dt'_0 v(t'_0) + (t_0 - \bar{t}_0)v \end{cases} \tag{7.1}$$

so that, as in eq. (6.2)

$$x_{0B}(t_0) - x_{0A}(t_0) = x_{0B}(0) - x_{0A}(0) = d_0 \tag{7.2}$$

Again, motion does not modify the distance ℓ_0 separating ***A*** and ***B***, as seen from S_0. The same distance seen from S (call it ℓ) instead *increased*, as the unit rod measuring it underwent a Lorentz contraction. One has

$$\ell = \frac{\ell_0}{R} \tag{7.3}$$

In fact the observer in S_0 will check: (i) that the distance d_0 between ***A*** and ***B*** remained the same after acceleration, in accordance with (7.2); (ii) that the unit rod at rest in S contracted by a factor R ; (iii) that the observer in S using his contracted rod to measure the ***A–B*** distance finds it larger by a factor $1/R$ than before departure. This measurement is an objective procedure and its result (= number of times the unit rod fits into the ***A–B*** distance) cannot depend on the subjective point of view of an observer. Therefore the observer in S finds indeed what the observer in S_0 sees him finding, namely an ***A–B*** distance given by eq. (7.3). By the way, this is essentially the Lorentz contraction of the distance separating the two spaceships, as d_0 is shorter than d by the usual factor R.

We can consider the trip of the two spaceships as an attempt to export to the system S the synchronization of distant clocks valid in S_0. We will see that the attempt fails. In fact we will show that the transformations relating S_0 and S are necessarily the inertial ones, if the synchronization actually transported by the two spaceships is adopted in S.

Motion is the same for ***A*** and ***B*** and all effects of motion will necessarily coincide. Therefore τ, the proper time marked by clocks aboard the two spaceships when the acceleration ends, is given by

$$\tau = \int_0^{\bar{t}_0} dt'_0 \sqrt{1 - v_0^2(t'_0)/c^2} \tag{7.4}$$

The clocks aboard ***A*** and ***B*** accumulate exactly the same delay with respect to the clocks at rest in S_0. Therefore two events simultaneous in S_0 will be such also in S, even if they take place in different points of space near which ***A*** and ***B*** happen to pass at some time $t_0 > \bar{t}_0$. Clearly we have here a case of absolute simultaneity and the condition $e_1 = 0$ must hold in S, reducing the equivalent transformations to the inertial ones.

We will now show, in agreement with the previous section, that the velocity, relative to S, of a light pulse traveling from ***A*** to ***B*** when the two spaceships are at rest in S is not c, but has the value demanded by the inertial transformations. Let a light signal leave ***A*** at time t_A and reach ***B*** at time t_B, these times being measured in S by the clocks aboard the two spaceships. For $t_0 \geq \bar{t}_0$ clock retardation in S is due to the constant velocity v. Therefore one has:

$$t_A = \tau + (t_{0A} - \bar{t}_0)R; \qquad t_B = \tau + (t_{0B} - \bar{t}_0)R \tag{7.5}$$

where t_{0A} and t_{0B} are the S_0 times of departure of the light signal from ***A*** and arrival at ***B***, respectively (we assume $0 < \bar{t}_0 < t_{0A} < t_{0B}$). By subtracting the first equation in eq. (7.5) from the second we get

$$t_B - t_A = R(t_{0B} - t_{0A}) \tag{7.6}$$

We will next calculate the velocity $\tilde{c}$ of the light signal in S, defined as

$$\tilde{c} = \frac{\ell}{t_B - t_A} \tag{7.7}$$

where ℓ is given by eq. (7.3).

The point $\tilde{x}_{0B}$ of S_0 where the light pulse is detected in ***B*** can be considered either from the point of view of pulse propagation, or of ***B*** motion. Thus it must satisfy two equations:

$$\tilde{x}_{0B} = x_{0A} + c(t_{0B} - t_{0A}); \qquad \tilde{x}_{0B} = x_{0B} + v(t_{0B} - t_{0A}) \tag{7.8}$$

where x_{0A} and x_{0B} are the positions of ***A*** and ***B***, respectively, at the time t_{0A} of emission of the light signal. In fact, while the light signal goes from x_{0A} to $\tilde{x}_{0B}$ with velocity c, ***B*** moves from x_{0B} to $\tilde{x}_{0B}$ with velocity v. From the equations in eq. (7.8) follows:

$$x_{0B} - x_{0A} = c(1 - \beta)(t_{0B} - t_{0A}) \tag{7.9}$$

where $\beta = v/c$. Remembering that eq. (7.2) holds for all times t_0 one gets

$$c(1-\beta)(t_{0B}-t_{0A}) = \ell_0 \tag{7.10}$$

By using eq. (7.3) for transforming the ***A–B*** distance and eq. (7.6) for transforming the time interval, we get from eq. (7.10)

$$c(1-\beta)\frac{t_B - t_A}{R} = R\ell \tag{7.11}$$

from which it is easy to obtain

$$\frac{t_B - t_A}{\ell} = \frac{1+\beta}{c} \tag{7.12}$$

Comparing with eq. (7.4) we get finally:

$$\tilde{c} = \frac{c}{1+\beta} \tag{7.13}$$

This is what one expects from the inertial transformations, since the straight line connecting ***A*** and ***B*** is parallel to their velocity ($\theta = 0$). A similar result with a negative sign in the denominator is found for the velocity relative to S of a light pulse traveling from ***B*** to ***A***, again in agreement with the inertial transformations ($\theta = \pi$). These results can perhaps be found surprising, as they are not in agreement with the Lorentz transformations of the TSR. In fact, the breakdown from the theory, we can now say, has definitely been introduced with eq. (7.5) where the transformations from S_0 to S of t_{0A} and t_{0B} do not contain the space coordinates. Replacing eq. (7.5) with the Lorentz transformations of time leads of course to $\tilde{c} = c$. Notice, however, that eq. (7.5) is a necessary consequence of the equal behavior of the two spaceships. Once more we can conclude that the spontaneously established synchronization of clocks during the acceleration of ***A*** and ***B*** is not compatible with the TSR. The result implies the inequivalence of the inertial systems, in the sense that the second postulate of the theory of relativity does not hold and the velocity of light relative to an inertial system depends on the absolute motion of the latter. This implies the existence of an objectively privileged system.

It should also be stressed that the concrete performance of the above experiment with the two accelerating spaceships would not really allow one to detect the privileged inertial system, as an arbitrary inertial system could be assumed to be isotropical and used as a starting frame of ***A*** and ***B***. In this sense the above considerations are compatible with the weak form of the relativity principle discussed in the last section.

Irreversible processes cannot be synchronized

We can show that the inertial transformations based on absolute simultaneity give the most natural description of the physical reality by considering irreversible processes on board the two spaceships. For example, a radioactive sample has an exponentially decreasing decay rate, which could be used to measure time. Such a "clock" cannot be synchronized and continues its running independently of human will. It is well known, however, that such a clock suffers the relativistic retardation, when in motion, just like man-made ordinary clocks.

A second example is provided by the ageing of biological structures. To develop this point, we will suppose that our spaceships ***A*** and ***B*** have two passengers P_A and P_B, who are twins. Of course in principle nothing can stop them from resynchronizing their reversible clocks once the acceleration has ended and the two spaceships are at rest in S. If they do so, however, they find that they have different biological ages at the same (resynchronized) S time, even if they started the space trip at the same S_0 time and with the same acceleration, as stipulated above. Moreover, reversible and irreversible clocks will not agree anymore.

To make everything as clear as possible, suppose the twins P_A and P_B were born in the point M of the x_0 axis of S_0 equidistant from the two points from which ***A*** and ***B*** start their trip. When the time of departure is near, P_A and P_B are brought from M to ***A*** and ***B***, respectively, with the same speed. As said above, at time $t_0 = 0$ the spaceships start accelerating in the $+x_0$ direction, and they do so in the same way, until at time $\bar{t}_0$ of S_0 they reach a preassigned velocity v parallel to $+x_0$. At $t_0 \geq \bar{t}_0$ the spaceships are at rest in the inertial system S moving with velocity v with respect to S_0. We have seen that the clocks of ***A*** and ***B*** are retarded in the same way and that the S-S_0 transformations must be the inertial ones. We can now add that also the ageing of the twins in ***A*** and ***B*** must be the same, given that they experience identical velocities. Therefore the twins have the same biological age when the times marked on their clocks are the same, if these were synchronized in S_0 before departure and never resynchronized afterwards. Ageing is an irreversible way of time keeping and cannot be modified. The twins can of course inform one another about their ages by exchanging pictures (e.g. via telefax) in which the time is recorded at which they were taken. The receiving twin will check in his records that at the time shown on his brother's picture he looked exactly the same, and therefore had the same age.

Naturally P_A and P_B can use a different synchronization of reversible clocks, if they wish, e.g. Einstein's synchronization leading to the Lorentz transformations between S and S_0. To do so they must send a light signal, e.g. from ***A*** to ***B***, and they must reset at least one clock. We can even suppose that every twin has two clocks and keeps the first one set on absolute time, while regulating the second one to show the Lorentz time. More exactly we assume that:

P_A	has a first clock	C_A	marking the absolute time	t_A
P_A	has a second clock	$\hat{C}_A$	marking the Lorentz time	$\hat{t}_A$
P_B	has a first clock	C_B	marking the absolute time	t_B
P_B	has a second clock	$\hat{C}_B$	marking the Lorentz time	$\hat{t}_B$

After the initial times $t_0 \leq \bar{t}_0$ of S_0 for which $t_A = \hat{t}_A = t_B = \hat{t}_B$ a resynchronization of $\hat{C}_B$ is carried out as follows. At some preestablished time a light signal is sent from ***A*** to ***B***. Of course it can be checked with C_B that this signal arrives at ***B*** a time

$$\frac{\ell(1+\beta)}{c} \tag{8.1}$$

later, as we saw that the velocity of light from ***A*** to ***B*** in S is given by eq. (7.13). Forcing the velocity of light to be c in his rest frame ("Einstein synchronization"), the observer in ***B*** will reset his clock $\hat{C}_B$ at a time earlier by

$$-\frac{\ell\beta}{c} \tag{8.2}$$

in such a way that eq. (8.1) is indeed replaced by ℓ/c, as required. If τ is the delay generated by velocity and acceleration up to time $\bar{t}_0$ (the same as in eq. (7.4)), at time t_0 of S_0 (after resynchronization of $\hat{C}_B$ so that $t_0 \geq \bar{t}_0$) the clocks mark the following times

C_A	marks	$t_A = \bar{t}_0 - \tau + (t_0 - \bar{t}_0)R$
$\hat{C}_A$	marks	$\hat{t}_A = t_A$
C_B	marks	$t_B = t_A$
$\hat{C}_B$	marks	$\hat{t}_B = t_B - \ell\beta/c$

Clearly, at the same physical time one has

$$t_B - t_A = 0 \tag{8.3}$$

but

$$\hat{t}_B - \hat{t}_A = -\frac{\ell\beta}{c} \tag{8.4}$$

The simultaneity of C_A and C_B is different from that of $\hat{C}_A$ and $\hat{C}_B$! If P_A and P_B exchange pictures in which the times marked by the clocks C and $\hat{C}$ are both shown, they discover that they had the same age at the same t, but different ages at the same $\hat{t}$. This fact gives a clear prevalence to the inertial transformations. Natural phenomena are once more better described by the inertial transformations when they take place in concretely produced inertial systems.

Export of absolute simultaneity succeeds

It will now be shown that, contrary to the case of relativistic simultaneity, the export of absolute simultaneity to a different inertial system is fully possible.

Once more, two identical spaceships ***A*** and ***B*** are initially at rest on the x axis of the inertial system S (this time not the privileged one!) at a distance ℓ from one another. The system S has velocity v relatively to the privileged frame S_0. The clocks at rest in S are synchronized in such a way that the inertial transformations hold between S_0 and S. The clocks in ***A*** and ***B*** are synchronized locally, before departure, with nearby clocks of S. At time $t = 0$ the spaceships start accelerating in the $+x$ direction, and they do so equally, having the same velocity $v(t)$ at any time t of S, until at $t = \bar{t}$ they reach a preassigned velocity $v' \equiv v(\bar{t})$ parallel to $+x$. At $t = \bar{t}$ acceleration ends. For $t \geq \bar{t}$ the spaceships are at rest in a different inertial system S' moving with velocity v'. With respect to S the positions of ***A*** and ***B*** at any time $t \geq \bar{t}$ are given by:

$$\begin{cases} x_A(t) = x_A(0) + \int_0^{\bar{t}} dt' v(t') + (t - \bar{t})v' \\ x_B(t) = x_B(0) + \int_0^{\bar{t}} dt' v(t') + (t - \bar{t})v' \end{cases} \tag{9.1}$$

so that

$$x_B(t) - x_A(t) = x_B(0) - x_A(0) = \ell \tag{9.2}$$

As eq. (9.2) shows, the motion of ***A*** and ***B*** does not modify their distance ℓ as seen from S. The theory of the inertial transformations shows that the velocity of S' (and then also of ***A*** and ***B*** after the end of acceleration) relative to S is given by

$$u = \frac{v' - v}{R^2} \tag{9.3}$$

After the end of the acceleration the distance between ***A*** and ***B*** in S' (call it ℓ') is different, as the unit rod measuring it undergoes a Lorentz contraction different from that applying to rods at rest in S. In this case one has from the inertial transformations

$$\ell' = \frac{R}{R'}\ell \tag{9.4}$$

where

$$R = \sqrt{1 - v^2/c^2}; \qquad R' = \sqrt{1 - v'^2/c^2} \tag{9.5}$$

It will next be shown that the transformations relating S and S' are necessarily the inertial ones, if no clock resynchronization is applied in S' correcting what nature itself generated during the acceleration of the two spaceships. In other words the inertial transformations are maintained when $\boldsymbol{A}$ and $\boldsymbol{B}$ pass from S to S' in the described way. Motion (relative and absolute) is the same for $\boldsymbol{A}$ and $\boldsymbol{B}$ and all effects of motion will necessarily coincide. Therefore the clocks aboard $\boldsymbol{A}$ and $\boldsymbol{B}$ will accumulate exactly the same delay with respect to the clocks at rest in S and two events simultaneous in S will be such also in S', even if they take place in different points of space. Clearly we have once more a case of absolute simultaneity and the condition $e'_1 = 0$ must hold in S'.

In order to develop the point further we check next that the velocity, relative to S', of a light pulse traveling from $\boldsymbol{A}$ to $\boldsymbol{B}$ when the two spaceships are at rest in S' (and move with velocity v with respect to S) has the value demanded by the inertial transformations. Let a light signal leave $\boldsymbol{A}$ at time t'_A and reach $\boldsymbol{B}$ at time t'_B, these times being measured in S' by the clocks aboard the two spaceships. The velocity $\tilde{c}'$ of the light signal in S' is, by definition

$$\tilde{c}' = \frac{\ell'}{t'_B - t'_A} \tag{9.6}$$

where ℓ' is given by eq. (9.4). Now define, with $0 < \bar{t} < t_A < t_B$:

- τ: proper time marked by clocks aboard $\boldsymbol{A}$ and $\boldsymbol{B}$ at the S time $\bar{t}$ at which acceleration ends;
- t_A: S time of departure of the light signal from $\boldsymbol{A}$;
- t_B: S time of arrival of the light signal at $\boldsymbol{B}$.

As for $t \geq \bar{t}$ clock retardation in S' is due to the constant velocity one has from the inertial transformations between two moving systems:

$$t'_A = \tau + \frac{R'}{R}(t_A - \bar{t}); \qquad t'_B = \tau + \frac{R'}{R}(t_B - \bar{t}) \tag{9.7}$$

By subtracting the first equation from the second one, one gets

$$t'_B - t'_A = \frac{R'}{R}(t_B - t_A) \tag{9.8}$$

We will next calculate $\tilde{c}'$ as defined by eq. (9.6). The point $\tilde{x}_B$ of S where the light pulse is absorbed in **B** must satisfy:

$$\tilde{x}_B = x_A(t_A) + \frac{c}{1+\beta}(t_B - t_A); \qquad \tilde{x}_B = x_B(t_A) + \frac{v' - v}{R^2}(t_B - t_A) \quad (9.9)$$

if $x_A(t_A)$ and $x_B(t_A)$ are the positions of the spaceships **A** and **B**, respectively, at time t_A of emission of the light signal. In fact, while the light signal goes from x_A to $\tilde{x}_B$ with velocity $c/(1+\beta)$, spaceship **B** moves from x_B to $\tilde{x}_B$ with velocity u given by eq. (9.3). By subtraction, from the equations in eq. (9.9) it follows:

$$x_B(t_A) - x_A(t_A) = \frac{c(1-\beta')}{R^2}(t_B - t_A) \quad (9.10)$$

Remembering that eq. (9.2) holds for all times t one gets

$$\frac{c(1-\beta')}{R^2}(t_B - t_A) = \ell \quad (9.11)$$

By using eq. (9.4) for transforming the **A**–**B** distance and eq. (9.8) for transforming the time interval, we get from eq. (9.11)

$$\frac{c(1-\beta')}{R^2}\frac{R}{R'}(t'_B - t'_A) = \frac{R'}{R}\ell' \quad (9.12)$$

from which it is easy to obtain

$$\frac{t'_B - t'_A}{\ell'} = \frac{1+\beta'}{c} \quad (9.13)$$

Comparing with eq. (9.4) we get finally:

$$\tilde{c}' = \frac{c}{1+\beta'} \quad (9.14)$$

This is what one expects from the inertial transformations, since the straight line connecting the spaceships **A** and **B** has been assumed parallel to their velocity ($\theta' = 0$). A similar result, with a negative sign in the denominator, is found for the velocity relative to S of a light pulse traveling from **B** to **A**, again in agreement with the inertial transformations ($\theta = \pi$). We can conclude that the export of the absolute synchronization of clocks during the motion with equal acceleration of **A** and **B** takes place spontaneously.

Resynchronization of clocks

The inertial transformations based on absolute simultaneity imply the existence of a single isotropic inertial reference system ("privileged system"). I have shown, however, that a resynchronization of clocks (ROC) in all inertial systems is possible leading to a different, arbitrarily chosen, isotropic "privileged" system (Selleri, 2005b: 325). Such a ROC does not modify any one of the empirical consequences of the theory, which is thus compatible with a formulation of the relativity principle weaker than adopted in Einstein's theory of special relativity.

Assuming the inertial transformations, the S_0 system is initially considered to be privileged, and the velocity of light relative to it isotropic. Other inertial systems (S, S', ...) are initially described as "moving" and relative to them the observers detect an anisotropic velocity of light given by equations like eq. (2.3). In Selleri (2005) I described a process of ROC showing that it is uniquely determined by the new inertial frame S chosen to replace S_0 as "privileged" and by the requirement that absolute simultaneity should be preserved.

The first assumption is that ROC should change the velocity of light relative to S from the value given by eq. (2.3) with $e_1 = 0$ to c. Therefore the time required by a point-like flash of light to cover the distance ℓ in a direction forming an angle θ with respect to the x axis should be modified in the following way

$$\frac{\ell}{c}\left(1 + \frac{v}{c} cos\theta\right) \rightarrow \frac{\ell}{c} \tag{10.1}$$

Given the homogeneity of space one can assume, without loss of generality, that the flash of light is emitted in the origin, so that $\ell cos\theta = x$. Therefore the recipe is very simple: subtract xv/c^2 to the time shown by the clock having a position with first coordinate x. Thus the new time $\tilde{t}$ that should replace t for the considered clock is

$$\tilde{t} = t - x\frac{v}{c^2} \tag{10.2}$$

If this is done systematically for all clocks at rest in S a new situation is obtained in which the speed of light appears to be isotropic and equal to c.

We will also say how ROC should be implemented in the other inertial frames in order to preserve the validity of absolute simultaneity. It is well known that the absolute simultaneity can be obtained simply by choosing one system S to be privileged, synchronizing clocks according to the Einstein procedure in this system, and then synchronizing clocks in all other systems moving past S by adjusting these clocks to $t' = 0$ whenever they fly past a clock in S which shows $t = 0$.

As far as the initially privileged frame S_0 is concerned one can show that if the post-ROC time $\tilde{t}_0$ is defined as follows

$$\tilde{t}_0 = t_0 - x_0 \frac{v}{c^2} \tag{10.3}$$

then the overlapping of clocks of S_0 and S takes place at $\tilde{t} = \tilde{t}_0 = 0$, as required. After dealing with the post-ROC and pre-ROC privileged systems, one should say how ROC should be in the most general inertial system in order to preserve the absolute simultaneity. It has been possible to show that the ROC in S' satisfies the following rule:

$$\tilde{t}' = t' - x' \frac{R'^2}{1 - vv'/c^2} \frac{v}{c^2} \tag{10.4}$$

where R' is given by eq. (9.5). Notice that eqs. (10.2) and (10.3) are particular cases of eq. (10.4) for $v' = 0$ and $v' = v$, respectively. Therefore the most general recipe for ROC is eq. (10.4).

It could be shown that the ROC given by eq. (10.4) maintains the validity of the inertial transformations with the new inertial system S replacing S_0 in the role of "privileged" system and S_0 becoming a regular "moving" inertial system also from the point of view of the new transformations. Furthermore any other system S' obtains transformations appropriate to the new roles of S and S_0.

As an example consider the inverse pre-ROC inertial transformations from S to S_0 given by

$$x_0 = R\left(x + \frac{v}{R^2} t\right); \qquad t_0 = \frac{1}{R} t \tag{10.5}$$

According to eq. (10.1) the old time t should be replaced by $\tilde{t} + xv/c^2$. If this is done in the first equation, a short calculation leads to

$$x_0 = \frac{1}{R}(x + v\tilde{t}) \tag{10.6}$$

Similarly, if t and t_0 as given by eqs. (10.1) and (10.2), respectively, are substituted in the pre-ROC transformation of time eq. (10.5), and eq. (10.6) is used, one easily gets

$$\tilde{t}_0 = R\tilde{t} \tag{10.7}$$

Clearly eqs. (10.6) and (10.7), together with the trivial transformations of y and z, are direct inertial transformations from an isotropic system S to an S_0 moving with velocity $-v$.

The possibility to choose freely the isotropic reference frame had been anticipated in the excellent paper by Rizzi, Ruggiero and Serafini (Rizzi *et al.* 2005: 1835), where one reads: " ... the 'privileged role' played by S_0 ... is a merely artificial element, S_0 being just the IRF [inertial reference frame] in which, by stipulation, Einstein synchronization has been performed: as a matter of fact, any IRF S can play the role of S_0 ".

Buonaura (2004: 627) used the inertial transformations to show that Maxwell's equations, passing from the isotropic inertial system S_0 (where they retain the usual form) to another inertial system S, acquire a generalized form depending on the velocity of S relative to S_0 ("absolute velocity"). Thus Maxwell's equations are affected by a change of reference frame given that the velocity of the frame to which these equations are referred is modified. In spite of this, the inertial transformations are compatible with a form of relativity ("weak relativity"), as the isotropic system can be chosen arbitrarily, meaning that nothing in the theory allows one to conceive an experiment leading to the discovery of the really isotropic system.

It is useful to distinguish two formulations of the relativity principle:

R1. *Strong relativity*, according to which the laws of physics are exactly the same in all inertial systems. This is Einstein's formulation.

R2. *Weak relativity*, stating merely the impossibility to measure the absolute velocity of the Earth. This principle does not demand necessarily the validity of the Lorentz transformations and opens a logical space for new theories, such as the one based on the inertial transformations. It is amusing to notice that this formulation is essentially the original one given by Galileo.

In spite of the impossibility to detect experimentally a privileged frame it is necessary to insist that all the statements in favor of absolute simultaneity made in the present paper are correct. Furthermore I repeat that the theory with the free e_1 applied to the rotating platform shows that only the choice $e_1 = 0$ gives a satisfactory explanation of the Sagnac effect. Similarly, only with $e_1 = 0$ one can obtain a reasonable description of aberration and of the differential retardation of separating and reuniting clocks. The paradoxes of the special theory of relativity disappear in a theory based on $e_1 = 0$. The growing evidence for the existence in nature of superluminal signals can easily be accommodated in the same theory, while it is incompatible with relativity due to the presence of a famous causal paradox.

The need to adopt $e_1 = 0$ is strong and clear. Nevertheless I must admit that this approach to the physics of space and time might seem somewhat contradictory. On the one hand it points to a theory of space and time in which such conceptions as absolute velocity, privileged frame and absolute simultaneity have a central role, while, on the other hand, relativism comes back in the arbitrariness of the choice of the "privileged" inertial reference frame. Both sides of the contradiction ($e_1 = 0$ and weak relativity) are absolutely necessary if the assumptions made in the second section are all correct.

As a final remark one can add that from the point of view of the inertial transformations the validity of weak relativity appears accidental, more than fundamental. It would be enough to discover a very small non-invariance of the two-way speed of light to make the whole game of ROC impossible.

References

Anandan, J. (1981) 'Sagnac Effect in Relativistic and Nonrelativistic Physics', *Phys. Rev.* D24, 338–46.

Bailey, J. *et al.* (1977) *Nature* 268, 301.

Builder, G. (1957) *Austr. J. Phys.* 10, 246.

Buonaura, B. (2004) *Found. Phys. Lett.* 17, 627.

Einstein, A. (1952) 'On the Electrodynamics of Moving Bodies', in A. Einstein, H.A. Lorentz, H. Weyl, and H. Minkowski, *The Principle of Relativity*, Dover, New York, pp. 37–65.

Ghosal, S.K., B. Raychaudhuri, A.K. Chowdhuri and M. Sarker (2004) *Found. Phys. Lett.* 17, 457.

Hasselbach, F. and M. Nicklaus (1993) 'Sagnac Experiment With Electrons: Observation of the Rotational Phase Shift of Electron Waves in Vacuum', *Phys. Rev.* A48, 143–51.

Jammer, M. (1979) 'Some Fundamental Problems in the Special Theory of Relativity', in G. Toraldo di Francia (ed.), *Problems in The Foundations of Physics*, Bologna and North Holland, Amsterdam: Società Italiana di Fisica, pp. 202–36.

Landau, L.D. and E.M. Lifshitz (1996) *The Classical Theory of Fields*, Oxford: Butterworth-Heinemann.

Langevin, P. (1921) *Compt. Rend.* 173, 831.

Langevin, P. (1937) *Compt. Rend.* 205, 304.

Mansouri, M. and R. Sexl (1977) *General Relat. Gravit.* 8, 497, 515, 809.

Poincaré, H. (1898) *Rev. Metaphys. Morale* 6, 1.

Post, E.J. (1967) 'Sagnac Effect', *Rev. Mod. Phys.* 39, 475–93.

Puccini, G. and F. Selleri (2002) 'Doppler Effect and Aberration of Light from the Point of View of Absolute Motion', *Nuovo Cim.B*, 117, 283.

Reichenbach, H. (1958) *The Philosophy of Space & Time*, Dover, New York.

Rizzi, G., M.L. Ruggiero and A. Serafini (2005) 'Synchronization Gauges and the Principles of Special Relativity', *Found. Phys.* 34, 1835–87.

Sagnac, M.G. (1913) *Compt. Rend.* 157, 708, 1410.

Sagnac, M.G. (1914) *J. de Phys.* 4, 177.

Selleri, F. (1995) 'Space, Time and Their Transformations', *Chin. J. Syst. Eng. Electr.* 6, 25.

Selleri, F. (1996a) 'Noninvariant One-Way Velocity of Light', *Found. Phys.* 26, 641.

Selleri, F. (1996b) 'Noninvariant One-Way Velocity of Light and Particle Collisions', *Found. Phys. Lett.* 9, 43.

Selleri, F. (1997) 'Noninvariant One-Way Speed of Light and Locally Equivalent Reference Frames', *Found. Phys. Lett.* 10, 73.

Selleri, F. (2002) 'Bell's Spaceships and Special Relativity', in R.A. Bertlmann and A. Zeilinger (eds.), *Quantum [Un]speakables*, Berlin, pp. 413–28.
Selleri, F. (2004) 'Sagnac Effect: End of the Mystery', in G. Rizzi and M.L. Ruggiero (eds.), *Relativity in Rotating Frames*, Dordrecht: Kluwer, pp. 57–77.
Selleri, F. (2005a) 'On the Differential Retardation Between Separating and Reuniting Clocks', in: O. Nawrot et al. (eds.), *Between Philosophy and Natural Sciences*, Essays for L. Kostro's 70th birthday, Gdansk University Press.
Selleri, F. (2005) 'The Inertial Transformations and the Relativity Principle', *Found. Phys. Lett.* 18, 325.
Selleri, F. (2006) 'Superluminal Signals and the resolution of the causal paradox', *Found. Phys.* 36, 443.

7 Global Positioning System and the twins' paradox

Tom Van Flandern

Abstract

In the Global Positioning System (GPS), all atomic clocks in all reference frames (in orbit and on the ground) are set once and stay synchronized. We can use this same trick to place a GPS-type clock aboard the spacecraft of a traveling twin. That clock will stay synchronized with Earth clocks, allowing a clear resolution of the twins' paradox in special relativity – why the traveler expects to come back younger, and why the stay-at-home twin is not entitled to the same expectation.

Background

About the Global Positioning System

The Global Positioning System (GPS) consists of a network of 24 satellites in roughly 12-hour orbits, each carrying atomic clocks on board. The orbital radius of the satellites is about four Earth-radii (26,600 km). The orbits are nearly circular, with a typical eccentricity of less than 1%. Orbital inclination to the Earth's equator is typically 55 degrees. The satellites have orbital speeds of about 3.9 km/s in a frame centered on the Earth and not rotating with respect to the distant stars. Nominally, the satellites occupy one of six equally spaced orbital planes. Four of them occupy each plane, spread at roughly 90-degree intervals around the Earth in that plane. The precise orbital periods of the satellites are close to 11 hours and 58 minutes so that the ground tracks of the satellites repeat day after day, because the Earth makes one rotation with respect to the stars about every 23 hours and 56 minutes. (Four extra minutes are required for a point on the Earth to return to a position directly under the Sun because the Sun advances about one degree per day with respect to the stars.)

The on-board atomic clocks are good to about 1 nanosecond (ns) in epoch, and about 1 ns/day in rate. Since the speed of light is about one foot per nanosecond, the system is capable of amazing accuracy in locating anything on Earth or in the near-Earth environment. For example, if the satellite clocks are fully synchronized with ground atomic clocks, and we know the time when a signal is sent from a satellite, then the time delay for that signal to reach a ground receiver immediately reveals the distance (to a potential accuracy of about one foot) between satellite and ground receiver. By using four satellites to triangulate and determine clock corrections, the position of a receiver at an unknown location can be determined with comparable precision.

Relativity in the GPS

General Relativity (GR) predicts that clocks in a stronger gravitational field will tick at a slower rate. Special Relativity (SR) predicts that moving clocks will appear to tick slower than non-moving ones. Remarkably, these two effects cancel each other for clocks located at sea level anywhere on Earth. So if a hypothetical clock at Earth's north or south pole is used as a reference, a clock at Earth's equator would tick slower because of its relative speed due to Earth's spin, but faster because of its greater distance from Earth's center of mass due to the flattening of the Earth. Because Earth's spin rate determines its shape, these two effects are not independent, and it is therefore not entirely coincidental that the effects exactly cancel. The cancellation is not general, however. Clocks at any altitude above sea level do tick faster than clocks at sea level; and clocks on rocket sleds do tick slower than stationary clocks.

For GPS satellites, GR predicts that the atomic clocks at GPS orbital altitudes will tick faster by about 45,900 ns/day because they are in a weaker gravitational field than atomic clocks on Earth's surface. SR predicts that atomic clocks moving at GPS orbital speeds will tick slower by about 7,200 ns/day than stationary ground clocks. Rather than have clocks with such large rate differences, the satellite clocks are reset in rate before launch to compensate for these predicted effects. In practice, simply changing the international definition of the number of atomic transitions that constitute a one-second interval accomplishes this goal. Therefore, we observe the clocks running at their offset rates before launch. Then we observe the clocks running after launch and compare their rates with the predictions of relativity, both GR and SR combined. If the predictions are right, we should see the clocks run again at nearly the same rates as ground clocks, despite using an offset definition for the length of one second.

We will refer to a clock whose natural ticking frequency has been corrected in this way as a "GPS clock". This will help in the discussion of SR effects such as the twins' paradox. A GPS clock is pre-corrected for relativistic rate changes so that it continues to tick at the same rate as Earth clocks even when traveling at high relative speeds. So a GPS clock carried by the traveling twin can be used to determine local time in the Earth's frame at any point along the journey – a great advantage for resolving paradoxes.

We note that this post-launch rate comparison is independent of frame or observer considerations. Since the ground tracks repeat day after day, the distance from satellite to ground remains essentially unchanged. Yet, any rate difference between satellite and ground clocks continues to build a larger and larger time reading difference as the days go by. Therefore, no confusion can arise due to the satellite clock being located some distance away from the ground clock when we compare their time readings. One only needs to wait long enough and the time difference due to a rate discrepancy will eventually exceed any imaginable error source or ambiguity in such comparisons.

Theory and observations compared

The highest precision GPS receiver data is collected continuously in two frequencies at 1.5-second intervals from all GPS satellites at five Air Force monitor stations distributed around the Earth. An in-depth discussion of the data and its analysis is beyond the scope of this paper (Alley and Van Flandern 1998). These data show that the on-board atomic clock rates do indeed agree with ground clock rates to the predicted extent, which varies slightly from nominal because the orbit actually achieved is not always precisely as planned. The accuracy of this comparison is limited mainly because atomic clocks change frequencies by small, semi-random amounts (of order 1 ns/day) at unpredictable times for reasons that are not fully understood. As a consequence, the long-term accuracy of these clocks is poorer than their short-term accuracy.

Therefore, we can assert with confidence that the predictions of relativity are confirmed to high accuracy over time periods of many days. In ground solutions with the data, new corrections for epoch offset and rate for each clock are determined anew typically once each day. These corrections differ by a few ns and a few ns/day, respectively, from similar corrections for other days in the same week. At much later times, unpredictable pseudo-random errors due to imperfections in the clocks build up, so comparisons with predictions become increasingly uncertain unless these empirical corrections are used. But within each day, the clock corrections remain stable to within about 1 ns in epoch and 1 ns/day in rate.

The initial clock rate errors just after launch would give the best indication of the absolute accuracy of the predictions of relativity because they would be least affected by accumulated random errors in clock rates over time. Unfortunately, these have not yet been studied. But if the errors were significantly greater than the rate variance among the 24 GPS satellites caused by orbit differences, which is less than 200 ns/day under normal circumstances, it would have been noticed even without a study. So we can state that the clock rate effect predicted by GR is confirmed to within no worse than $\pm 200/45{,}900$ or about 0.7%, and that predicted by SR is confirmed to within $\pm 200/7{,}200$ or about 3%. This is a very conservative estimate. In an actual study, most of that maximum 200 ns/day variance would almost certainly be accounted for by differences between planned and achieved orbits, and the predictions of relativity would be confirmed with much better precision.

Twelve-hour variations (the orbital period) in clock rates due to small changes in the orbital altitude and speed of the satellites, caused by the small eccentricity of their orbits, are also detected. These are observed to be of the expected size for each GPS satellite's own orbit. For example, for an orbital eccentricity of 0.01, the amplitude of this 12-hour term is 23 ns. Contributions from both altitude and speed changes, while not separable, are clearly both present because the observed amplitude equals the sum of the two predicted amplitudes.

Speed of light constant?

Other studies using GPS data have placed far more stringent limits than we will here. But our goal here is not to set the most stringent limit on possible variations in the speed of light, but rather to determine what the maximum possible variation might be that can remain consistent with the data, allowing for systematic errors. The GPS operates by sending atomic clock signals from orbital altitudes to the ground. This takes a mere 0.08 seconds from our human perspective, but a very long (although equivalent) 80,000,000 ns from the perspective of an atomic clock. Because of this precision, the system has shown that the speed of radio signals (identical to the "speed of light") is the same from all satellites to all ground stations at all times of day and in all directions to within ± 12 meters per second (m/s). The same numerical value for the speed of light works equally well at any season of the year.

Technical note: measuring the one-way speed of light requires two clocks, one on each end of the path. If the separation of the clocks is known, then the separation divided by the time interval between transmission and reception is the one-way speed of the signal. But measuring the time interval requires synchronizing the clocks first. If the Einstein prescription for synchronizing clocks is used, then the measured speed must be the speed of light by definition of the Einstein prescription (which assumes the speed of light is the same in all inertial frames). If some other non-equivalent synchronization method is used, then the measured speed of the signal will not be the speed of light. Clearly, the measured signal speed and the synchronization method are intimately connected. *It is impossible in principle to measure the speed of light independent of theory using electromagnetic or slower phenomena.* However, we can look for changes in the time interval required for a radio signal to travel a known distance, such as from a satellite in a circular orbit to the ground at different times of day or seasons of the year, to determine if the signal speed varies. And the preceding result shows that it does not vary by an amount comparable to Earth's orbital motions around the Sun or the Galaxy or relative to the microwave "background," so "aether wind" theories are thereby falsified. Either SR is correct and the speed of light is invariant, or the density of any aether field that carries light must be entrained by the local gravity field.

Lorentzian relativity

As history buffs may know, the Lorentz Ether Theory (LET) (Lorentz 1931: 208–11)[1] appeared a year before Einstein's 1905 publication of SR. Of course, LET incorporated both the relativity principle (taken from Poincare, but it was first formulated about a generation earlier) and the Lorentz transformations that bear his name. The essential new element introduced by Einstein the following year was the equivalence of all inertial frames, thereby eliminating the need for the luminiferous ether. This first *postulate* of SR makes the Lorentz transformations reciprocal; i.e. they work equally well from any inertial frame to any other, then back again; so it has no meaning to ask which of two identical clocks in different frames is ticking slower in any absolute sense. The second *postulate* of SR makes the speed of light independent of not only the speed of the source (which is also true generally for waves in any medium, including luminiferous ether), but also independent of the speed of the observer (which is a feature unique to SR).

Both SR and LET explain all existing electromagnetic-based experiments and, in that sense, would remain viable theories of the relativity of motion. But the difference between them is much more than aesthetic. In addition to a great difference in practicality for use in systems such as the GPS (in favor of LET), the two theories differ about whether or not material bodies can exceed the speed of light in forward time. In SR, that is proved impossible because time ceases to advance for any entity traveling at the speed of light. By contrast, in LR, no speed limit for material bodies exists. It is true that speed relative to the preferred frame causes electromagnetic-type clocks (which include all ordinary mechanical, biological, and atomic clocks) to slow by the relativistic factor $\gamma = 1/\sqrt{1 - v^2/c^2}$ just as in SR. But in LR, time itself is not affected. (Here, v is the relative speed of the body and c is the speed of light.) So the question of which theory better represents nature is of major importance to the future of physics, which is presently invested in the belief that speeds faster than light in forward time are not possible.

Today, our concepts of the "luminiferous ether" are considerably different than they were in Lorentz's day. It is now widely recognized that the local gravity field serves as the "preferred frame" of LET. With this alteration from Lorentz's original concept but without any change in the math or structure of the theory, LET has now become known simply as "Lorentzian relativity" (LR). Although LR has no intrinsic speed limit, it recognizes the innate difficulty of material bodies composed in part of electrons, while propagating in a luminiferous ether, being able to exceed the wave speed of that ether, the speed of light. LR treats this as analogous to a propeller-driven aircraft exceeding the speed of sound without any outside assists, such as from gravity. A force that cannot itself propagate faster than light cannot propel material bodies faster than light.

Of critical importance to choosing the model that best represents nature, none of the eleven independent experiments testing SR verify frame reciprocity or distinguish SR from LR. In fact, historically, de Sitter, Sagnac, Michelson, and Ives concluded from their respective experiments that SR was falsified in favor of the Lorentz theory. De Sitter argued that the forward displacement of starlight (aberration) depended on absolute, not relative, speeds because both components of a double star, each with some unique velocity, had the same aberration. Sagnac argued that the fringe shifts expected but not seen in the Michelson-Morley experiment *are* seen if the experiment is done on a rotating platform. Michelson argued in the 1925 Michelson-Gale experiment that the Earth was just such a rotating platform. Ives argued that ions radiated at frequencies determined by absolute, not relative, motion because they had to pick a specific frequency to radiate at. In each case, a complex-but-now-familiar SR explanation could account for the same observed results (Table 7.1).

Indeed, the GPS itself is a practical realization of Lorentz's "universal time," wherein all clocks ("GPS clocks") remain synchronized despite being in many different frames with high relative speeds. However, subsequent reinterpretation of SR allowed that theory to survive these objections. This "magic" is envisioned to happen by virtue of each clock in the system being synchronized to an imaginary clock in the Earth-centered inertial (ECI) frame, instantaneously co-located with the moving clock, and assumed to be in a gravitational potential equal to that at sea level at Earth's poles. (Note the "coincidence" that the magic makes use of the Lorentzian preferred frame, the local gravity field.) This trick makes the clock rates all the same as they would have been if they were at rest in the ECI frame and in a constant potential field. This is all very nice, but hardly what Einstein envisioned when speaking of two clocks in relative motion, one at a station and one on a passing train. How simple special relativity would have become all these years if physicists had realized that all they had to do was reset the clock rates so they all ticked at the same rate as the reference clock in the local gravity field!

The converse is also true. Suppose we did not change the clock rates before launch, but instead let them tick at their design rates in accord with whatever speed and potential they experienced in orbit. Now, suppose we tried to Einstein-synchronize the system of clocks. Satellite and ground clocks would tick at different rates. And if we tried to work in any local, instantaneously co-moving inertial frame, the corrections needed to synchronize with each orbiting clock would be unique to that observer's frame and different from moment to moment because both clocks are accelerating. The practical difficulties of operating the system would be virtually insurmountable. What we would gain by doing that is constancy of the measured speed of light in all inertial frames. But because all clocks are now re-synchronized to just the ECI frame in the GPS, the speed of light is constant in that one frame, and the invariance of the speed of light in other inertial frames is of no practical value.

Both Lorentz in 1904 and Einstein in 1905 chose to adopt the principle of relativity discussed by Poincare in 1899, which apparently originated some years earlier in the nineteenth century. Lorentz also popularized the famous transformations that bear his name, later used by Einstein. However, Lorentz's relativity theory assumed an aether, a preferred frame, and a universal time. Einstein did away with the need for these. But it is important to realize that none of the eleven independent experiments said to confirm the validity of SR experimentally distinguish it from LR – at least not in Einstein's favor.

Several of the experiments in 7.1 and 7.2 bearing on various aspects of SR gave results consistent with both SR and LR. But Sagnac in 1913, Michelson following the Michelson-Gale confirmation of the Sagnac effect for the rotating Earth in 1925 (not an independent experiment, so not listed in the tables), and Ives in 1941, all claimed at the time they published that their results were experimental contradictions of Einstein SR because they implied a preferred frame. In hindsight, it can be argued that most of the experiments contain some aspect that makes their interpretation simpler in a preferred frame, consistent with LR. In modern discussions of LR, the preferred frame is not universal, but rather coincides with the local gravity field. Yet, none of these experiments are impossible for SR to explain (Table 7.2)

For example, Fresnel showed that light is partially dragged by the local medium, which suggests a certain amount of frame-dependence. Airy found that aberration did not change for a water-filled telescope, and therefore did not arise in the telescope tube. That suggests it must arise elsewhere locally.

Table 7.1 Independent experiments bearing on Special Relativity – descriptions and years

Experiment	*Description*	*Year*
Bradley	Discovery of aberration of light	1728
Fresnel	Light suffers drag from the local medium	1817
Airy	Aberration independent of the local medium	1871
Michelson-Morley	Speed of light independent of Earth's orbital motion	1881
De Sitter	Speed of light independent of speed of source	1913
Sagnac	Speed of light depends on rotational speed	1913
Kennedy-Thorndike	Measured time also affected by motion	1932
Ives-Stilwell	Ions radiate at frequencies affected by their motion	1941
Frisch-Smith	Radioactive decay of mesons is slowed by motion	1963
Hafele-Keating	Atomic clock changes depend on Earth's rotation	1972
GPS	Clocks in all frames continuously synchronized	1997

Michelson-Morley expected the Earth's velocity to affect the speed of light because it affected aberration. But it didn't. If these experimenters had realized that the aether was not a single entity but changed with the local gravity field, they would not have been surprised. It might have helped their understanding to realize that Earth's own Moon does not experience aberration as the distant stars do, but only the much smaller amount appropriate to its small speed through the Earth's gravity field.

Another clue came from De Sitter in 1913, elaborated by Phipps (Phipps 1989: 549–51), both of whom reminded us that double star components with high relative velocities nonetheless both have the same stellar aberration. This meant that the relative velocity between a light source and an observer was not relevant to stellar aberration. Rather, the relative velocity between local and distant gravity fields determined aberration. In the same year, Sagnac showed non-null results for a Michelson-Morley experiment done on a rotating platform. In the simplest interpretation, this demonstrated that speeds relative to the local gravity field do add to or subtract from the speed of light in the experiment, since the fringes do shift. The Michelson-Gale experiment in 1925 confirmed that the Sagnac result holds true when the rotating platform is the entire Earth's surface.

When Ives and Stilwell showed in 1941 that the frequencies of radiating ions depended on their motion, Ives thought he had disposed once and for all of the notion that only relative velocity mattered. After all, the ions emitted at a particular frequency no matter what frame they were observed from. He was unmoved by arguments to show that SR could explain this too because it seemed clear that nature still needed a preferred frame, the motion relative to which would determine the ion frequencies. Otherwise, how would the ions know how often to radiate? Answers to Ive's dilemma exist, but not with a comparable simplicity.

Richard Keating was surprised in 1972 that two atomic clocks traveling in opposite directions around the world, when compared with a third that stayed at home, showed slowing that depended on their absolute speed through space – the vector sum of the Earth's rotation and airplane speeds – rather on the relative velocities of the clocks. But he quickly accepted that astronomers always use the Earth's frame for local phenomena, and the solar system barycentric frame for other planetary system phenomena, to get results that agreed with the predictions of relativity. Being unaware of LR, he did not question the interpretation at any deeper level.

Table 7.2 summarizes what the various experiments have so say about a preferred frame. These experiments confirm the original aether-formulated relativity principle to high precision. However, the issue of the need for a preferred frame in nature is, charitably, not yet settled. Certainly, experts do not yet agree on its resolution. But of those who have compared both LR and SR to the experiments, most seem convinced that LR more easily explains the behavior of nature.

Table 7.2 Independent experiments bearing on Special Relativity – type and notes on reciprocity

Experiment	*Type*	*Notes on reciprocity*
Bradley	Aberration	Moon exempt
Fresnel	Fresnel drag	Existence of aether
Airy	Existence of aether	Water in scope ignored
Michelson-Morley	No universal aether	Aether "entrained"?
De Sitter	c independent of source	Double star aberration
Sagnac	c depends on rotation	Local gravity field non-rotating
Kennedy-Thorndike	Clocks slow	Motion w.r.t. local gravity field
Ives-Stilwell	Ions slow	Motion w.r.t. local gravity field
Frisch-Smith	Mesons live longer	Motion w.r.t. local gravity field
Hafele-Keating	Clocks depend on rotation	Preferred frame indicated
GPS	Universal synchronization	Preferred frame = local gravity

In LR, speed relative to the preferred frame (the local gravity field) causes clocks to slow. Electromagnetic-based forces become increasingly less efficient with increasing speed relative to the preferred frame, and approach zero efficiency as speed approaches c. There are natural, physical reasons why these things should be so (Van Flandern 1993). The frame of the local gravity field acts as a preferred frame. Universal time and remote simultaneity exist.

The single most important difference is that, in SR, nothing can propagate faster than c in forward time. In LR, electromagnetic-based forces and clocks would cease to operate at speeds of c or higher. But no problem in principle exists in attaining any speed in forward time using forces such as gravity that retain their efficiency at high speeds.

In a recent article, Ashby (Ashby 2002: 41–47) claimed that the clock-epoch correction term (also called "time slippage" term) in the Lorentz transformations, vX/c^2 (see eq. (1) below), can be dropped even when its value is large, but he is very vague about why. However, this particular term is the only difference of consequence between Einstein synchronization of clocks in different inertial frames and Lorentz synchronization of clocks to an underlying "universal time". And the GPS system has been designed to use Lorentz synchronization, for which one frame, the local gravity field or ECI, is special; not Einstein synchronization, wherein clocks tick at their natural rates and all inertial frames are equivalent. By itself, this does not prove LR "right" or SR "wrong". But the practical difficulties for GPS of not changing the natural rates of clocks pre-launch, or with the use of SR for any frame but the Lorentzian preferred frame, are very great. If a ring of satellites (A, B, C, . . ., Y, Z) circled the Earth in a common orbit, and each satellite tried to Einstein-synchronize with the next in sequence, then when Z tried to complete the circuit by Einstein-synchronizing with A, the corrections required would lead to time readings for A different from the starting readings, making closure impossible.

Introducing the twins

The "twins' paradox" is an illustration of the complexity of SR's interpretations of nature. Suppose two identical twins start out at some common instant. One remains on Earth. The other (the "traveler") is on a spacecraft headed for Alpha Centauri (AC) four light-years away at 99% of the speed of light, for which speed, the time dilation factor is $\gamma \approx 7$. (We choose a large value of γ so that the effects of the relativity of motion will be large and obvious, not subtle.) Upon arrival at AC, the traveler turns back to Earth at the same speed (or is replaced by a traveler already headed toward Earth of identical biological age at the moment they pass, to avoid need for an acceleration). The round trip requires slightly over 8 years Earth time; let's say 98 months to be specific. This is path 1 in Figure 7.1. When the twins are reunited, the Earth-bound twin is 98 months old, and the traveler is 14 months old (a factor of 7 less) (Figure 7.1).

That much is a clear prediction of SR. Note especially that no accelerations need actually occur at the beginning or end of the journey, nor even in the middle if we do the "twin replacement" trick, or by noting that half the age discrepancy for the full journey has already occurred by the mid-point before any turn-around event. That is consistent with cyclotron experiments showing that accelerations as such, even as great as 10^{19} g (where g is the acceleration of gravity at sea level), have no effect on clocks (Bailey *et al.* 1997: 301–5). At each stage of the journey where an event occurs, comparisons can be made without ambiguity between adjacent points, one in each of the inertial frames containing the clocks or twins to be compared. Despite the fact that many textbooks discussing the twins' paradox treat accelerations as essential, that is illusory. Accelerations are unbounded in size, and in principle can be done in an instant, allowing no local time to elapse in any relevant frame. Accelerations do not change local clocks, clock rates, or biological aging.

Now we come to the paradox part: why isn't the traveler entitled to claim that the spacecraft remained at rest and the Earth traveled away at 99% of the speed of light, then turned around and came back? From that perspective, the original traveler would argue that the Earth-bound twin should be the younger one. We will examine the rather different answers to that question offered by SR and by LR.

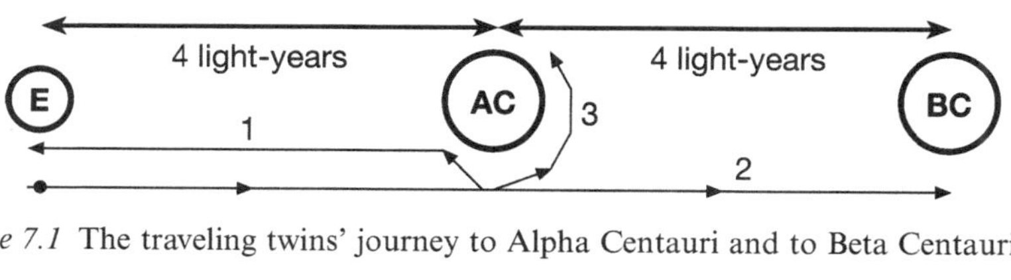

Figure 7.1 The traveling twins' journey to Alpha Centauri and to Beta Centauri.

The traveler takes along a GPS clock

Let's more closely follow what is happening to our two twins. Imagine that the inertial frame of reference containing Earth and AC is filled everywhere with synchronized clocks at rest, so that any traveler can always look out a window and read what time Earth-frame people think it is (see Figure 7.2). And let the traveler take a "GPS clock" along on his journey, along with an unadjusted "normal" clock. The GPS clock is preset in rate before the journey so that, once placed aboard the spacecraft, it will remain synchronized in epoch and rate with clocks on Earth, just as real GPS clocks do. Then the on-board GPS clock will always give readings identical to the nearest Earth-frame clock visible outside the spacecraft window. This instant ability to compare the traveler's time in his own frame with time everywhere in the Earth-AC frame will prevent paradoxes from arising (Figure 7.2).

Now let's examine the journey details. When the traveler's journey begins, the on-board native clock ticks slower than the GPS clock by a factor of 7. But isn't that already an asymmetry present at a stage where there is simply a relative motion, and no way to decide which twin should be aging slower? LR answers simply "yes" because the frame of the local gravity field is the preferred frame in which clocks tick fastest, and time in all other relatively moving frames passes more slowly. But SR offers the opposite answer. And understanding that answer is the key to understanding SR.

SR is a mathematical theory built around the Lorentz transformations. Let the time T be the reading on a clock fixed in the Earth frame; and let X be the relative location in the Earth frame of a clock fixed in the spacecraft frame moving at speed v relative to the Earth frame. Let t be the time reading on the natural clock in the traveling spacecraft, and x be the relative location of Earth in the spacecraft's frame. In that frame, Earth passes the spacecraft with a speed $-v$. As before, the Lorentz time dilation/length contraction parameter is γ, having a value of 7 when $v = 0.99c$. So in general, the relation between the Earth-frame clock and the spacecraft-frame clock is:

$$\begin{aligned} t &= \gamma\left(T - vX/c^2\right) \\ x &= \gamma(X - vT) \end{aligned} \tag{1}$$

Because the relationships are reciprocal (all inertial frames are equivalent in SR), the inverse relations must also hold:

$$\begin{aligned} T &= \gamma\left(t - vx/c^2\right) \\ X &= \gamma(x - vt) \end{aligned} \tag{2}$$

Now let's compare time in the two frames. First, let an observer at a clock fixed on Earth watch time on the spacecraft clock. Then the observer is looking at a point $X = vT$. Substituting that into eq. (1), we get the unsurprising result that $x = 0$, meaning that the spacecraft clock remains fixed at the origin of its own coordinate system. And from the time transformation, using the definition of γ, we get $t = T/\gamma$, which restates the well-known prediction of SR we cited above that the spacecraft clock will appear to the Earth observer to be ticking $\gamma = 7$ times slower than the Earth-fixed clock.

Figure 7.2 Experiment and theory both agree that a traveling twin will come back younger than a stay-at-home twin. Artwork copyright©2002 by Boris Starosta.

That much is routine. But before we leave the Earth frame, let's do one more calculation. Suppose the spacecraft frame is likewise filled with clocks everywhere, Einstein-synchronized with each other. Because all such clocks are at rest relative to one another, they can simply exchange light signals and assume that the light takes equal time for the uplink and downlink portions of its round trip. Then the average of the transmission and reception times on one clock must agree with the signal reflection time on the other clock by definition of "Einstein synchronization". Now, from the Earth frame, let's peer into the spacecraft frame at the point $X = 0$, the fixed location of the Earth observer. From eq. (1), we now have $t = \gamma T$. The meaning of this relation is that the Earth observer sees a succession of spacecraft-frame clocks parading by, and time on the succession of clocks goes by $\gamma = 7$ times *faster* than Earth time.

You read that right. According to SR, at the same time that each and every clock in the spacecraft frame is seen by the Earth frame to tick $\gamma = 7$ times *slower*, time in the spacecraft frame on a succession of passing clocks is passing by $\gamma = 7$ times *faster* than Earth time. To repeat this essential point, in any inertial frame with a relative motion, all individual clocks tick slower but overall frame time moves forward at a faster rate. Such is the effect of the vX/c^2 term in the transformation. The time difference between clocks in different frames is a function of the different rates they tick at *and* of the "time slippage" effect, whereby time is a function of location in any relatively moving frame. In SR, this would remain true even if we used the GPS trick and eliminated the rate differences between clocks. In general, the time slippage effect dominates the effect of a changed clock rate for any clock in a frame with relative motion. (N.B. time is everywhere the same when viewed from within an inertial frame. But it is everywhere different in that same frame from the perspective of any other frame with a relative motion because of time slippage.)

It is worth studying the points in the preceding two paragraphs because, while mathematically permissible, they defy our intuitions and are what makes relativity such an unintuitive theory.

The traveler looks back at the Earth twin

Now let's switch to the spacecraft frame and look back at the Earth-frame clock at $x = vt$ using eq. (2). Then we get $X = 0$ and $T = t/\gamma$. So the spacecraft clock sees the Earth clock ticking $\gamma = 7$ times slower than itself. Substituting $x = 0$, we get $T = \gamma t$: the spacecraft twin sees Earth-frame clocks streaming by outside his window with time elapsing on a succession of them at $\gamma = 7$ time faster. So both effects, relative clock rates and frame time from time slippage, are reciprocal and symmetric between the two frames.

And that is SR's answer to the symmetry challenge we posed at the outset of the spacecraft's journey. Both LR and SR predict that the spacecraft's clocks will appear to tick slower than all Earth-frame clocks as viewed by

Earth-frame observers. But SR (only) predicts that the situation looks reciprocal and symmetric for observers in the spacecraft frame looking at clocks in the Earth frame. The traveler is therefore not surprised by the behavior of the GPS clock on board, which correctly records a combination slower clock rate plus a fast time slippage for the Earth frame, and stays in agreement with the Earth-frame clocks passing by just outside the spacecraft window.

Now consider the traveler's arrival at AC, when the spacecraft turns around or is replaced by a spacecraft heading Earthward at the same speed. Neither the natural clock nor the on-board GPS clock changes rate or reading significantly during the turn-around. However, before the turn-around, the Earth-frame clocks say 4 years (actually, 49 months) of Earth time have elapsed since the journey began, and the spacecraft natural clock says 7 months of spacecraft time have elapsed. But the spacecraft infers that only a single month has elapsed back on Earth because a single clock in another frame is affected only by clock slowing, not time slippage; whereas the spacecraft agrees that 49 total months have elapsed at AC because the journey began with a 48-month time slippage for AC and added one more month during the journey. As usual, time slippage corrections (where applicable) dominate clock readings in other frames.

We now come to the crux of the resolution of the paradox, which seems paradoxical only in SR. The traveler has inferred that only one month has elapsed back on Earth since his journey began, so by spacecraft-frame reckoning, Earth time is just one month later than the actual departure time. For example, if the journey commenced in 2000 January, when the traveler arrives at AC, the on-board GPS clock reads 2004 February; but the traveler infers that Earth clocks still read 2000 February. Because of the finite speed of light, the traveler can see Earth only as it was, not as it is now, and therefore cannot check this inference by direct observation. According to SR, all these clock-reading inferences are not just illusions, but reflect the real, physical time for each frame involved. So at the same time and place that an AC resident infers that Earth time is 2004 February, the spacecraft traveler infers it is 2000 February – a 4-year difference; and both are correct for their respective frames.

Then the spacecraft turns around. Nothing changes locally. But inferences about remote time change greatly because of time slippage, which now has the opposite sign. Now the traveler infers that Earth time is 2008 February – 4 years into the future instead of the past. As a consequence, the traveler will again infer that only one month of Earth time will elapse during the return journey, and all participants agree that Earth time upon the traveler's return will be 2008 March. The traveler arrives back younger (14 months old) than the stay-at-home twin (98 months – over 8 years old). According to LR, this is because the traveler had a high speed relative to the preferred frame, the local gravity field; and there never was any symmetry between the two frames. But according to SR, the elapsed time is a combination of slowed aging and time slippage effects, the latter changing discontinuously when the direction of the traveler changes. (We note in passing that the effect that SR expects accelerations or frame changes to have on remote clocks would constitute an instantaneous action at a distance, a violation of the causality principle.)

The Earth twin thinks the traveler should naturally age slower because his clocks run slower. And the traveler thinks the Earth twin should naturally age faster because the time slippage effect dominated the slowed aging effect when that twin turned around. Note that the traveler could have taken any path whatever as long as the spacecraft speed relative to the Earth frame remained 0.99 c. Then at every instant along the journey, the traveler's biological age will be a factor $\gamma = 7$ less than the elapsed time on the on-board GPS clock, which will agree with the Earth clocks upon return.

Indeed, the spacecraft could have simply continued in a straight line past AC and on to Beta Centauri (BC), say (for purposes of this example only), 8 light-years from Earth. Then there was no turn-around event, but the traveler is still just 14 months old on arrival, and twins born on BC at the same time the Earth twin was born (according to Earth-frame clocks) will still be 98 months old. And the traveler will infer that the BC twins started out 96 months old at the journey's beginning, and aged just 2 months during the journey. So clearly, neither the turn-around event nor any acceleration is essential to the result; and the SR resolution of the paradox retains its symmetry and the equivalence of all frames.

Hence, from the traveler's perspective, wherever the spacecraft goes in any direction without changing speed relative to the Earth frame, it will encounter Earth-frame clocks with *more* elapsed time than on the natural clock aboard the spacecraft (but with the same elapsed time as for the on-board GPS clock). But the traveler will infer that all clocks in the Earth frame are always ticking slower than the natural spacecraft clock, and beings in the Earth frame are always aging more slowly than the traveler. Everything encountered can be explained by a combination of clock rate changes and time slippage. The traveler cannot infer that the Earth-bound twin will be the younger one upon the spacecraft's return because Earth experienced a time slippage event (whether sudden or gradual), rapidly aging everyone on Earth during the traveler's journey. That time slippage event is no different in character than the one that a hypothetical twin on BC would experience if the traveler continued on past AC without a turn-around event.

The traveler takes different paths

What we have just described are careful and correct inferences of SR as applied to the twins' paradox. This also shows the essentially mathematical nature of the theory, because it does violence to what we fondly call "common sense". The most important point to note carefully is that the theory is internally consistent, and no mathematical contradictions can be found no matter how the transformation equations are manipulated, or how many frames or twins are introduced. The next important point to note is that SR makes demands on our credulity that LR does not. Let's examine why.

At the point of turn-around on the original journey from Earth to AC, the traveler's inferences about time on Earth changed suddenly. Instead of the physically unrealistic instant turn-around, let's assume the spacecraft "orbits" around AC to perform the turn-around. (To stay at a safe distance from the star at the same speed, this would require propulsion, not just gravity.) This can still take a time short enough to be neglected, especially at such a high relative speed. So the traveler's spacecraft changes from headed away from Earth and inferring the Earth year is 2000, to traveling toward Earth and inferring the Earth year is 2008. Again, SR says this is real, physical time, and not an illusion.

So before commencing a journey back to Earth, let's suppose the traveler orbits AC several times. Then each time the traveler heads away from Earth in that orbit, Earth time drops back to 2000; and each time the traveler heads toward Earth, inferred Earth time becomes 2008. The Earth-time is intermediate for intermediate orbital positions. Now the significance of repeating this situation several times is that, as Earth time goes to 2008, many people will have died and others will be born. And on each occasion that Earth time reverts to 2000, some of the dead will be resurrected and some living young children in 2008 will cease to exist in 2000. Note that while all this is happening according to SR, the on-board GPS clock representing LR's "universal time" continues to insist that Earth time is the same as spacecraft time and AC time: 2004 everywhere. In SR, effects of this type are never observable because they "lie outside the observer's light cone," hidden from direct view by the finite speed of light. Nonetheless, SR insists that such changes affect real, physical time and are not mere illusions, because the viewpoint of each inertial frame is just as valid as that from any other frame.

In LR, one reference frame (the local gravity field) is preferred; and speed cannot affect time, but only the rate of ticking of mechanical, electromagnetic, or biological clocks. However, just as we do not assume that time has been affected when the temperature rises and causes a pendulum clock to slow down, LR says that changes in clock rates are changes in the rates of physical processes, and do not affect space or time. So by carrying an on-board GPS clock on the spacecraft, we are offered a clear choice between models: Earth time can be what SR infers it is, or it can be what the GPS clock says it is. In the former case, SR works, but leads to heavy-duty complexities and fantastic inferences about the nature of time at remote locations. Moreover, the proof that nothing can travel faster than light in forward time stands intact. In the latter case, LR works with great simplicity and in full accord with our intuitions about the universality of the instant "now". And the speed of light is no longer a universal speed limit because time itself is never affected either by motion or by gravity.

Aside from these practical difficulties with the use of SR in the GPS, Einstein's special relativity is also under challenge in a more serious way from the "speed of gravity" issue, because the proven existence of anything propagating faster than light in forward time (as all experiments indicate is the case for gravity) would falsify SR outright (Van Flandern 1998: 1–11; Van Flandern and Vigier 2002: 1031–68). So it is entirely possible that reality is Lorentzian, not Einsteinian, with respect to the relativity of motion. In that case, physics may have no speed limit when the driving forces are gravitational or electrodynamic rather than electromagnetic in nature. And that may be the most important thing that the GPS has helped us to appreciate.

Notes

1 This contains a summary of and citation to the original 1904 paper.

References

Alley, C.O. and T. Van Flandern (1998) 'Absolute GPS to Better Than One Meter', Available online at: http://metaresearch.org/solar%20system/gps/absolute-gps-1meter.ASP

Ashby, N. (2002) 'Relativity and the Global Positioning System', *Phys. Today*, May, 41–47.

Bailey, J., K. Borer, F. Combley, H. Drumm, F.J.M. Farley, J.H. Field, *et al.* (1997) 'Measurements of Relativistic Time Dilation for Positive and Negative Muons in a Circular Orbit', *Nature* 268, 301–5.

Lorentz, H.A. (1931) 'Lectures on Theoretical Physics', Vol. III, *The Principle of Relativity For Uniform Translations*, London: Macmillan & Co., pp. 208–11.

Phipps, T. (1989) 'Relativity and Aberration', *Amer. J. Phys.* 57, 549–51.

Van Flandern, T. (1993) *Dark Matter, Missing Planets and New Comets*, Berkeley, CA: North Atlantic Books.

Van Flandern, T. (1998) 'The Speed of Gravity – What the Experiments Say', *Phys. Lett. A* 250, pp. 1–11.

Van Flandern, T. and J.P. Vigier (2002) 'Experimental Repeal of the Speed Limit for Gravitational, Electrodynamic, and Quantum Field Interactions', *Found. Phys.* 32 (7), 1031–68.

8 A defense of absolute simultaneity

Michael Tooley

The basic thesis that I shall be defending in this paper is that events in the actual world do stand in the relation of absolute simultaneity, and in support of that thesis I shall offer two arguments. The first will be a metaphysical argument that focuses upon causation and space-time, while the second argument will appeal to quantum mechanics.[1]

The metaphysical argument: causation and space-time

This first argument involves four main steps. First, it will be argued that there are good reasons for rejecting a relationalist (or reductionist) view of space-time in favor of a realist account. Secondly, it will then be argued that once a realist view of space-time is accepted, the continued existence of space-time becomes a fact that stands in need of an explanation – and, specifically, a causal explanation. Thirdly, I shall then argue that the most plausible causal explanation is one that postulates a conservation principle for space-time itself, and, in particular, one that postulates non-branching causal processes. Finally, I shall then show that such a conservation principle enables one to define the notion of absolute location in space-time, and, thereby, the relation of absolute simultaneity.

One important feature of this argument, I think, is that it does not involve any special postulate whose only function is either to guarantee the existence of the relation of sameness of location, or to guarantee that events stand in the relation of absolute simultaneity. For the existence of both of these relations follows simply from the idea that parts of space-time are causally connected, together with the most plausible hypothesis concerning the conservation law that governs that connection.

Let us now turn to the argument. The first stage involves the claim that there are good reasons for accepting a realist view of space-time. This issue of the choice between an absolute, or realist, view of space-time, and a reductionist, or relational one raises, of course, some difficult questions. However philosophers of physics who reject the idea of absolute simultaneity do not in general rest their rejection of it upon a rejection of a realist view of space-time: they maintain, rather, that regardless of whether one accepts an absolute view of space-time or a relational one, there is no reason to accept the idea that events in our world stand in the relation of absolute simultaneity.

Because of this, I shall not attempt to offer a detailed defense of realism with regard to space-time. I shall, however, briefly set out what seems to me one of the strongest arguments for a non-relational view, and which runs as follows. First, consider the fact that there are locations where there *could*, at a given time, have been a physical object, even though, as a matter of fact, that was not the case. The possibility involved here is not that of a mere logical possibility: our everyday experience of the motion of objects, for example, makes it reasonable to believe that there are locations where there are, as a matter of fact, no physical objects, but where the existence of such objects is not only logically possible, but *empirically* possible.

Secondly, I think it is plausible that statements of empirical possibility – in contrast to statements of logical possibility – require truth makers in the world, and that the truth makers in question cannot consist of ontologically primitive, modal facts: any empirical possibilities must instead be grounded in non-modal, categorical states of affairs – together, presumably, with laws of nature.

If this is right, then consider any spatio-temporal location where there is, as a matter of fact, no physical object, but where it is empirically possible that there could have been such an object. What is the relevant categorical basis of the empirical possibility in question?

In response, it might be suggested that answers are certainly possible that do not involve a realist view of space-time. Perhaps, for example, it is the presence of some physical field in an otherwise empty location that is the ground of the empirical possibility of there being an object at that location. Or perhaps the empirical possibility is somehow grounded in the categorical properties of physical objects, events, or fields that exist at other locations than the one in question. I think, however, that it is fair to say that both of these hypotheses are extremely speculative: there is no evidence either for the view that what makes it possible for there to be an object at a given location is that there is already some field at that location, or for the view that the possibility is, instead, dependent upon fields or events or objects at other locations. But if these accounts are rejected, it is hard to see what plausible alternative there is to the view that the categorical properties that are the basis of the empirical possibilities in question are properties of the space-time points themselves, and, thus, properties that are in no way dependent upon the properties of physical objects, or events, or fields – there or elsewhere. If this is right, and there are no plausible alternatives, then the postulation of absolute, or substantival space-time is justified as providing a categorical ground for empirical possibilities concerning alternative locations of physical objects.

Let us now move on to the second stage in the argument. It involves the claim that, once one accepts a realist view of space-time, it is very reasonable to hold that there are causal relations between temporal slices of space-time. Why should this be so? One argument is as follows. Assume that a realist account of space-time is correct. Given that assumption, what external relations must obtained between different spatio-temporal regions? One external relation that must obtain between some spatio-temporal regions is that of temporal priority. Let us ask, then, what account is to be given of that relation.

One answer, which has been defended by a number of philosophers, is that temporal priority is to be analyzed in causal terms (Mehlberg 1935: 119–258; 1937: 111–231; Mellor 1981: Chapters 9 and 10; 1998: Chapters 10–12; Tooley 1997: Chapter 9). But if such an account is right, and if one accepts a realist view of space-time, then it seems that one may be able, by focusing upon the possibility that space-time might be completely empty, to argue that in order for different parts of space-time to be temporally related to one another, there must be causal relations between them.

But is this conclusion justified? An important challenge to it involves the point that causal theories of temporal priority have often been formulated, not in terms of actual causal *connections* between states of affairs, but in terms of possible causal connections, in terms of causal *connectibility*. Moreover, it may well appear, initially, that there is a very good reason for opting for a modal formulation, since there would seem to be possible worlds where the *actual* causal relations that obtain between events may not be sufficient to fix all of the temporal relations that obtain between events. So, for example, it may be that event *C* causally gives rise to two events – *A* and *B* – but that there is no causal process leading from *A* to *B*, or from *B* to *A*. The actual causal relations in such a case do not settle how *A* and *B* are temporally related. And given such possibilities, it is natural, if one is attracted to a causal account of temporal priority, to respond by appealing to possible causal connections, and to say, for example, that if *A* is temporally prior to *B* then it must be the case that there could have been an event in the immediate vicinity of *A* that would have given rise to an event in the immediate vicinity of *B*.[2]

If an account of temporal priority in terms of causal connectibility is correct, the argument mentioned above is undercut, for what makes it the case that one part of space-time is temporally prior to another need not be a matter of causal connections between the two parts: it could instead be a matter of causal connections that would have been obtained had there been events in the relevant parts of space-time.

In response, however, I think that it can be shown that modal formulations of causal accounts of temporal priority are open to a decisive objection – an objection that becomes evident once one focuses upon the concept of causal connectibility. For that concept presumably has to be analyzed in terms of counterfactuals, and then one is faced with the question of the truth makers for the relevant subjunctive conditionals: what makes it true that, although event *A* is not causally connected to event *B*, there could have been a causal process connecting those events, or, at least, connecting alternative events in the same spatio-temporal locations? It is hard to see what answer there can be, other than one that appeals not only to relevant laws of nature, and to the intrinsic properties of the events in question, but also to the temporal relation between those events. But if there is no other answer, then causal approaches to temporal priority that appeal to causal connectibility are unable, on pain of circularity, to specify truth makers for the relevant subjunctive conditionals.

The upshot is that the idea of focusing upon the possibility of a space-time that is completely empty, and then of appealing to a causal account of temporal priority to show that there must be causal relations between spatio-temporal regions themselves, cannot be undercut by arguing that a causal analysis of temporal priority should be formulated in terms of possible causal connections, rather than actual ones.

At this point, two other responses are possible. First, it might be suggested that there is simply no good reason to accept a causal account of temporal priority. But it is doubtful that that contention can be sustained, given that alternative accounts of temporal priority seem very problematic. Thus, on the one hand, other reductionist accounts of the direction of time fail to generate any direction at all in the case of some simple universes – such as a universe containing a single particle, or a universe containing two uncharged particles rotating about one another. They also generate the wrong direction in the case of universes where, because of unusual – but perfectly possible – initial conditions, the direction defined by such features as the direction of increase in entropy, or the direction of the expansion of the universe, or the direction of open forks, etc., are all opposite to that of the direction of time.[3]

On the other hand, if one opts instead for a non-reductionist account of the direction of time, then one encounters the problem that such an account cannot, apparently, provide any explanation of why it is that the things that we normally take as evidence of the direction of time are in fact evidence, and this in turn means that it becomes rather mysterious, given a realist account of temporal priority, how one can ever know what the direction of time is.

By contrast, these difficulties that confront alternative accounts do not pose any problems for a causal approach. Thus it can be argued, for example, that any simple universe that one would naturally classify as a temporal universe also contains causal relations. Nor is there any problem concerning the epistemology of our temporal judgments, since it can be argued that the facts that we take as evidence for the direction of time are evidence for the direction of causation. And finally, provided that one adopts a realist account of causation, one can handle without difficulty cases where, because of extremely improbable initial conditions, the direction of time is opposite to that indicated by various processes taking place in time – such as increase in entropy.

There are, then, a number of considerations that provide strong *prima facie* support for a causal account of temporal priority. But that still leaves a final response that needs to be addressed – namely, that whatever advantages such an approach may have, it must still be rejected, since it is exposed to decisive objections.

One of the most detailed and careful statements of the case against a causal approach to temporal priority has been given by J. J. C. Smart in his article "Causal Theories of Time" (Smart 1971: 61–77), where he argues that causal theories of time are exposed to a number of very strong objections. I believe that it can be shown, however – as I have argued elsewhere (Tooley 1997: Chapter 9) – that while Smart's objections do tell against causal theories of time that are conjoined with relational theories of space-time, the objections have no force if a causal theory of temporal priority is embedded in an absolute theory of space-time.

In view of the above, I believe that there are good reasons for thinking that one is justified in appealing to a causal account of temporal priority as part of an argument for the conclusion that, given a realist account of space-time, some parts of space-time must stand in causal relations to one another. But I also want to argue that one need not, at this point, rest everything upon an appeal to a causal theory of time, since there are at least two other reasons that can be offered for holding that there are causal relations between spatio-temporal regions. Thus, another line of argument is as follows. First, if space-time is a continuant, an account is needed of its identity over time, of what it is that makes spatio-temporal regions, existing at different times, part of a single space-time. Secondly, any satisfactory account of identity over time for ordinary entities must involve causal relations between temporal slices (Armstrong 1980: 67–78). Hence, thirdly, if space-time is a continuant, its unification into a single persisting entity should, presumably, be explicable in parallel fashion: different spatio-temporal regions are part of a single, persisting space-time because of causal connections between them. Therefore, it is reasonable to hold that there are causal relations between different temporal parts of space-time.

A final consideration that also supports the idea of causal connections between spatio-temporal regions is this. It is not a necessary truth that space-time continues to exist. How, then, is this contingent state of affairs to be explained? If one accepted a relational account of space-time, then the continued existence of space-time would be reducible to the continued existence of objects in space-time, and that, in turn, could be explained by relevant conservation laws. But this type of explanation is ruled out if one adopts a realist account, since then one is saying that space-time could exist even if it were totally empty. So what explanation can be offered, given a realist view of space-time? It is hard to see what alternative there is to the idea that the continued existence of space-time is to be explained by the hypothesis that spatio-temporal regions existing at one moment causally give rise to spatio-temporal regions existing at later moments.

But might not one perhaps respond as follows at this point: "why think that there has to be *any* explanation at all for the continued existence of space-time? After all, explanation has to stop somewhere."

This attempt to blunt the present argument involves, however, a confusion – namely, that between explanations of laws, and explanations of temporally located states of affairs. In the case of laws, the idea that one will ultimately arrive at basic laws that cannot be further explained – rather than there being an infinite regress of more and more basic laws – is very plausible. But when it is a question of the existence of some temporally located state of affairs, one does not appeal to the idea that explanation has to stop somewhere. In that case, one attempts, whenever possible, to explain the existence of a state of affairs at one time in terms of the existence of a state of affairs at an earlier time, and so on, in principle, back forever, if time has no beginning.

Notice, in addition, what is involved in the suggestion that one should make no attempt to explain the continued existence of space-time. It is very different from saying, for example, that perhaps some small event that is highly localized in space-time has no explanation. For if one holds that there is no explanation of the continued existence of space-time, then one is accepting an accident of cosmic proportions, since it involves a non-denumerably infinite sequence of states of affairs of the same type. If there were no explanation for such a sequence of states of affairs, the sequence would be massively improbable – indeed, its probability would be infinitesimally close to zero. To treat the continued existence of space-time as not in need of any explanation is, accordingly, not a rationally defensible position.

What account can be offered, then, of the continued existence of space-time? If one accepted a relational account of space-time, then, as noted earlier, the continued existence of space-time would be reducible to the continued existence of objects in space-time, and that, in turn, could be explained by relevant conservation laws. But this type of explanation is ruled out if one adopts a realist account, since then one is saying that space-time could exist even if it were totally empty. So what explanation can be offered, given a realist view of space-time? It is hard to see what alternative there is to the idea that the continued existence of space-time is to be explained by the hypothesis that spatio-temporal regions existing at one moment causally gives rise to spatio-temporal regions existing at later moments. This hypothesis, of course, does not generate very exciting predictions; however the same is true of conservation laws in general. But it does generate predictions that, although unexciting, are massively confirmed, and so, unless some better explanation of the continued existence of space-time can be advanced, there would seem to be excellent reason for accepting this hypothesis.

To sum up this second stage in my argument: we have seen that there are at least three substantial reasons for holding that parts of space-time stand in causal relations to one another. One of the arguments does involve the claim – which many philosophers would either question or reject – that temporal priority is to be analyzed in causal terms. In response, I indicated briefly some reasons for thinking that a causal account of temporal priority deserves serious consideration. In any case, the other two arguments for the claim that parts of space-time stand in causal relations to one another did not involve any particularly controversial assumptions.

The third and final step in the argument involves asking how, exactly, different parts of space-time are causally related to one another. This question cannot be answered, of course, by experimenting with space-time to determine what causal connections are present: answering the question will be a matter, rather, of determining what hypothesis provides, overall, the best explanation of the continued existence of space-time.

What hypothesis, then, does so? The candidate that I wish to propose is the following: "The Principle of the Parallel, Non-Branching Conservation of Space-Time":

(1) Every space-time point is such that its existence is a cause of the existence of at least one other space-time point.
(2) Every space-time point is such that its existence is caused by the existence of at least one other space-time point.
(3) If *P* and *Q* are any two causally connected space-time points and *S* is any other space-time point on the line determined by *P* and *Q*, then the existence of *S* is causally connected both with the existence of *P* and with the existence of *Q*.
(4) If *P*, *Q*, *R*, and *S* are any four space-time points such that the existence of *P* causes the existence of *Q*, and the existence of *R* causes the existence of *S*, then the line defined by *P* and *Q* is parallel to the line defined by *R* and *S*.

How do the different clauses in this conservation principle function? The answer is that clauses (1) and (2) capture the general idea that space-time is a persisting entity, while clause (3) imposes the requirement that all space-time points on any line containing two causally connected space-time points must also be causally connected, with the result that there is no causal gappiness. Finally, clause (4) ensures that appropriate betweenness and congruence relations are preserved, and that the causal relations involved in the continuing existence of space-time are non-branching ones, with the result that what one has is a strict *conservation* of space-time.

But what grounds can be offered for thinking that this particular principle provides the best explanation of the continued existence of space-time? Is it clear that there are no alternative hypotheses that are preferable?

What alternatives need to be considered? First of all, in view of the fact that it is an absolutely central principle within the Theory of Special Relativity that nothing can travel faster than the speed of light in a vacuum, any conservation principle for space-time that allowed the existence of one space-time to be caused by the existence of a space-time point to be that was not in the past light cone of that space-time point, or, equivalently, that allowed the existence of a space-time point to cause the existence of a space-time point that was not in its future light cone, should presumably be excluded.

Secondly, once one departs from the conservation principle formulated above, and allows non-parallel causal processes, and possibly also branching causal processes, the only non-arbitrary way of allowing the existence of a space-time point to be causally connected with the existence of some, but not all, of the space-time points within its past and future light cones, is via principles that distinguish between space-time points on the surface of the light cones, and space-time points in the interior of the light cones. It would seem, then, that, as regards alternatives to the Principle of the Parallel, Non-Branching Conservation of Space-Time, there are only three possibilities that are serious candidates, since there are only three alternatives that neither entail the existence of causal processes that are faster than those involved in the transmission of light, nor suffer from arbitrariness. First, there is an expansive, "conservation" principle according to which the existence of a space-time point is causally connected to the existence of every space-time point to which it is connected via optical lines: "Expansive Conservation of Space: Principle I":

> Every space-time point is such that its existence is a cause, via a straight-line causal process, of the existence of every space-time point on the surface of its future light cone.

Secondly, there is an expansive, "conservation" principle according to which the existence of a space-time point is causally connected to the existence of all and only space-time points that lie on the inside of its future light cone: "Expansive Conservation of Space: Principle II":

> Every space-time point is such that its existence is a cause, via a straight-line causal process, of the existence of every space-time point in the interior of its future light cone.

Thirdly, there is the expansive, "conservation" principle that postulates both the causal connections involved in Principle I, and those involved in Principle II: "Expansive Conservation of Space: Principle III":

> Every space-time point is such that its existence is a cause, via a straight-line causal process, of the existence of every space-time point in its future light cone.

How does the Principle of the Parallel, Non-Branching Conservation of Space-Time compare with these three alternatives? It seems to me that it is superior, for at least three reasons. In the first place, that principle is *simpler* with respect to the number of states of affairs that it postulates. For whereas the Principle of the Parallel, Non-Branching Conservation of Space-Time does not entail anything beyond the existence of *non-branching* causal chains connecting some space-time points, all three of the expansive conservation principles entail that every space-time point is causally connected, via *non-denumerably many* causal chains, with *every* space-time point on the inside of the relevant forward and backward light cones. In addition, each of those causal chains branches *infinitely often at every point*, whereas, according to the Parallel, Non-Branching Conservation Principle, the causal chains connecting space-time points never branch. The latter theory is therefore vastly simpler with respect to the actual causal connections that it postulates.

Secondly, and as an immediate consequence of the preceding point, the three expansive conservation principles involve massive *causal overdetermination*, since, for every space-time point, there are non-denumerably many, causally unconnected space-time points such that the existence of any one of those points is a sufficient cause of the existence of the space-time point in question. By contrast, no such overdetermination exists in the case of the Parallel, Non-Branching Conservation Principle.

Finally, in adding new laws to a scientific theory, it seems reasonable, other things being equal, to assume that new laws will be similar in mathematical and causal form to known laws that deal with comparable phenomena. Thus, in particular, if the goal is to arrive at a law dealing with the continued existence of space-time, a plausible initial hypothesis is that the law in question will be parallel in form to laws dealing with the conservation of matter and energy, momentum, electrical charge, etc. The latter, however, are *conservation* laws. It is therefore likely that the law governing the continued existence of space-time will also be a conservation law. The Principle of the Parallel, Non-Branching Conservation of Space-Time, in postulating non-branching causal connections between space-time points, entails that space-time, rather than increasing or decreasing, is strictly conserved, and so it is a law of the relevant form. By contrast, the other three proposals, in entailing the existence of branching causal processes – indeed, infinitely branching ones – are not conservation laws, and therefore are radically different in form from the laws dealing with the continued existence of matter, energy, momentum, and electrical charge.

The conclusion, in short, is as follows. First, the Principle of the Parallel, Non-Branching Conservation of Space-Time both predicts and causally explains the continued existence of space-time, so that, in the absence of a superior hypothesis, its acceptance can be justified via an inference to the best explanation. Secondly, it is superior to the three alternative hypotheses with regard to simplicity, both because it postulates far fewer causal processes, and because it avoids massive causal overdetermination. Thirdly, it is a hypothesis that is parallel in form to other central, conservation principles of physics, whereas the other three alternatives are not. There would seem to be excellent grounds, therefore, for regarding that principle as more plausible than its competitors.

Given the Principle of the Parallel, Non-Branching Conservation of Space-Time, it is possible to give an account of sameness of location. The idea is the very natural one of offering the same sort of analysis of the identity of locations over time that can be offered of the identity of physical objects over time. The latter, however, can be analyzed – and very plausibly – in terms of causal relations between temporal parts. Paralleling such an analysis for the case of sameness of locations relative to substantival space-time gives one the following analysis:

Two distinct space-time points *P* and *Q* are part of one and the same spatial location at different times

means the same as:

The Principle of the Parallel, Non-Branching Conservation of Space-Time is true, and *P* and *Q* are space-time points such that either the existence of *P* causes the existence of *Q*, or the existence of *Q* causes the existence of *P*.

The relation of sameness of location need not, then, be taken as a primitive relation, since it can be analyzed in terms of the ontologically fundamental relation of causation. Nor is there any need for an independent postulate that deals with sameness of location, in order to ensure that there are space-time points that are in the same location at different times, since given the above analysis of sameness of location in terms of causation, the existence of space-time points that are in the same location at different times will be entailed by a very plausible principle concerning the conservation of space-time.

Given an account of sameness of location, together with the idea of simultaneity relative to a frame of reference, the idea of absolute simultaneity can be defined very simply as follows:

> Two events, *E* and *F*, are *absolutely simultaneous*

means the same as:

> *E* and *F* are simultaneous relative to some frame of reference that is at rest with respect to absolute space-time – that is, which is such that no part of it is ever in different spatial locations at different times.

The upshot is that there is no need to postulate a relation of absolute simultaneity, since the existence of that relation follows very quickly from the existence of the relation of simultaneity relative to a frame of reference, together with the Principle of the Parallel, Non-Branching Conservation of Space-Time.

The scientific argument: quantum mechanics and absolute simultaneity

The second, and very different argument for the thesis that events in our world do stand in the relation of absolute simultaneity involves an appeal to quantum mechanics – and specifically, to an idea that is absolutely central in quantum mechanics, namely, that of the collapse of a wave packet.

Intuitively, the argument can be put as follows. Consider an electron that has been fired towards the screen of a cathode ray tube. According to quantum mechanics, the initial conditions do not causally determine what point on the screen the electron will strike. All that follows from the laws of quantum mechanics is that there are various non-zero probabilities of the electron's striking different parts of the screen. But once the electron strikes the screen, the probabilities of it hitting other parts of the screen must change from having non-zero values to being equal to zero. But then must it not be the case that all of these changes are absolutely simultaneous with the event that consists of the electron's hitting the screen in a certain location?

The idea of the collapse of the wave packet thus seems very clearly to involve the notion of absolute simultaneity. Put in such a general way, however, the force of this point may seem open to question, since one might well respond that, if quantum mechanics is inconsistent with the Special Theory of Relativity, then perhaps one should conclude, "So much the worse for quantum mechanics." How does one know that it is the Special Theory of Relativity, rather than quantum mechanics, that needs to be modified?

This is a perfectly reasonable response to the argument as I have formulated it above. However, what we shall now see is that we can recast the argument in terms of a very famous theorem of quantum mechanics, and certain crucial experiments associated with it, and that, when we do so, it becomes clear that it is the Special Theory of Relativity that must be rejected, rather than quantum mechanics.

The story begins with a famous argument advanced by Albert Einstein, Boris Podolsky, and Nathan Rosen (Einstein *et al.* 1935: 770–80). Their argument involves a thought experiment designed to show that quantum mechanics does not provide a complete description of the world. In outline, the structure of the argument is this. On the one hand, quantum physics does not attribute any properties to an object until a relevant measurement has been made. So an electron, for example, does not have any definite position until a measurement of its position is made, nor any definite momentum until a measurement of its momentum is made, and so on. But, on the other hand, an experimental situation can be described in which there will be a correlation that can be satisfactorily explained *only* on the assumption that objects do have properties before any relevant measurement on them has been carried out. So the description of the physical world provided by quantum physics must be incomplete.

What experimental situation did Einstein, Podolsky, and Rosen have in mind, as posing a problem for quantum mechanics? It was one based upon the fact that it is possible to create a pair of particles that are *correlated* in the sense that there is some determinable property such that, if the value of that property is fixed for one of the particles, it must also be fixed for the other. So, for example, a pair of electrons can be created in such a way that the total spin of the two electrons along a given axis must have a certain value, and, if this is the case, then, once the spin of either of the electrons, along the given axis, is measured, and so has a determinate value, the spin of the other electron, along the same axis, must also have a determinate value. Suppose, then, that such a pair of electrons is created, that they are then separated, and that at time t a measurement is made of the spin, along a given axis, of one of the electrons. Then, *at least* from that time onward, the spin of that electron, along the given axis, has a determinate value. But what about the other electron? When does it have a determinate spin, along the axis in question?

Only two views seem possible. The one is that neither electron *acquires* a determinate spin as the result of measurement, since both electrons have a determinate spin, along the axis in question, from the time that they are created. Measurement merely enables us to know what those determinate values are; it does not bring those determinate values into existence.

This is the view favored by Einstein, Podolsky, and Rosen. It implies, of course, that the description of the electrons given by quantum mechanics is incomplete, since quantum mechanics does not attribute determinate spins to the electrons prior to a measurement being made.

The other view is that the two electrons acquire a determinate spin, along any given axis, only when a measurement is made on at least one of the electrons. But if this is right, a crucial question arises – namely, what is the relation between the time of the acquisition of a determinate spin by the electron on which the measurement is carried out, and the time at which the other electron acquires a determinate spin?

Since the second electron surely does not acquire a determinate spin at an earlier time, either it comes to have a determinate spin only later, or else it acquires a determinate spin at the same time. If it has a determinate spin only later, there are two possibilities. One is that there is a temporal gap between the time at which the first electron acquires a determinate spin, and every moment at which the second electron has a determinate spin. But surely this cannot be possible, since it would imply that, if a measurement were made on the second electron during the temporal gap in question, then it would be possible for the measurement to yield a value for the spin along the axis in question that, in conjunction with the value of the spin of the first electron, would imply a violation of the principle of conservation of spin. So there cannot be any temporal gap.

Alternatively, it may be that, while the second electron does not acquire a determinate spin when the first electron does, it has a determinate spin at every time throughout some immediately following, open temporal interval. This possibility is best addressed, however, after we have considered the third possibility – namely, that the second electron acquires a determinate spin at precisely the same time as the first electron. The question that immediately arises, of course, is what is meant by "at the same time" in this context. In particular, can it be interpreted merely as simultaneity relative to some inertial frame, *F*? It would seem that it cannot, since if it were true merely that the acquisition of a determinate spin by the second electron was simultaneous with the acquisition of a determinate spin by the first electron relative to some inertial frame, *F*, then there would be another inertial frame, *F**, such that, relative to that frame, the second electron acquires a determinate spin *after* the first electron does, and with an intervening gap, so that we would be confronted, once again, with the possibility of a measurement being made on the second electron during the gap between the moment when the first electron acquires a determinate spin and the moment when the second electron would otherwise acquire a determinate spin, in virtue of the measurement made on the first electron. Thus, if simultaneity is merely relative simultaneity, there will be inertial frames where there is a temporal gap, and so, from the perspective of those inertial frames, there will be the possibility of pairs of measurements generating values that violate conservation principles. Only if simultaneity is absolute simultaneity will such a possibility be excluded.

Finally, what about the possibility that, while the second electron does not have a determinate spin at the time when the first electron acquires one, it has a determinate spin at every time throughout some immediately following, open temporal interval? The answer is that precisely the same problem arises as in the case when the two electrons acquire a determinate spin at the same time, for, in characterizing the relevant open interval, one has to make use of the idea that the instant that immediately precedes that interval is simultaneous with the instant at which the first electron acquires a determinate spin, and, once again, it will not do to interpret the simultaneity as simultaneity relative to a frame of reference: only absolute simultaneity will do.

What the Einstein-Podolsky-Rosen thought experiment shows, in short, is this. Either particles have determinate states prior to measurement, in which case quantum mechanics does not provide a complete description of physical reality, or else correlated particles must acquire corresponding determinate properties simultaneously, in the absolute sense – or at least without there being an intervening temporal gap – in which case the Special Theory of Relativity does not provide a complete description of the spatio-temporal relations between events.

Given the argument to this point, I think it is fair to say that it would be plausible to back the Special Theory of Relativity, and so to conclude that quantum mechanics fails to provide a complete description of reality. But there is more to the story. In 1964 a physicist, John S. Bell, published a paper in which he showed that the quantitative predictions generated by quantum mechanics logically *preclude* there being properties that make it the case, for example, that an electron possesses determinate spins along various possible axes before any measurements are made (Bell 1964: 195–200).[4] So the thrust of the Einstein-Podolsky-Rosen thought experiment is no longer merely that either the Special Theory of Relativity, or else quantum mechanics, is incomplete. It is rather that either the Special Theory of Relativity is incomplete, or quantum mechanics is *false*.

When it was a matter merely of the incompleteness of quantum physics, that option was not too difficult to accept. But with Bell's theorem on board, it became rather more difficult to hold that the Special Theory of Relativity provides a complete description of the spatio-temporal relations between events, since one could do so only by holding that quantum mechanics, which is very well confirmed indeed, is false.

The crucial point, however, is that, with the proof of Bell's theorem, it becomes possible to put the question to an experimental test. It is no longer a matter for speculation which view should be adopted. What happens, then, when the relevant experiments are performed? The answer is that the predictions of quantum mechanics in this area are correct (Aspect *et al.* 1981: 460–67; 1982: 91–94). But this means, as we have seen, that it will not do to say, for example, that the electron on which the measurement is not made has a determinate spin from the time that the two electrons are created. It must be the case, rather, either that there is a time at which it *acquires* a determinate spin, and which, for the reason indicated earlier, must be absolutely simultaneous with the time at which the other electron acquires a determinate spin, or else it has a determinate spin at every moment after a time that is absolutely simultaneous with the time at which the first electron acquires a determinate spin. Both possibilities presuppose absolute simultaneity. The conclusion, accordingly, is that there are experimental grounds for holding that absolute simultaneity is a relation that obtains between events in our world.

Summing up

The question of whether the relation of absolute simultaneity is a relation that is present in our world is a metaphysical question of considerable interest, especially given that tensed or dynamic accounts of the nature of time typically involve the idea of a present that is universe-wide, and that is not relative in any way to frames of reference. What I have tried to do here is to show that there are good reasons to believe that absolute simultaneity is a relation that holds between events in our world. For that view can be supported by at least two arguments, one which is a metaphysical argument that focuses upon the best explanation of the continued existence of space-time, realistically conceived, and the other of which turns upon appeal to quantum mechanics, and, in particular, to Bell's Theorem, and the experimental results connected with it.

Notes

1 These two arguments were first set out in Chapter 11 of Tooley (1997). The metaphysical argument is there embedded, however, in a modified version of the Theory of Special Relativity. Here I am setting it out as an independent argument.
2 Compare the view advanced by Robb (1914).
3 For additional discussion, see Tooley (1987): especially Section 7.4, pp. 221–28.
4 For clear formulations of Bell's theorem which do not presuppose any specialist background, see d'Espagnat (1979: 158–81), and Mermin (1981: 397–408; 1985: 38–47).

References

Armstrong, David M. (1980) 'Identity Through Time', in Peter van Inwagen (ed.), *Time and Cause*, Dordrecht: D. Reidel, pp. 67–78.

Aspect, A., P. Grangier, and G. Roger (1981) 'Experimental Tests of Realistic Local Theories via Bell's Theorem', *Physical Review Letters* 47, 460–67.

Aspect, A., P. Grangier, and G. Roger (1982) 'Experimental Realization of Einstein-Podolsky-Rosen-Bohm *Gedankenexperiment*: A New Violation of Bell's Inequalities', *Physical Review Letters* 48, 91–94.

Bell, John S. (1964) 'On the Einstein-Podolsky-Rosen Paradox', *Physics* 1, 195–200.

d'Espagnat, B. (1979) 'The Quantum Theory and Reality', *Scientific American* 241/5, 158–81.

Einstein, Albert, Boris Podolsky, and Nathan Rosen (1935) 'Can Quantum-Mechanical Description of Physical Reality be Considered Complete?', *Physical Review* 47, 770–80.

Mehlberg, Henryk (1935) 'Essai sur la théorie causale du temps', *Studia Philosophica* 1, 119–258 and (1937) 2, 111–231.

Mellor, D. H. (1981) *Real Time*, Cambridge: Cambridge University Press.

Mellor, D. H. (1998) *Real Time II*, London: Routledge.

Mermin, David N. (1981) 'Quantum Mysteries for Anyone', *Journal of Philosophy* 78, 397–408.

Mermin, David N. (1985) 'Is the Moon there When No One Looks? Reality and the Quantum Theory', *Physics Today* April, 38–47

Robb, Alfred A. (1914) *A Theory of Time and Space*, Cambridge: Cambridge University Press.

Smart, J. J. C. (1971) 'Causal Theories of Time', in Eugene Freeman and Wilfrid Sellars (eds.), *Basic Issues in the Philosophy of Time*, La Salle, Illinois: Open Court, pp. 61–77.

Tooley, Michael (1987) *Causation – A Realist Approach*, Oxford: Oxford University Press.

Tooley, Michael (1997) *Time, Tense, and Causation*, Oxford: Oxford University Press.

9 Cosmic simultaneity[1]

Richard Swinburne

Einstein claimed that various phenomena synthesized by his Special Theory of Relativity had the consequence that

> we cannot attach any *absolute* signification to the concept of simultaneity, but that two events which, viewed from a system of coordinates, are simultaneous, can no longer be looked upon as simultaneous events when envisaged from a system which is in motion relatively to that system.
>
> (Einstein 1923: 42–43)

I argue in this paper that Einstein gave good, but not compelling grounds for supposing that the concept of absolute simultaneity would have no application in a universe empty of matter and so governed entirely by the laws of Special Relativity. But I claim that there is absolute simultaneity in our homogenous and isotropic universe of galaxies receding from each other with a metric described by the Robertson-Walker line element. "Cosmic Time" provides a correct standard of absolute simultaneity.

Part 1

It is important to distinguish between an analysis of the nature of a concept, and primary and secondary tests for its application. An atom of (uncharged) hydrogen just is an atom consisting of one proton and one electron. But sometimes concepts are too basic to be analyzed; "cause," "property" (in the philosophical sense), and "color" are like this. One can merely illustrate by examples, cases of one thing or event causing another and show some of the entailments thereof (e.g. that if A causes B, then B does not cause A). The primary tests for the application of a concept are the best tests anyone could ever use – given the physical laws by which the universe is governed. Where the concept can be analyzed, the primary tests are constituted by the analysis. The test that an uncharged gas is hydrogen is that it consists of atoms each consisting of one proton and one electron. But where the concept cannot be analyzed, there are still primary tests.

Φ is a primary test for the value of some property, e.g. the distance between two points, if it is a member of a set of tests Φ, γ, χ, etc., such that if the use of all those tests gives one definite value for the property, then the value given by those tests is the best possible estimate of the value of the property. No different tests can upset the common result of all the primary tests. α is a secondary test for the value of a property, if we only use it because we have good reason to believe that it would give the same result as one or more primary tests if we were able to use the latter. If a secondary test gives a result different from that given by the primary tests, the former is considered wrong and the latter right. If different primary tests give in some particular situation different results from each other, then we cannot estimate reliably the value of the property. But we must distinguish between two cases – one where the primary tests give an analysis of the concept of the value of the property, and the other where they do not but are still the best tests anyone could ever have for that value. In the former case the different results given by the different primary tests show that the concept of that property having a definite value does not apply in this situation; in the latter case the different results show only that we can never know what the value of the property is in this situation. And if with respect to all situations different primary tests give different results for the value of some property this may be either because the concept of it having a value has no application, or because there are limits to the possibility of human knowledge of what that value is.

Thus in my view (for which I do not argue here) the spatial distance between two points can be analyzed in terms of how many rigid rods which coincide with each other at one place, can be laid between the two points; and the rigidity of a rod can be analyzed in terms of it continuing to coincide with the members of the only family of rods which preserve coincidence relations with each other. There is no content to the concept of a distance between two points apart from what would be measured by such rods. Hence in this case the primary tests for the application of the concept are provided by an analysis of the concept. But for distances of any magnitude it is not in practice possible to lay down rigid rods, and so we have to use the secondary tests of stellar parallax, apparent luminosity of Cepheid stars, red shift etc.

Part 2

With these points in mind, and before coming explicitly to discuss the concept of simultaneity at different places, I need to consider the criteria for measuring temporal interval at any one place, the "proper time" at that place as it is often called. Places are places on a frame of reference, that is, an extended system of physical objects (normally material objects) stationary relative to each other. Frames of reference move relative to each other, and an object stationary in one frame may be in motion relative to another. A book on a car seat will be stationary relative to the car, but may be in motion relative to the Earth. Scientists seek the simplest laws of nature yielding the most accurate predictions. Any law of nature can be formulated in various different ways, simple or complicated; but the simplicity of a law of nature depends on the simplicity of the simplest way of formulating it. Laws of nature involve assertions about forces arising from distances between objects; and they can be expressed most simply if these distances are measured relative to some frames of reference rather than others. A frame of reference relative to which the laws of nature can be formulated, most simply, I shall call a basic frame. There might in principle be either many such equally basic frames (equibasic frames), or just one unique one.

In a space empty of matter there is (or can be constructed) a class of equibasic frames called "inertial frames" which are such that when measurements of distance are made relative to these frames Newton's laws of motion hold with perfect accuracy. Every different inertial frame moves with uniform velocity relative to every other inertial frame. Most small regions of the Earth's surface are near approximations to an inertial frame.

How then is temporal interval to be measured on an inertial frame such as (approximately is) the surface of the earth? To say that an event occurred *n* units of time after another event at the same place is to say that if a true clock recorded a certain reading at the instant of the occurrence of the first event, then it would record a reading greater by *n* units at the instant of the occurrence of the second event. Statements about temporal interval are statements about what would have been or would be recorded by true clocks. A clock measuring intervals in terms of one kind of unit, say hours, will measure time on the same time scale as another clock measuring intervals in terms of another kind of unit, say minutes, if all measurements of intervals between any two events by the two clocks are linearly related to each other; that is, each measurement by one clock in terms of its units is a constant multiple of the measurement of the same interval by the other clock in terms of its unit; a clock with only a minute-hand and a clock with only an hour-hand measure time on the same time scale.

But what is it for a clock to be a true clock? Any actual recurrent process – the rotation of the Earth, its revolution around the Sun – forms a clock, and yields a potential time scale whereby we can measure intervals. So too does any mechanical device which could be built – such as a pendulum clock – in which there is a recurrent process. Every occurrence of the process or some function thereof is assumed to occur after the same period of time, and we call this interval between any two occurrences – provisionally – a time unit. The rotation of the Earth takes a day. This gives us units in terms of which we can measure processes. If we assume that the Earth rotates at a regular rate against the background of the "fixed" stars, we can divide the day into hours and minutes and seconds. On this basis we may postulate laws of nature describing how bodies move under the influence of forces of various kinds.

There are two primary tests for a potential time scale to be a true time scale (that is for the measurement made by a certain clock to measure the passage of time correctly). The first test is that judgments of temporal interval yielded by that scale between events which form part of our history and in particular of our personal experience must not diverge radically from those which we ordinarily make by means of the day and year scales which have been in use for so many thousands of years. Two events on Earth on different days which last from when the Sun is at its highest point in the sky one day to when it is at its highest point the next day must be said to last approximately the same period of time as each other. So too must two events in different years from when the Sun is in Capricorn to when it is in Capricorn the next year. For if the adoption of a proposed time scale meant that the last sidereal year was a thousand times longer than the previous sidereal year, that time scale could not be adopted – however simple the resulting laws of physics. This is so because such expressions as "equal time interval," "much longer time," etc. are given their meanings by the circumstances in which ordinary users of language judge it appropriate to use them. We only know what these expressions mean because we have been taught what these circumstances are. If physics radically contradicted all such obvious common-sense judgments about time intervals, we would have to conclude that it did not use "time" in the same sense as did ordinary language. Certainly some law of physics may be simpler if the "*t*" in it is said to be a thousand times greater for the last sidereal year than for the preceding one. But if we are to say that "*t*" denotes time, our judgments of t-interval must correspond largely to the judgments of time interval which would be made in ordinary circumstances of personal experience and familiar human history by an ordinary user of language. Otherwise why say that the physicists' "*t*" denotes time?

The first test for a clock measuring the true time scale is a criterion which clocks need satisfy only approximately. Subject to it the second primary test is that the true time scale at any place is that which will give the simplest form to the laws of nature. Our ground for not choosing a typical medieval town clock as the true timepiece is that, if we used it, we would have to postulate highly complex laws of nature. These laws would have the consequence that days were of uneven length, and so the Earth would be rotating at a significantly ever-varying rate. Bodies would attract each other with forces varying from instant to instant. The principle that the scientist should choose from among proposed laws compatible with observed phenomena which are the simplest possible, leads to the rejection of such a time scale. The reason why – in my view – scientists choose the simplest proposed laws yielding the most accurate predictions, is that such laws are more probably the true laws than are more complex laws; and – in my view – they are right to think so (Swinburne 2001: Chap. 4). Hence the time scale used in them is that most probably true. If we choose the scale given by the daily rotation of the Earth relative to the "fixed stars" we find that bodies obey

Newton's three highly simple laws of motion and the simple inverse square laws of gravitation and electrostatic and magnetostatic attraction to a high degree of accuracy; and obey the comparatively simple laws which give better predictions of phenomena, such as Einstein's field equations and the laws of Quantum Theory – to an even higher degree of accuracy.

Once upon a time, the "daily rotation" scale provided the official scale for measuring temporal interval. This was then replaced by the "annual revolution" scale, and in 1966 the "annual revolution" scale was replaced by a scale based on the natural frequency of radiation of a line of cesium.[2] The role of each new definition has been marginally to correct and not to replace the previous method of measuring temporal interval; and the grounds for adopting each new definition were that it had the consequence that what we reasonably believe to be the true laws of nature predict even more accurately. The true scale is that which would be used in the true laws of nature. The time scale which is most probably (on our present evidence) the true scale is that which is used in what are most probably (on our present evidence) the true laws of nature. That the frequency of radiation of an atom of a given kind is constant is most probably (on our present evidence– but see note 2) such a true law.

My account of our grounds for adopting the particular time scale which we do are grounds arising from the behavior of material bodies and of electromagnetic radiation on Earth, and are therefore grounds for adopting it for judging proper time on Earth. But since there is good reason (which I shall discuss briefly later) to suppose that the same fundamental laws of nature operate at all other places in the universe as operate on places on earth, time at any place in the universe is to be measured by a clock based on the same scale – e.g. the radiation frequency of cesium there. But do these two primary tests analyze the concept of temporal interval, or are they simply the best tests we can have (given the actual laws of nature) for the applicability of a concept too basic to be exhausted by the results of our human tests? We can see the answer to that by asking ourselves what we would say if there was not at any place any unique simplest time scale such that if measurements were made by it all laws of nature would predict with perfect accuracy everywhere. E. A. Milne once suggested that electromagnetic phenomena could best be dealt with by his t-scale, and dynamical processes by his τ-scale, and that these scales were related asymptotically (Milne 1948, passim and especially p. 244f). Fortunately science has found no need to adopt his suggestion. But if we found that Milne's suggestion was correct, would we say that there was a true temporal interval between two events, but we could never know what it was; or would we say that the only fact of the matter was that between any two events there was one interval measured by one scale and a different interval measured by another scale? Would we say, for example, that the interval between t_1 and t_2 was a certain number of times as long as the interval between t_0 and t_1 but we could never know how many; or that it was of the same length as the latter interval by one scale and twice as long by another scale and that there was no one true scale? For myself I cannot give any content to the former suggestion. I cannot understand what would be meant by there being a true time scale independent of what instruments showed; and so I am inclined to say that the primary texts for temporal interval at one place analyze the nature of the concept. A major reason why I am inclined to say this is that it is not merely physically impossible, but logically impossible to have a better way of judging whether two non-overlapping temporal intervals are of equal length than by seeing whether (subject to satisfaction of the first primary test) that result follows from using clocks which give to the true laws of nature their simplest form.[3] It is not logically possible to move one of these intervals adjacent to the other and check whether they coincide! But since all the evidence is that at each place there is the same unique best way of measuring temporal interval, it shows that there is in fact a unique correct true time scale. So I suggest that we agree with Newton that there is an "absolute, true, and mathematical time"; but I suggest that this is a contingent truth and, contrary to Newton, that time does not "of itself and from its own nature, [flow] equably without relation to anything external" (*Principia*, Scholium to Definition, viii). Conventionalism about temporal interval at any one place[4] is false, but only contingently so.

The primary tests for temporal interval give rise to secondary tests such as those which ascertain the age of objects from the proportions of radioactive substances contained in them. We find the proportion of atoms of substances which decay in some period of time, as measured by our best clocks, namely by primary tests; and so assuming the same rate of decay to operate at all temporal periods we find the half-life of such substances, namely the expected value of the period of time in which half of some mass of a substance will decay. We find that two isotopes of Uranium, U^{238} and U^{235} decay, respectively, into isotopes of lead Pb^{206} and Pb^{207} with half-lives of 4510 and 707 million years. We then infer how rocks containing uranium are formed and so infer the relative proportion of these isotopes of uranium and lead to be expected in the Earth's rocks on their first formation. Then from the proportion of the different isotopes of lead to the different isotopes of uranium currently found in rocks, we calculate their age. This method utilizes primary tests to measure the rates of decay over a few years, and generalizes that the same rates of decay will hold over millions of years where we cannot in practice use the primary tests, because then there were no observers who could have measured the passage of time by clocks. Having illustrated, for temporal interval at one place, what constitutes a secondary as opposed to a primary test, I shall not bother to discuss secondary tests for the main temporal concept which I am about to discuss, simultaneity at different places, since it should be clear by now what the contrast between primary and secondary tests amounts to. My interest is solely in primary tests.

Part 3

So much for time at one place. What are the criteria, the primary tests, for an event at one place occurring at the same time as a certain event at another place? How, that is, can we synchronize a clock at one place so that it records a certain time at the exact instant when a clock at another place records that time? We can synchronize clocks at the same place simply by adjusting them to record the same temporal instant, as they touch each other. Synchronization at the same place I shall term direct synchronization. Synchronization at two distant places I shall term indirect synchronization. Intuitively, before we introduce any modern science, there would seem two possible methods of synchronizing clocks at distant places *P* and *Q*, the clock-transport method and the signal method. The clock-transport method is to have two clocks at *P* synchronized with each other directly at t_0 and then move one of them to Q; when it reaches Q at say t_1 you synchronize the clock at Q with it directly. The signal method is to send a signal at t_0 from *P* to *Q*. It arrives at *Q* at t_1. As soon as it arrives the signal is sent back from *Q* to *P* where it arrives at t_2. Now t_1 must lie between t_0 and t_2 by the principle that causes necessarily precede their effects. We then make the simplest hypothesis about the velocity of the signal coherent with the rest of physics. This hypothesis will yield a solution about where t_1 lies between t_0 and t_2, and hence provide a procedure for synchronizing clocks at *P* and *Q*. The observer at *P* can then tell the observer at *Q* that the signal was sent at "t_0" and received back at "t_2" by *P*'s clock, and the observer at *Q* can then set his clock so that by that clock the signal would have arrived at *Q* at the time intermediate between t_0 and t_2 dictated by that simplest hypothesis.

It would seem that, whether or not they analyze the concept of simultaneity, one or other of these methods or perhaps both of these methods taken together are the primary tests for synchronizing clocks at different places. A clock at *P* and a clock at *Q* will show the same time if a clock transported from *P* to *Q*, showing the same time as *P*'s clock on departure, shows the same time as *Q*'s clock on arrival; and if a signal sent from *P* to *Q* and back again arrives at *Q* at the instant on *P*'s clock, predicted by the simplest hypothesis coherent with the rest of physics about the velocity of the signal and information about the instants at which, by *P*'s clock, it left *P* and returned to *P*. (Time at each place is to be measured by true clocks, identifiable as such by the methods described in the previous section.)

Unfortunately, it is a physical fact that different applications of each method conflict with each other and with applications of the other method in some circumstances. First, take the signal method. A signal is sent from *P* to *Q* and back again. The time of its arrival at *Q*, t_1 on *Q*'s clock, must lie between t_0, the time of its departure from *P*, and t_2 the time of its return to *P*. If we had a signal which returned to *P* in less than any finite period of time and so had infinite velocity, we could uniquely identify t_1. For then (if we ignore any infinitesimal intervals) $t_1 = t_0 = t_2$. No such signal is known to science. In the absence of one, it is clearly desirable to choose the signal with the fastest two-way velocity, that is a signal whose total journey there and back is the fastest. For such a signal will narrow down the possible values of t_1 as much as possible. The fastest signal known to physics is that of light and all other electromagnetic radiation, which has – at any rate within the region of our galaxy – a mean or two-way velocity of approximately 300,000 km/sec (denoted by c). (We cannot of course measure the one-way velocity of light, i.e. its velocity relative to P on its journey from P to Q, without having first synchronized a clock at Q with one at P.) Now light (and all other electromagnetic radiation) has the peculiar property not shared by material objects that its mean velocity has the same value in all inertial frames. By this is meant that when a light signal is sent between any two points P and Q and back again (stationary or in motion relative to each other), and that passage is marked as a passage on an inertial frame moving relative to P or Q with any velocity, then the mean velocity of the signal will be the same in all such frames.

That the mean velocity of light has the same value (c) in all inertial frames is the most natural extrapolation from a very large number of experimental results, the most celebrated of which is the Michelson-Morley experiment. The most natural simplification of this extrapolation is that the signal has in every inertial frame its mean velocity as its one-way velocity relative to the source in every direction, namely, $t_1 = \frac{t_0+t_2}{2}$. Since, as we have seen, science aims to formulate the simplest laws of nature compatible with phenomena, the assumption must be adopted – unless it leads to complexities elsewhere. It does not initially lead to any such complexities.

Given this assumption that the one-way velocity of light is the same as its two-way velocity (c) in all inertial frames, Einstein derived as the fundamental hypothesis of Special Relativity the "Lorenz transformations" which relate measurements of time and distance in different inertial frames to each other. If an inertial frame F' is moving with uniform velocity v along the x-axis relative to inertial frame F, then an event at (x, y, z, t) in the frame F will have coordinates (x', y', z', t') in F' (the first three coordinates being Cartesian spatial coordinates and the last coordinate the temporal coordinate, spatial and temporal origin points being the same for measurements in both frames) such that:

$$x' = \beta(x - vC)$$
$$y' = y$$
$$z' = z$$

$$t' = \beta\left(t - \frac{vx}{c^2}\right)$$

where

$$\beta = \frac{1}{\sqrt{1 - \frac{v2}{c2}}}$$

Spatial distances are measured by rods in the way briefly described earlier. So x' is the distance measured in this way by rods in F' between two events whose distance apart measured in F in the same kind of way is x. t is the temporal interval between these events as measured by the signal method and true clocks in F, whereas t' is the temporal interval between the same events as measured by the signal method and true clocks in F' – all measurements being made on the assumption that the one-way velocity of light is the same in all frames. For clocks on the same frame of reference (that is, for clocks stationary among themselves), these two temporal values (t and t') will be the same. The signal method provides an unambiguous way of synchronizing clocks at different places in the same frame; and using it provides confirmation that natural processes run at the same rate at each place on any one frame. These processes include, as well as the radiation frequency of a line of cesium, biological growth and decay.

Yet for events on frames moving relative to each other, the signal method gives different results for the interval between two events, according to the frame on which the measuring clocks are situated. Consider two events E_1 and E_2, say flashes of lightning. It is a consequence of the Lorenz transformations that if E_1 and E_2 are simultaneous as judged by the light signal method used in F, they will not be simultaneous as judged by the light signal method used in F', and conversely. For, if the time interval between E_1 and E_2, as judged in F, is $t = 0$, then the time interval between them as judged in F' will be $t' = -\frac{\beta vx}{c^2}$. Conversely if t', the time interval between E_1 and E_2 as judged in $F' = 0$, $t = \frac{vx'}{\beta c^2}$. Consequently, the signal method for synchronizing the clocks leads to inconsistency. For if it is valid, you find of two events that they both are and are not simultaneous. The relativity of simultaneity, as judged by the signal method, is, as we have seen, predicted by the Special Theory; but the formula $t' = \beta\left(t - \frac{vx}{c^2}\right)$ has been well confirmed independently.

Exactly the same problems arise with the other method for synchronizing clocks – the clock transport method. It follows from the Lorenz transformations that a clock taken from P to Q and then brought back to P will have slowed down by P's clock, the difference between the two clocks being greater, the faster the moving clock is moved (relative to P). But if P and Q are on the same reference frame (that is, stationary relative to each other), and the moving clock is moved slowly, the difference between the clocks is going to be very small. We can work out the limit of this process – that is, to talk slightly metaphorically, what the clock would read if it were moved infinitely slowly; and then directly synchronize the clock at Q with the moving clock corrected for what it would have shown if its motion had been infinitely slow. We then move the moving clock back from Q to P, and again correct its reading for what it would have read if it had been moved infinitely slowly. That reading will exactly coincide with the reading on the clock which stayed at P. The simplest hypothesis is to suppose that the clock ticks at the same rate when moved from P to Q as when moved in the other direction. With this assumption, the moving clock corrected for infinitely slow motion will yield a correct measure of simultaneity for clocks on the same inertial frame (see Ellis and Bowman 1967: 116–36).[5] But again if *P* and *Q* are on different inertial frames in relative motion the clock transport method will give different results for when clocks are synchronized, depending on which is the frame in which the velocity of the clock is measured. This is for the simple reason that a clock moving with a velocity which is infinitely slow relative to *P* will not be moving with a velocity which is infinitely slow relative to *Q*, and vice versa. So each method of synchronization gives different results for when two clocks on different inertial frames are synchronized, varying with the frame in which the measurements are made.

Faced with this problem, there are two possible reactions. The first reaction is to say that the two primary tests for simultaneity do not provide an analysis of the concept, but are merely the best tests we can use for its application. There is a truth about which events at one place are (absolutely) simultaneous with which events at other places, but what the Special Theory has shown is that we cannot discover which these events are. That is, there is a true frame of reference *F* relative to which alone the signal and clock-transport methods of synchronization give unique true results. To hold this involves holding that it is only motion infinitely slow relative to *F* which does not slow down a clock; and that only in *F* is the velocity of a reflected light signal constant in both directions. In all other inertial frames it will have some velocity j (different for each frame) in a certain direction and $\frac{cj}{2j-c}$ in the opposite direction, with intermediate velocities in intermediate directions. This will mean that its mean velocity for a two-way trip will be the same in all inertial frames, as observable. The reinterpretation of physics leading to a unique measure of simultaneity can be carried through perfectly consistently and with identical complexity for any choice of preferred inertial frame,[6] (and given the applicability of the Special Theory, any non-inertial frame would be a less basic frame and so one which science ought not to use). But the trouble is, if we confine ourselves to the Universe of Special Relativity, namely a Universe empty of matter, that there are no grounds for choosing one inertial frame rather than any other as the preferred frame, since – as we have seen – all inertial frames are equibasic. So unless we have grounds for adopting a scientific theory other than Special Relativity, and on the basis of it for judging that some inertial frame is more basic than others, or that some different frame of reference is more basic than inertial frames, we have no grounds for making our measurements of simultaneity relative to one frame of reference rather than any other.

The alternative reaction adopted by Einstein and expounded in text books on the Special Theory of Relativity is to say that the ordinary concept of simultaneity could only be applied if two events simultaneous by the light signal method (and the clock method, although this was not considered by Einstein and most others) in one frame were simultaneous in all inertial frames, and so we must substitute for it the concept of simultaneity in a frame. Two events are simultaneous in a certain frame if the clocks in that frame synchronized by the light signal (and clock transport) method show the same reading when the two events occur. So E_1 and E_2 may be simultaneous in *F*, but not in F^1. But no contradiction arises, since it does not follow from E_1 and E_2 being simultaneous in *F* that they will be simultaneous in *F′* or any other frame.

Now there are, I think, grounds for supposing that our concept of (absolute) simultaneity is more basic than the primary tests for its application; whereas by contrast, I claimed that there was no more to the concept of equal temporal interval at a place than was contained in the tests for its application. For it is entailed by an event at *P* being simultaneous with an event at *Q* that if a signal of infinite velocity were sent from *P* to *Q* and reflected back to *P* (and so arrived back at *P* after an interval less than any finite interval), the time of its arrival at *Q* would be (by *P*'s clock and by any other clock, and ignoring infinitesimal differences) the same time as the time of its emission from *P*. It is merely the physical impossibility of there being such a signal which prevents our using this method. But that does not show any logical impossibility in the sending of such a signal; and so there is content to the notion of absolute simultaneity – that is, what would be shown by the use of such a signal. Special Relativity, it would seem, simply reveals a limit to what we can discover, not to what is the case. By contrast there is no logically possible thought experiment in the case of measuring equal intervals of time at one place which would give a better result than the primary tests described in Part 2.

However the view that in a universe of Special Relativity there is an undiscoverable time frame of reference involves postulating not merely that there is some truth physically impossible for humans to discover, but that nature as it were conspires against our discovering it. Nature ensures that light behaves in such a peculiar way that the evidence will always support equally each of an infinite number of hypotheses about which is the true frame. We would have to suppose that "Nature contains both deep asymmetries and compensatory factors which exactly nullify these asymmetries" (Zahar 1983: 39) in the sense of preventing us from ever discovering them. Scientific laws will have simpler formulations if we do not have to postulate such a coincidental balancing arrangement. That suggests that in such a universe, incomprehensible as it may seem, there would be no absolute truth about when on *P*'s clock the signal arrives at *Q*. While the existence of absolute simultaneity in the universe of Special Relativity is certainly logically possible, considerations of simplicity suggest that it would probably not exist there.

Part 4

However our universe is not on the large scale the Universe of Special Relativity. It consists, we now know, of galaxies of stars, grouped in clusters and superclusters. An observer on Earth notes that, on the evidence of their red shift and other secondary tests for distance, all other galaxies (apart from those very close to our own galaxy) are in recession from our galaxy; and that (when allowance has been made for some random clumping of galaxies) there are the same number of galaxies at any given distance in any direction from ourselves. All other such galaxies at any given distance in whatever direction have, when allowance is made for small random velocities, the same velocity of recession from our galaxy, a velocity which increases with distance (possibly uniformly but probably at an accelerating rate); that is namely the observed region of the Universe is, relative to our galaxy, approximately isotropic. The isotropy of the Universe relative to ourselves is shown further by the fact that the temperature of the cosmic microwave background radiation (the presumed red-shifted relic of the Big Bang) is to a high degree of approximation the same in all directions. Now is it only relative to our galaxy at this instant on its clocks that this isotropic recession of observed galaxies occurs, or does it occur relative to all other galaxies (observed and as yet unobserved) at all instants on the clocks of each? The principle of simplicity tells us (in the absence of counter-evidence) to postulate that for all galaxies at all instants on the clocks of each galaxy every part of the Universe is approximately isotropic (namely galaxies at any given distance from any other galaxy – apart from any very close to it – are distributed uniformly and recede from it with approximately the same velocity). We can avoid the awkwardness that the isotropy is only approximate by postulating an imaginary frame of reference in the vicinity of each galaxy, relative to which the isotropy is exact – all other such frames at the same distance from it are distributed isotropically and recede from it with uniformly the same velocity. Such a frame I shall term a fundamental frame; and I shall follow the terminology of some of the literature of cosmology in terming an observer imagined as situated on such a frame a fundamental observer. A galaxy may rotate relative to its fundamental frame but any velocity of recession of the galaxy from the frame will be small, temporary, and equally likely to be in any direction.

The supposition that we can postulate associated with each galaxy a fundamental frame such that at any instant relative to it the Universe of mutually receding fundamental frames is isotropic, may be called the Principle of Isotropy. That it holds is an empirical postulate which might turn out to be false. For example, radio messages might one day be received from astronomers on a distant galaxy reporting that relative to them the observable Universe is very far from isotropic. But until such messages are received or other counter evidence is obtained, it seems correct to adopt the principle on grounds of that simplicity which is evidence of truth.

If then the observable Universe is always isotropic relative to each fundamental observer, it seems a further simplification to postulate that the laws of physics as formulated by an observer on each fundamental frame, using his measures of distance and clocks, are the same; and so that they are the same on each galaxy, if we make allowances for any random motion or rotation of the galaxy. This principle we shall call the Principle of Equivalent Laws. By this principle, if a clock of a certain construction is a true clock on some galaxy, then a clock of qualitatively the same construction will be a true clock on any other galaxy. There is detailed observable evidence that it holds on distant observable galaxies. We find that light spectra of elements on such galaxies have the same shapes as do the spectra of elements known to us shifted to the red, a shift which can be accounted for most simply by the recession of those galaxies from ourselves. It follows that the spectra of elements on such galaxies have the same shapes relative to each other as spectra of elements known to us, and hence an atomic clock graduated by the natural frequency of a line of one element there would ensure such simple laws of nature as the constancy over time of the frequency of light emitted by the other elements. And then the very regular behavior of pulsars (spinning neutron stars emitting pulses of light which reach us at precise regular intervals, e.g. every 15 milliseconds) can be explained simply by supposing that the same laws of nature hold in their region as measured by its clocks, as hold in our region measured by our clocks; and so that pulsars are true local clocks. That all this continues to hold for galaxies too distant to be observed is the simplest extrapolation from the data.

It follows from the Principle of Equivalent Laws that the fundamental frame in the vicinity of each galaxy is for the laws of cosmology an equibasic frame, viz. the laws of cosmology acquire their simplest form relative to any fundamental frame, but no fundamental frame is more basic than any other one. Hence *either* each process of a certain kind on each fundamental frame, including the ticking of clocks (such as the rotation of pulsars), occurs at a rate generally different from a process of the same kind on another fundamental frame, while on each frame a process of one kind occurs at the same rate relative to each process of another kind, *or* each process of the same kind occurs at the same rate on each fundamental frame. For example, *either* on every frame the natural frequency of a certain line of a cesium atom has the same value relative to the radiation frequencies of all other lines of atomic spectra on that frame but different from the frequency of that line on other fundamental frames, *or* the frequency is the same on all fundamental frames.

Clearly the simplest supposition to make and hence the one which is most probable in the absence of counter-reasons is the latter. This principle I shall call the Principle of Similar Clocks. If there are no objections to adopting the principle, then there will be a unique method of ascertaining the instant by the clocks on any galaxy at which a distant event occurred. For by the Principle of Isotropy either the mutual recession of galaxies has been going on since an instant by the clocks of our galaxy when all galaxies were very close together (an instant soon after the Big Bang) or it has been going on since an instant by the clocks of our galaxy when all galaxies were approximately stationary relative to each other. But in either of these cases clocks on each galaxy could have been synchronized at the instants referred to. In the first case clocks would have been very close together and so could have synchronized directly by comparison at what would be approximately the same place. In the second case, when all galaxies were approximately stationary relative to each other, all fundamental frames would have formed the same one most basic frame and hence there would have been a preferred frame. Clocks could then have been synchronized on this preferred frame by either method of indirect synchronization, making the simplest supposition about the applicability of that method compatible with the rest of physics. The simplest supposition (to be adopted in the absence of counter-reasons) about the velocity of light on it (even though it would not have been an inertial frame) is, as on any one inertial frame, that it has the same velocity in both directions. The simplest supposition about the effect of motion on clock transport on this frame is, as on any one inertial frame, that the effect is non-existent for infinitely slow transport; and again in the absence of counter-reasons we should make this supposition. Since there are no such counter-reasons we should suppose that if these two tests had been done when the fundamental frames were close or stationary relative to each other, they would have yielded a unique simplest result – a fundamental observer on any frame would identify the same set of instants on all frames as simultaneous with each other. Hence there is a unique instant, a unique plane of simultaneity, by all the clocks on all frames at which the expansion of the universe began.

Given the Principle of Similar Clocks, an observer on each galaxy can retrodict how many years ago by his clocks this early instant occurred; and his present instant will be simultaneous (absolutely) with the instant on the clock of each other galaxy when an observer on that galaxy judges that the same number of years has elapsed since that early instant. Every true clock kept stationary in its fundamental frame (that is, moved with the mean velocity of the matter in its vicinity) will have ticked at the same rate (relative to the true clocks on all other galaxies).

The Principle of Similar Clocks does not hold in the Universe of Special Relativity. There the equibasic frames are the inertial frames. Now consider three inertial frames *A*, *B*, and *C* all moving with uniform rectilinear velocity relative to each other. At the initial instant of the experiment *B* passes *A*, and observers on each synchronize clocks directly; later *B* passes *C* which is moving towards *A*, and observers on *C* synchronize their clock directly with the clock on B. It is a consequence of the Special Theory that when *C* passes *A* the clock on *C* will record a smaller passage of time than the clock on *A*. For although processes of the same kind occur at the same rate on all three frames as measured by local clocks, processes on *B* and/or *C* occur more slowly than processes on *A* as measured by clocks on *A*. Consequently the Principle of Similar Clocks is false for a world of inertial frames. But in the actual universe, described by modern cosmology, the equibasic frames are not inertial frames, but galaxies (or more precisely the fundamental frames as defined earlier), and the possible relative motions of these are very different from the possible relative motions of equibasic frames on Special Theory. For the only instants, according to modern cosmology, at which a fundamental frame meets another fundamental frame are, because of the principle of isotropy, the instants, if any, at which it meets all fundamental frames. Hence an experiment of the type described above could not be performed for *A*, *B*, and *C* as fundamental frames. But if the universe were to contract and all galaxies came close to each other again, it follows from the Principle of Similar Clocks that a true clock on each galaxy whose reading could then be discovered directly or by light signal or clock transport methods by an observer on some other galaxy would be found to measure the same passage of time since the beginning of the expansion.

There could be evidence of a kind more readily obtainable about the truth or falsity of the Principle of Similar Clocks if messages were received (by light signal) from inhabitants of distant galaxies. They could report that at the instant of the sending of the message more years had elapsed since the instant of the Big Bang (or the instant at which all galaxies were stationary relative to each other) than we judge by our clocks at the instant of receiving the message. This would falsify the principle. Alternatively, such a message could report that less years had elapsed than we judge, and this would be consistent with the principle, since light would have taken some time to reach us.

The Principle of Similar Clocks gives us a cosmic time, a unique true clock for measuring the time of occurrence of events on the cosmic scale (e.g. the instant at which galaxies are a certain distance apart). It is clearly simpler to make measurements on a local scale by the same method as we should use on a cosmic scale, that is by clocks stationary in their fundamental frame. And there is an additional reason for regarding the local fundamental frame as the one true frame of measurement in which clocks should be stationary when they are used to make judgments of local simultaneity. This is that it is a consequence of our three principles (isotropy, equivalent laws, and similar clocks) and the assumption that light has its one-way velocity (c) relative to each fundamental frame through which it passes, that the universe is homogeneous and its geometry is everywhere uniform, and so that a formula known as the Robertson-Walker line element will hold. This asserts that[7] for any fundamental frame taken as origin an interval ds^2 is invariant:

$$ds^2 = c^2dt^2 - R^2(t)\left\{\frac{du^2}{l - ku^2} + u^2d\theta^2 + u^2sw^2\theta d\varphi^2\right\}$$

ds is the spatio-temporal interval between two neighboring events E_1 and E_2. $ds = 0$ when these events are points on the path of a ray of light. E_1 occurs on a fundamental frame located by spherical polar coordinates (u, θ, φ) at cosmic instant t, and E_2 occurs at $(u + du, \theta + d\theta, \varphi + d\varphi)$ at cosmic instant $t + dt$. θ and φ are the normal angular coordinates. u is a "co-moving" radial coordinate, such that if a frame has it at one instant it has it at all instants. It is related to the frame's "coordinate distance" r by $u = r/R(t)$. $R(t)$ is a function of time, known as the scale factor of the Universe, and measures the extent to which frames are spread out. $R'(t)$ is the rate of change of $R(t)$. The greater $R'(t)$, the greater (for given $R(t)$) is the velocity with which galaxies are receding from each other. k is a constant having values $k = 0$ for Euclidean Space, $k = +1$ for elliptic space (and also for spherical space), $k = -1$ for hyperbolic space. From the Robertson-Walker metric cosmologists develop various more specific "models," that is cosmological theories, by giving a particular value to k and making some particular supposition about the equation governing the value of $R(t)$.

The Robertson-Walker metric has the consequence that the one-way velocity of light on distant galaxies is *not* the same relative to all fundamental observers at all instants. The proper distance from a galaxy G_a at the origin, of a photon, the particle of light, emitted from a galaxy G_b having u-coordinate u_b at cosmic instant t_1 in the direction of G_a, is at cosmic instant t_2:

$$l = R(t_2)\left\{\sigma(\mathrm{v_b}) - \int_{t_1}^{t_2} \frac{\mathrm{cdt},}{\mathrm{R(t)}}\right\}$$

$$where\ \sigma = (u_i) = \int_0^{u_i} \frac{du}{\sqrt{1 - kv^2}}$$

Hence the velocity of the photon at t relative to G_a, marking velocity of approach by a negative sign, will be

$$\frac{dl}{dt} = R'(t_2)\left\{\sigma(\mathrm{u_b}) \int_{\mathrm{t_i}}^{\mathrm{t_2}} \frac{\mathrm{cdt}}{\mathrm{R(t)}}\right\} - c$$

So, given an expanding (or contracting) universe (that is, $R'(t) \neq 0$), the (one-way) velocity of light passing through a distant galaxy must be judged by each observer to vary with his distance from that galaxy. Of course a distant observer cannot measure that velocity at that instant, but he needs to make a hypothesis about what that velocity is in order to explain the subsequent behavior of the photon (e.g. when it will reach him) and the hypothesis which will most simply explain that behavior is given by the above formula. One consequence of that formula is that, with a rapid rate of expansion of the universe, photons omitted from some distant galaxies will never reach us. (Such galaxies are said to lie beyond our "event horizon".) Yet such galaxies are a finite distance away from us, and so if light had the same one-way velocity on such a galaxy relative to our galaxy as it does on our galaxy, any such photon would reach us at some later time. The situation in the actual universe as described by modern cosmology is quite different from that in the universe of Special Relativity.

As we saw in Part 3, the local laws of physics acquire their simplest form if we suppose that the velocity of light locally is the same in all directions relative to the local frame. But that, we have now seen, has the consequence that it will not be the same in all directions relative to any other fundamental frame. I conclude that cosmic time provides a unique simplest and so probably correct standard of simultaneity not merely on the cosmic scale, but also on the local scale (by selecting a unique frame of reference, that of the fundamental frame associated with the local galaxy, relative to which local measurements should be made).

Notes

1 This paper is a new version of material originally published as Chapter 11 of my *Space and Time*, second edition, The Macmillan Press, 1981. I am most grateful to Dr Pedro Ferreira for reading an earlier version of this paper and helping me to correct it in the light of the most recent cosmological developments.

2 There are prospects for replacing the cesium scale by a different scale also based on radiation frequency, an optical frequency of a line of either mercury or strontium, which being much greater than the cesium frequency would allow for more precise measurements of temporal interval. See *Science*, 19 November, 2004. This new scale would be merely a more precise form of the previous scale; it would not yield measurements different from those of the old scale, merely ones more precise. However it may well become established that the value of the fine structure constant (α), and so the radiation spectra of elements, are changing with time. (See for example Barrow 2003: Chapter 12.) In that case physicists would need to introduce a time scale based on fundamental laws in which that constant does not appear but from which it follows how that constant changes with time.

3 I am not claiming that in all cases nothing could be ø when the best (logically possible) tests show that it is not ø, or conversely; merely that for any given ø, there is some reason for supposing that whether or not something is ø is whatever is shown by the best logically possible tests. I suggest, however, that reflection on the concept of "equal temporal interval" suggests that necessarily two temporal intervals are of equal length if the best (logically possible) primary tests indicate that they are of equal length.

4 For examples of such conventionalism, see Reichenbach 1958: 117, "All definitions … [of temporal interval] … are equally admissible"; and Carnap 1966: 83: "We cannot say that the pendulum is the 'right' choice as the basis for our time unit and my pulse beat the 'wrong' choice."

5 This paper led to a panel discussion Grünbaum *et al.* 1969: 1–81; to which Ellis replied in 1971: 177–203.

6 For the first stages of such a reinterpretation and demonstration of its consistency see Grünbaum 1964: Chap. 12, Section B.

7 The mathematical results which follow are due largely to Rindler. See Rindler 1956: 662–77.

References

Barrow, J. D. (2003) *The Constants ofNature*, Vintage, Chap. 12.

Carnap, Rudolf (1966) *Philosophical Foundations of Physics*, Basic Books.

Einstein, Albert (1923) 'On the Electrodynamics of Moving Bodies' (originally published 1905), in (trans.) W. Bennett and G. B. Jeffrey, *The Principle of Relativity*, Methuen.

Ellis, Brian (1971) 'On Conventionality and Simultaneity', *Australasian Journal of Philosophy* 49, 177–203.

Ellis, Brian and Peter Bowman (1967) 'Conventionality in Distant Simultaneity', *Philosophy of Science* 34, 116–36.

Grünbaum, Adolf (1964) *Philosophical Problems of Space and Time*, Routledge and Kegan Paul.

Grünbaum, Adolf, Wesley C. Salmon, Bas C. van Fraassen and Allen I. Janis (1969) 'A Panel Discussion of Simultaneity by Slow Clock – Transport in the Special and General Theories of Relativity', *Philosophy of Science* 36, 1–81.

Milne, E. A. (1948) *Kinematic Relativity*, Clarendon Press, passim and especially pp. 244f.

Newton, I. (1934) *Principia*, Scholium to Definition, viii.

Reichenbach, Hans (1958) *The Philosophy of Space and Time* (originally published 1928) (trans. M. Reichenbach and J. Freund), Dover Publications.

Rindler, W. (1956) 'Visual Horizons in World Models', *Monthly Notes of the Royal Astronomical Society* 116, 662–77.

Swinburne, Richard (2001) *Epistemic Justification*, Clarendon Press.

Zahar, Elie (1983) 'Absoluteness and Conspiracy' in Richard Swinburne (ed.), *Space, Time and Causality*, D. Reidel.

10 Presentism, eternalism and relativity physics

Thomas M. Crisp

Eternalism, roughly, is the view that our most inclusive quantifiers range over past, present and future entities; its opposite is presentism, the view that our most inclusive quantifiers range only over present entities. Many have argued against presentism on the grounds that presentism is incompatible with the theory of relativity.[1] Relativity, goes the argument, is a paradigmatically successful scientific theory; presentism contravenes it; so much the worse for presentism. I shall urge in what follows that this line of reasoning is inconclusive at best. I grant that orthodox relativity theory favors an ontology of eternalism. But I shall draw on some recent work by Julian Barbour and collaborators to argue that there are presentist-friendly variations on orthodox relativity theory on all fours with the orthodox approach in terms of empirical adequacy and theoretical virtue. Current physics, I shall urge, gives us no good reason to prefer orthodoxy to these unorthodox variations. The upshot, I shall suggest, is that the incompatibility of presentism and orthodox relativity theory implies nothing very interesting about how to resolve the presentism/eternalism debate.

I begin by stating more precisely what I shall mean by "eternalism" and "presentism," then show why relativity theory is commonly thought to imply the former.

Presentism, eternalism and orthodox relativity

Eternalists think of reality as spread out in time as well as space, comprising past, present and future entities. Presentists disagree, holding that every spatiotemporal thing whatsoever is a present thing. Less roughly, I shall think of eternalism as the thesis that the spatiotemporal world – the totality of spatiotemporal entities – is embedded in a four-dimensional manifold M of point-locations such that (a) M is structured by a primitive distance relation – the space-time interval; and (b) M is isomorphic to at least one model of general relativity. And presentism, I shall say, is the thesis that the spatiotemporal world – the totality of spatiotemporal entities – is embedded in an enduring, *three*-dimensional manifold of point-locations structured by a primitive spatial distance relation.

So far, eternalism and presentism. We look next at why relativity theory is typically thought to imply the former. On their usual construals, special and general relativity are *space-time* theories: roughly, theories that attempt to predict and explain physical phenomena in terms of the geometrical properties of a space-time manifold, a four-dimensional, differentiable manifold on which geometrical objects are defined at every point. We can think of a space-time theory *T* as having two parts: following van Fraassen (1987), its *theoretical structure* and its *theoretical hypotheses.* The theoretical structure of *T* is a family of mathematical space-time models, where each model is an n-tuple $< M, \Phi_1, \ldots, \Phi_{n-1} >$ such that *M* is a four-dimensional, differentiable point manifold and $\Phi_1, \ldots, \Phi_{n-1}$ are *geometrical objects* defined everywhere on *M* (roughly, set-theoretical mappings defined on *M* representing its geometrical structure and matter-energy distribution). The theoretical hypotheses of *T* describe the relationship between its space-time models and the empirical world. On the orthodox approach, special and general relativity include what we might call an *eternalist* theoretical hypothesis,[2] a proposition to the effect that one or more of the models of the theory is an isomorphic representation of *space-time*, the four-dimensional space of locations-at-a-time in which all physical goings-on – past, present and future – are embedded.[3]

Plainly enough, then, orthodox relativity theory implies eternalism since it includes a theoretical hypothesis to the effect that eternalism is true. One might well wonder, though, whether there's any deep incompatibility between presentism and relativity theory. For can't we simply replace the eternalist theoretical hypothesis of orthodox formulations of the theory with a presentist alternative? As follows: instead of thinking of some relativistic space-time model as an isomorphic representation of a four-dimensional spatiotemporal world, we suppose that only a *part* of some model represents by isomorphism. We suppose a foliation or slicing of this model's manifold into a series of three-dimensional space-like hypersurfaces. We then construe *one* member of this series as an isomorph of the *three*-dimensional world; other members of the series are construed as representations of past and future states of this 3-world. The entire series represents the evolution of the 3-world through time.

To be sure, this isn't the usual way of proceeding, but is it a *substantive* departure from orthodox relativity? Many physicists, I think, would answer this question in the affirmative. Adding a presentist theoretical hypothesis to relativity requires privileging a foliation of one or more of its models: for some one of its models, recall, we specify a particular slicing of the model into a series of space-like hyperspaces and think of the resulting series of 3-spaces as uniquely representing the evolution of our 3-world through time. But many would say that adding a preferred foliation to relativity in this way means rejecting both the letter and the spirit of the theory. One of Einstein's most important insights, it is commonly thought, is the idea that there *is* no privileged foliation of space-time (or our models thereof), no foliation such that it alone tells the correct story of how the cosmos evolves over time. Adding a presentist theoretical hypothesis to relativity means rejecting this central tenet of modern physics.

The upshot: presentism conflicts with both the letter and the spirit of orthodox relativity theory. Eternalism doesn't. Wherefore, says conventional wisdom, relativistic physics favors eternalism over presentism.

Well and good: relativistic physics, on its orthodox construal, favors eternalism over presentism. The interesting question is whether current physics gives us good reason to prefer the orthodox approach over the unorthodox, presentist-friendly approach sketched above. I shall argue that it doesn't. My tack will be to sketch a particular implementation of this unorthodox approach in the context of general relativity (GR) and then argue that current physics gives us no reason to prefer orthodox GR to this unorthodox variant. Though there are other ways of implementing a presentist approach to GR, I focus on one that is closely connected to some recent work on gravitation by Julian Barbour and collaborators. I'll close by saying something about the implications of my approach to GR for special relativity (SR).

A presentist-friendly variation on GR

As we've seen, a space-time theory has two parts, its theoretical structure (some class of mathematical space-time models) and its theoretical hypotheses (a set of propositions describing the relationship between the models of the theory and the empirical world). My presentist-friendly variation on GR – henceforth, presentist GR, "PGR" for short – has the following theoretical structure: its models are all and only the CMC-foliable models of general relativity. (There is a large subclass of the class of general relativistic models whose members admit of a unique foliation into surfaces of constant mean curvature (CMC). For an introduction to the notion of CMC slicing, see Gordon Belot and John Earman (2001: 239–40).)

PGR's theoretical hypotheses mimic those of orthodox GR except that it replaces orthodoxy's eternalist theoretical hypothesis[4] with a presentist theoretical hypothesis. The latter hypothesis consists of the following proposition:

> (PTH) At least one model of PGR represents the evolution of Space over time.

I shall now explain this proposition. To start with, it presupposes that the entirety of the physical world is embedded in Space, an enduring, three-dimensional object made up of enduring point-locations and structured by a primitive spatial distance relation. (It presupposes, then, that the spatio-temporal world is three- as opposed to four-dimensional.) It also presupposes that Space "evolves" over time. Something *x* evolves over time, let us say, if there is some property F such that x bears the *having* or *instantiation* relation – which I suppose to be a two-term connection of things to their properties – to F, but *WAS* (x doesn't bear the *having* relation to F) or *WILL* (x doesn't bear the *having* relation to F).

(PTH) says that at least one model of PGR "represents" the evolution of Space over time. This claim may be understood as follows. At least one model of PGR represents the evolution of Space over time, we shall say, iff there is at least one model $< M, g, T >$ of PGR that admits of a CMC foliation into a sequence S^* of space-like hypersurfaces such that there is a one-one map from a *history* H of Space onto S^* that takes each *instantaneous state* of H to a *corresponding* member of S^*. The key terms in the latter sentence may be explained as follows.

An *instantaneous state* is a triple $< \Sigma, h_{ij}, \phi >$ such that Σ is a three-dimensional point manifold, h_{ij} is a Riemannian metric defined everywhere on Σ, and ϕ is a family of fields $\phi_1, \ldots, \phi_n$ such that $\phi_1, \ldots, \phi_n$ are scalar, vector and/or tensor fields defined everywhere on Σ that, intuitively, represent the distribution of matter and energy across Σ. $< \Sigma, h_{ij}, \phi >$ is an instantaneous state *of* Space, we shall say, iff there is a one-one mapping Φ : Space $\rightarrow$ Σ such that (i) Φ maps spatial distances between space points p and q onto equal distances induced by h_{ij} between $\Phi(p)$ and $\Phi(q)$; and (ii), roughly, the quantity of mass-energy at a space point p is coded by the "values" of ϕ at $\Phi(p)$.

A *history* of Space, then, is any series S of instantaneous states such that (i) S is ordered by an irreflexive, asymmetric, and transitive relation R such that, for any $x, y \in S$, $R(x,y)$ iff *IS, WAS, OR WILL BE* (y is an instantaneous state of Space and *WAS* (x is an instantaneous state of Space)); and (ii) one member of S is presently an instantaneous state of Space.

Finally, an instantaneous state $< \Sigma, h_{ij}, \phi >$ *corresponds* to a hypersurface Σ' of $< M, g, T >$ iff there's a diffeomorphic embedding $\Phi : \Sigma \rightarrow M$ such that $\Phi(\Sigma) = \Sigma'$, $\Phi^*(h_{ij})$ is the Riemannian metric on Σ' induced by g, and $\Phi^*(\phi)$ codes the three-dimensional matter-energy distribution on Σ' induced by T.

In summary, according to (PTH), the physical world is embedded in an evolving 3-space whose history is given isomorphic representation by at least one CMC-foliable general relativistic space-time model. Such is our presentist-friendly variation on GR, roughly construed. The main difference between it and standard GR, the difference that matters for metaphysics anyway, is this: whereas the latter is typically construed as a theory describing the large-scale geometrical structure of a four-dimensional space-time, our presentist variation is a dynamical theory describing the evolution over time of a curved, three-dimensional space and its contents.

This variation on GR is closely related to an approach to gravitation recently advocated by Julian Barbour and collaborators.[5] Their view takes its start from the so-called "3+1" approach to general relativity, also known as *geometrodynamics*.[6] On the standard formulation of GR, the basic equations of the theory are Einstein's field equations, which describe the distribution of metric and matter fields across a four-dimensional space-time. With geometrodynamical GR, the basic equations are different. On Barbour's version, the basic equation is a Jacobi action principle that determines a class of geodesics through an infinite dimensional configuration space. Each point in the configuration space is an instantaneous 3-geometry (and perhaps an instantaneous matter-energy distribution, though this is typically omitted for simplicity). The geodesics through the configuration space fixed by the action principle are sequences of 3-geometries and may be thought of as representing dynamically possible histories of an evolving 3-space, each point on the curve being an instant of time in the 3-space's history. The connection with the standard space-time approach to GR is straightforward: the physically possible sequences of 3-spaces fixed by the action principle will each correspond to some sequence of space-like hypersurfaces induced by some foliation of a general relativistic space-time model.

Thus far the general contours of Barbour's approach to geometrodynamics. Recently, he and collaborators[7] have extended this basic approach to obtain a class of theories defined on a configuration space called *conformal* superspace, where this is the space one arrives at by quotienting the space of Riemannian metrics defined on some 3-manifold Σ– called *Riem*(Σ) – by the group of all diffeomorphisms of Σ and the group of all conformal transformations on *Riem*(Σ).[8] As above, the basic equation on their approach is an action principle that determines a class of geodesics through conformal superspace, each of which may be thought of as representing a dynamically possible history of an evolving 3-space. What's of interest to our project is that, on one of the theories they develop,[9] the class of curves through conformal superspace yielded by their action principle is precisely the class of interest to the proponent of PGR: the class of curves corresponding to the CMC foliations of CMC-foliable models of GR.

A word about presentism and absolute time

There are various ways of filling in the above sketch of our presentist variation on GR. For instance, I omitted from the above discussion any mention of a temporal metric. I suggested that the history of Space and its contents is given by a series of instantaneous states ordered by an analogue of the B-theoretic earlier/later relation. One might wonder what the metric structure of this series is. Pick any two "instants" in the series and there's this question: how much time elapses between the them? Note that to suppose that there's a definite answer to this question is to suppose that time has an *absolute* metric, in the sense that, for any two past, present or future instantaneous states of Space, there's a well-defined temporal distance between them, an *amount* of time such that it is the amount of time that lapses between the states in question.

Now, one can certainly fill in the details of our sketch of a presentist theoretical hypothesis by adding an absolute temporal metric into the sketch. I want to suggest, though, that the presentist need not do so. I want to suggest, that is, that the presentist is free to view the temporal structure of her theory as metrically amorphous, or as it's sometimes (misleadingly) put, as "timeless".

Barbour has argued in a fascinating series of papers (and books) for what he calls a "timeless" view of physics.[10] For present purposes, we needn't go into the full details of his program. I want instead to focus on his claim that GR can be interpreted as a timeless theory. As I'll try to show, presentism can be thought of as a timeless theory of time in just the same way that GR can be thought of as a timeless theory of gravity.

As we've already seen, Barbour's interpretation of GR takes its start from the so-called geometrodynamical approach to GR. What makes Barbour's version of geometrodynamics "timeless" is this.[11] The action principle at the heart of his theory – the principle that fixes the class of curves through superspace representing dynamically possible histories – has no time parameter. Points along the curves through superspace are labeled by an arbitrary parameter λ – arbitrary because any monotonically increasing parameterization will do – and the action principle describes the geodesic trajectories through superspace in terms of this arbitrary parameter. The dynamics of his theory, then, can be fully spelled out without mention of a time parameter and without postulating a primitive temporal metric. His approach is "timeless," then, in this sense: it postulates no primitive temporal metric.

What then of GR's usual notion of proper time? Barbour shows, in effect, that it is definable from other quantities invoked by his dynamics. Very roughly, we can think of his action principle as yielding a set of equations that describe dynamically possible sequences of 3-geometries in terms of arbitrary, monotonically increasing parameterizations of the sequences. It turns out that certain ways of parameterizing a sequence S greatly simplify the equations describing it. Barbour shows that these simplifying parameterizations correspond to proper time along a special class of time-like curves through the general relativistic space-time model corresponding to S. Proper time along all other time-like curves through the model is then definable in terms of the proper time along these curves. The upshot: GR's usual notion of proper time is definable from the dynamics of his theory, dynamics which invoke no primitive temporal metric.

Now, it's no part of my project to comment on the merits of Barbour's approach. I bring it up because it suggests an analogous "timeless" approach to presentism, one I think will be attractive to some presentists. Go back to our series of instantaneous states representing the history of Space and its contents. We inquired above into the metric structure of the series. Pick any two "times" in the series and there's this question: how much time elapses between the instantiation of one time and the other? One might follow Barbour in denying that the question is well-formed. On this approach, our ersatz history of Space has no intrinsic temporal metric. We can arbitrarily specify a metric: e.g. we could say that the time lapse between times t_1 and t_2 is given by the number of times the earth goes round the sun in the interval between t_1 and t_2. Or we could say that it's given by the number of oscillations of some particle in that interval. Or we could refuse to specify a single metric, but give instead trajectory relative metrics, the ones suggested by Barbour, for instance, in the previous paragraph. Some choices, note, will yield simpler laws of motion than others. The ones that yield the simplest laws will likely be the ones we work with in physics. But in each case, we're merely *stipulating* a metric; the series in itself is metrically amorphous.

I find this sort of "presentism-cum-conventionalism-about-time" attractive. I shall think of our presentist theoretical hypothesis as postulating this sort of presentism.

I turn now to the main question of this paper: does current physics give us some reason to prefer orthodox GR to the presentist variation on GR just sketched? Call this the Main Question.

Clarifying the Main Question

What we're looking for, then, is some reason to prefer orthodox GR over the unorthodox variation sketched above. Some will have broadly philosophical reasons for preferring orthodoxy. So, for example, some will argue that presentism is metaphysically impossible or somehow incoherent; others will argue that metaphysics must always start with what scientists believe and since scientists aren't presentists, presentistic GR is a non-starter. But I wish to bracket these kinds of considerations. (I shall assume that presentism isn't obviously impossible or incoherent and that one can sensibly do metaphysics without starting from the deliverances of current science.) Instead, I want to focus on a set of narrower questions: is there some *empirical* reason to prefer orthodox GR? Does PGR conflict with some deliverance of experimental physics? If not, then is there some *theoretical* reason to prefer orthodox GR – some truth-indicating theoretical virtue or set thereof displayed by orthodoxy and not by our presentist-friendly variant? So, for instance, if our presentist GR involved elaborate conspiracy on the part of nature to hide her true structure, a conspiracy not found in the orthodox picture, this would count against it. Or if presentist GR involved a myriad of unexplained coincidences not found in its orthodox cousin, or if it postulated explanatorily superfluous entities in a way that its orthodox cousin didn't, these would be theoretical costs not shared by orthodox GR and would constitute theoretical grounds for preferring the latter. So the Main Question comes to this: are there either empirical or truth-indicating theoretical grounds of the sort displayed in the last few sentences for preferring orthodoxy?

Empirical reason to prefer orthodoxy?

Here we're wondering whether there is observational evidence inconsistent with presentist GR. So is there some deliverance of astronomy, say, or some branch of experimental physics that conflicts with presentist GR? The answer to this question would be "yes" if we knew from astronomical observations, say, that the space-time models isomorphic to our universe aren't CMC foliable. It'd also be "yes" if we knew of observable phenomena which couldn't be *embedded* into a CMC-foliable model of GR. In very general terms, an observable phenomenon is embeddable into a model of GR if it can be represented by an n-tuple $< M, \Phi_1, \ldots, \Phi_{n-1} >$ such that (a) M is a point manifold representing some bit of space-time, or some bit of space through time, and $\Phi_1 \ldots \Phi_{n-1}$ are geometrical objects representing M's geometry and, e.g. observable particle trajectories or patterns of field intensity; and (b) $< M, \Phi_1, \ldots, \Phi_{n-1} >$ is a *submodel* of some general relativistic space-time model.

So what is there to say? Is there some deliverance of astronomy or experimental physics that conflicts in one of these ways with presentist GR? Perhaps so: the deliverances of quantum physics would seem not to be embeddable into the models of classical GR – so the search for a quantum theory of gravity. But set this problem aside: this is as much a problem for orthodoxy as it is for presentist GR. Is there some deliverance of observational astronomy or experimental physics – *modulo* the deliverances of quantum physics – that conflicts in one of these ways with presentist GR? The answer here, I think most would agree, is "no".

Theoretical reasons to prefer orthodoxy?

Presentist GR violates a central tenet of modern physics: no privileged foliations

Presentist GR, unlike its orthodox cousin, postulates a preferred foliation on its space-time models. On presentist GR, one way of foliating space-time is physically significant in a way that other ways of foliating aren't: it alone carves our space-time models into a sequence of 3-slices that accurately represents the evolution of Space over time; other ways of foliating induce sequences of 3-slices that *mis*represent the history of Space. But this violates a central tenet of modern physics according to which there is no preferred foliation of space-time (or our models thereof), no one correct way of carving space-time into spaces and times so that it and it alone tells the story of how the cosmos evolves over time.

True enough, presentist GR violates this tenet of modern physics. But in the present dialectical context, this doesn't amount to much of an objection since the question we're considering is whether we have some reason for preferring theories that comport with this tenet over theories that don't.

Egalitarianism about reference frames has led to some highly successful physics

Classical relativistic physics, one of the most successful research programs in physics to date, was built on the assumption that there are no privileged reference frames. Doesn't this strongly suggest that there aren't any?

No. An equally plausible explanation for the success of classical relativistic physics is that the world is housed in a 3-space whose evolution is described as above.

Orthodox GR is simpler than presentist GR

Like orthodox GR (and for that matter, most of any space-time theory), presentist GR can be formulated in generally covariant form. But its generally covariant formulation will be slightly more complicated than orthodoxy's on account of its more restrictive class of models. Some might take this as indication that orthodoxy is a simpler theory and thus more likely to be true.

But this strikes me as misguided. The class of PGR models isn't in any obvious sense more miscellaneous or gerrymandered than the full class of general relativistic space-time models, nor are the theoretical hypotheses of PGR in any obvious sense more complicated than those of orthodoxy. Since, on the view of theories we're working with, the so-called semantic view, a theory just *is* a class of models and a class of theoretical hypotheses, the claim that orthodoxy is a simpler theory looks wrong to me.

Maybe the suggestion, rather, is that, *ceteris paribus*, space-time theories that admit of simpler generally covariant formulation are more likely to be true. But why think this? One main reason for thinking this is that simplicity of generally covariant formulation often goes together with economy of postulated space-time structure (in general, the more structure you postulate, the more complicated your field equations get), and thus economy of postulated ontology. Since economy of the latter sort is arguably a truth-indicating theoretical virtue, to the extent that it explains the simplicity of a theory's generally covariant formulation, the latter will also be a truth-indicating theoretical virtue.

This all seems plausible enough, but in the present case, it's not clear that orthodoxy's simpler generally covariant formulation is a function of it postulating less ontology than presentist GR. This is because orthodoxy in fact postulates *more* ontology than presentist GR: it postulates a vast realm of past and future entities not postulated by the presentist.

Are there other reasons for thinking that orthodoxy's simpler generally covariant formulation is truth indicating? Not any obvious ones.

Presentist GR is ad hoc in a way that orthodoxy isn't

Presentist GR arbitrarily restricts the class of general relativistic models to the CMC-foliable models. Since there's no physical motivation for this restriction, presentist GR has a whiff of *ad-hoc-ness* about it: it looks as if the only motivation for the restriction is the desire to produce a theory friendly to presentism.

A few points in reply. First, it is notoriously difficult to spell out what the charge of *ad-hoc-ness* comes to exactly. I rather suspect that accusing a theory of *ad hoc-ery* has more to do with expressing one's distaste for the theory than with substantive criticism. Be that as it may, secondly, it's not true that there's no physical motivation for restricting the class of general relativistic models to the CMC-foliable models. For instance, Barbour and collaborators have recently argued that (a) there are good reasons for thinking the evolution of the cosmos over time to be governed by a variational principle like that at the heart of their theory; and (b), that, if it is, then CMC evolution of 3-geometry over time follows as a consequence.[12] It's an open question, I gather, whether they're right about this, but this much seems clear: it's going too far to say there's no physical motivation for thinking CMC slicing fundamental. Some very good physicists seem to think there's physical motivation for it.[13]

Thirdly, and most importantly, the objection presupposes that a physical theory is somehow untoward if a motivation for introducing certain of its features is the desire to produce a theory that comports well with one's background metaphysic. But this is surely wrong. GR, I take it, wasn't objectionably *ad hoc* when first put forward by Einstein, though many of its features were motivated by his background metaphysical assumptions (e.g. his assumptions about Leibnizian relationalism and Machian accounts of inertial effects).

Presentist GR postulates a conspiracy on the part of nature to conceal her true structure

This sort of complaint is sometimes lodgers against the so-called "neo-Lorentzian" approach to special relativity (see e.g. Balashov *et al.* 2003: 327–46). On this approach, instead of postulating a Minkowski manifold and analyzing the usual special relativistic effects (length contraction, time dilation, relativity of simultaneity, etc.) in terms of the Lorentzian geometry of this manifold, one postulates Newtonian space-time and classical electrodynamics and tries to account for the special relativistic effects by appeal to motion-induced deformations in our measuring equipment. The idea here is that, though the underlying metrical structure of space and time is the classical Newtonian structure – replete with absolute spatial and temporal distances, absolute velocity and absolute rest – systematic deformations in our measuring equipment hide this structure from us.

Given this sort of theory, one can see why one might complain of conspiracy.[14] But in the case of presentist GR, it's harder to see how the objection would go. PGR does not postulate an absolute temporal metric hidden from observation by slowing clocks: it postulates, rather, that there is no temporal metric intrinsic to the world. It does not postulate a Euclidean spatial metric hidden from us by distorted measuring equipment, and it does not imply that any object has a well-defined absolute velocity: it says that for any two instants of time, there's no one correct answer to the question how much time elapses between them. Thus it implies that there's no one right answer to the question how fast an object moves with respect to Space.

Still, one might complain, it *is* committed to absolute simultaneity and rest. First, it's committed to a well-defined simultaneity relation that holds independently of reference frame. That relation may be defined as follows:

> events x and y are absolutely simultaneous $=_{\text{df.}}$ for some events u and v, $u = x$ and $v = y$.

And second, it is committed to a well-defined notion of absolute rest or sameness of place that holds independently of reference frame: roughly, an object exhibits absolute sameness of place over an interval of time if, over the course of that interval, it overlaps one and the same region of Space.

And, goes the complaint, insofar as it's committed to these, it's committed to an untoward conspiracy. After more than a century of trying, no one has been able to detect a preferred simultaneity frame or distinguished state of rest. If there are such things, nature is doing a good job of hiding them.

Supposing there are such things, though, what is the sense, exactly, in which nature is "hiding" them? Let the skeptic speak: "Well, there's the fact that, where the effects of gravity can be ignored, relative to local Lorentz frames, the laws governing non-gravitational interactions appear to be Lorentz invariant. And doesn't this fact constitute a kind of conspiracy on the part of nature to hide absolute rest and simultaneity from us if there are such things?" No, not in any obvious sense. Is there some reason to *expect* that, were presentism true, the local laws governing non-gravitational interactions – the laws holding on a small neighborhood of any space point – would single out the rest frames? Not that I can see. Consequently, I can't see any reason for thinking that the defender of PGR who grants that the local laws governing non-gravitational interactions are locally Lorentz invariant is *ipso facto* committed to an untoward conspiracy to hide what we would otherwise expect to see.

Maybe the worry is this. Whereas the proponent of orthodoxy has an *explanation* for the fact that local Lorentz invariance holds – namely, the local approximate Minkowski geometry of space-time – the presentist must chalk this up to unexplained coincidence. So Michelle Janssen and Yuri Balashov:

> In the neo-Lorentzian interpretation it is, in the final analysis, an unexplained coincidence that the laws effectively governing different

> sorts of matter all share the property of Lorentz invariance, which originally appeared to be nothing but a peculiarity of the laws governing electromagnetic fields. In the space-time interpretation this coincidence is explained by tracing the Lorentz invariance of all these different laws to a common origin: the space-time structure posited in this interpretation.
>
> (Balashov and Janssen 2003: 341–42)

In brief: defenders of orthodoxy can appeal to the local Minkowski geometry of space-time to explain why the laws governing different sorts of non-gravitational interactions are all locally Lorentz invariant; PGR must chalk this up to fantastic luck. This looks bad for PGR.

But, by way of reply, it's very difficult to see why the local Minkowski geometry of space-time postulated by orthodox GR should count as an *explanation* of the fact that the laws governing non-gravitational interactions are locally Lorentz invariant. (Here I follow recent arguments to this effect by Harvey Brown and Oliver Pooley 2001: 270–71). It certainly seems *possible* – in the broadly logical sense – that matter and energy should be spread across a space-time whose background geometry is Minkowskian, but the laws governing the matter and energy don't satisfy the Lorentzian symmetries. (The scenario envisaged is analogous to the scenario envisaged by the neo-Lorentzian in which one has matter and energy spread across a space-time whose background geometry is Newtonian, but the laws governing the matter and energy don't satisfy the Galilean symmetries.) Since a Minkowskian background geometry would seem to be compatible with non-Lorentz-invariant laws, it's hard to see why the existence of local Minkowski geometry counts as an *explanation* of local Lorentz invariance.

Brown and Pooley argue that local Lorentz invariance is not something that can be derived from GR's postulation of local Minkowski geometry, but must be independently *assumed* (Brown and Pooley 2001: 270). If they're right, then the proponent of orthodoxy and the proponent of PGR look to be in the same boat: each postulates local Lorentz invariance; neither proposes to explain it.

(Before I move on, a recent claim by Barbour and collaborators is relevant here (Barbour *et al.* 2002). They claim to be able to derive local Lorentz invariance from the action principle at the heart of the Baierlein-Sharp-Wheeler (BSW) approach to geometrodynamics. If they're right, then they've shown that local Lorentz invariance isn't among the fundamental postulates of GR, but a consequence of them. What's interesting about this for our purposes is that, if they're right, the presentist and the proponent of orthodoxy are still in the same boat: for the presentist can grant with equanimity that the dynamical evolution of Space and its contents over time is governed by Barbour's BSW Lagrangian.)

What about unfoliable space-times?

It is well known that not all models of GR can be foliated into global space-like hypersurfaces. Gödel (1949), for instance, proposed a widely discussed model of general relativity that cannot be foliated. Doesn't this make trouble for the foregoing presentist variation on GR?

No. Why would it? If PGR is right, the physically possible models of GR are just the CMC-foliable ones, and unfoliable space-times like Gödel's represent interesting but physically impossible scenarios.

The leading candidates for a theory of quantum gravity don't involve a fixed foliation

Because it has proven impossible to marry classical general relativity to quantum theory, it is generally assumed that classical general relativity is, strictly speaking, false, and that the true theory of gravity is some yet-to-be-worked-out quantum theory of gravity.

So far, there is no consensus about exactly how this theory will go, but at present, the leading approaches make no use of a preferred foliation.[15] Does this give us reason to prefer orthodoxy? (Here we wouldn't be thinking of orthodoxy as preferable to PGR in the sense that the former but not the latter is *true*. Rather, we'd be thinking that the former is a better approximation to reality than the latter.)

I don't think it gives us any reason at all. For this to constitute reason to prefer orthodoxy, we'd also need reason to think that, even if we were to eventually settle on a theory lacking a preferred foliation, there wouldn't be an alternative, empirically adequate fixed foliation theory of comparable theoretical virtue. But, I submit, we've currently no reason to think the latter.

I'm out of ideas now about why orthodoxy might be theoretically preferable to our unorthodox, presentist-friendly variant. I take the upshot of this discussion to be that the answer to the Main Question is "no": at present, anyway, we have no good reason (from physics) for thinking that orthodox GR is more likely to be true than the above-sketched presentist-friendly variant. Note that I do *not* claim that we have reason for thinking that the latter is more likely to be true than the former. I claim only that current physics gives no reason to prefer one over the other. If I'm right about this, my argument leads us to this conclusion: presentism's incompatibility with orthodox GR tells us very little about whether we should or shouldn't be presentists.

On presentism and special relativity

Thus far, my discussion has focused exclusively on GR. If what I've said is on target, presentism's incompatibility with orthodox GR sheds little light on the presentism/eternalism debate. But most discussion in the literature on presentism and relativity physics focuses on the incompatibility of presentism with *special* relativity (SR). I close, then, by considering the question whether presentism's incompatibility with SR sheds any further light on the presentism/eternalism debate.

One thing to note here is that the above sort of strategy for formulating a presentist-friendly variation on GR doesn't carry over to SR. My approach to GR took its start from the fact that there is a large class of general relativistic space-time models – viz., the class of CMC-foliable models – for which there is a natural definition of a global time function. For obvious reasons, this sort of approach isn't available in the case of SR: Minkowski space-time admits of no non-arbitrary partitioning into space-like slices. So, you might object, even if I'm right that there are physically viable variations on GR friendly to presentism, the same can't be said of SR. So much the worse for presentism, then, since SR is a paradigmatically successful physical theory.

But this sort of argument strikes me as misguided. Suppose I'm a proponent of PGR. You point out that my theory conflicts with SR. (Perhaps as follows: my theory implies presentism; presentism conflicts with SR; so my theory conflicts with SR.) True enough, I reply, my theory conflicts with SR, but then again, so does orthodox GR: SR says that space-time is flat; orthodox GR, together with the fact that the universe contains matter, implies that it isn't. So PGR conflicts with SR, but so does orthodox GR. So far, anyway, my theory is no worse off than orthodox GR.

Reply: well, given orthodox GR, SR is at least locally correct: where the effects of gravity can be ignored, relative to local Lorentz frames, the laws governing non-gravitational interactions take their special relativistic forms. But so too with PGR: the proponent of PGR grants that, where the effects of gravity can be ignored, the laws governing non-gravitational interactions take their standard special relativistic forms.

Reply: yes, but it's a fundamental principle of SR that there are no preferred slicings of space-time. PGR conflicts with this principle and thereby incurs a steep cost. True enough, PGR conflicts with SR's "no privileged slicing" principle. But why think this a steep cost? Is this it? Is it that failure to comport with the principle costs PGR in empirical adequacy or truth-indicating theoretical virtue? It doesn't seem so. If the above arguments are on target, PGR and orthodox GR are on par in these respects. Since orthodox GR is about as good as it gets when it comes to empirical adequacy and theoretical virtue, it would seem that PGR pays no steep price in either for failing to comport with SR's "no privileged slicing" principle. Are there other reasons for thinking that PGR's failure to comport with the principle tells against the theory? No obvious ones.

Now I'm out of ideas about why we should think PGR's conflict with SR makes trouble for it. As best I can tell, SR gives us no good reason at all for thinking PGR false. But PGR entails presentism. So SR gives us no good reason for thinking presentism false.

I take the upshot of all of this to be that presentism's incompatibility with SR and GR implies nothing very interesting about how to resolve the presentism/eternalism debate.[16]

Notes

1 For discussion, see Callender (2000: S587–S599); Godfrey-Smith (1979: 233–44); Hinchliff (1996: 119–36; 2000: S575–S586); Maxwell (1985: 23–43); Monton (*forthcoming*); Putnam (1967: 240–47); Rea (1998: 225–60; 2003: 246–80); Rietdijk (1966: 341–44; 1976: 598–609); Saunders (2002: 277–92); Savitt (1994: 463–74; 2000: S563-S574); Sider (2001); Sklar (1974; 1981: 129–42); Stein (1968: 5–23; 1970: 289–94; 1991: 147–67).

2 This term comes from Monton (*forthcoming*).

3 I am characterizing the orthodox approach to SR and GR in terms of substantivalism about space-time. I realize, however, that it's a matter of controversy whether relativity requires commitment to substantivalism, and if so, what sort of substantivalism it requires. Since, so far as I can tell, everything I say in the sequel can be re-cast in terms of some version or other of relationalism, I shall ignore this controversy and press on as if orthodox relativity was a theory about a substantival space-time.

4 Where this, again, is a proposition to the effect that at least one general relativistic space-time model is an isomorph of our four-dimensional world.

5 See Anderson *et al.* (2003: 1571–1604); Barbour (2003: 1543–70); Barbour *et al.* (1999; 2002: 3217–48). For an introduction to Barbour's approach to relativity, see Barbour (1994: 2853–73) and Barbour (1999). For introduction and discussion of Barbour's approach aimed at philosophers rather than physicists, see Pooley (2001) and Butterfield (*forthcoming*).

6 So-called "geometrodynamical" GR traces back to work in the late 1950s by Paul Dirac, Richard Arnowitt, Stanley Deser and Charles Misner. See e.g. Arnowitt *et al.* (1962).

7 See Anderson *et al.* (2003); Barbour (2003) and Barbour *et al.* (1999).

8 Some of the theories they develop are defined on a slightly more complicated configuration space. I shall ignore this complication.

9 I am speaking here of their "CS+V" theory. See Anderson *et al.* (2003: 1586ff).

10 See Barbour (1999) for a popular exposition of the view.

11 For discussion, see Barbour (1994), Butterfield (*forthcoming*) and Pooley (2001).

12 See Anderson *et al.* (2003) and Barbour *et al.* (2002).

13 For discussion of other motivations, see Qadir *et al.* (1985); Tipler (1988: 222), and Valentini (1996: 45–66).

14 Though for an extended defense of the neo-Lorentzian approach, see Craig (2001).

15 For discussion, see Monton (*forthcoming*).

16 Thanks to William Lane Craig, Hans Halverson, Brad Monton, Brian Pitts and an anonymous referee for helpful comments on earlier versions of this paper.

References

Anderson, Edward, Julian Barbour, Brendan Foster, and Niall Ó Murchadha (2003) 'Scale-Invariant Gravity: Geometrodynamics', *Classical and Quantum Gravity* XX, 1571–1604.

Arnowitt, R., S. Deser and C.W. Misner (1962) 'The Dynamics of General Relativity', in Louis Witten (ed.), *Gravitation: An Introduction to Current Research*, New York: Wiley, pp. 227–65.

Baerlein, Ralph F., David H. Sharp and John A. Wheeler (1962) 'Three-Dimensional Geometry as Carrier of Information about Time', *Physical Review* 126, 1864–65.

Balashov, Yuri and Michelle Janssen (2003) 'Presentism and Relativity', *British Journal for the Philosophy of Science*, LIV, 327–46.

Barbour, Julian (1994) 'The Timelessness of Quantum Gravity: The Evidence from the Classical Theory', *Classical and Quantum Gravity*, XI, 2853–73.

Barbour, Julian (1999) *The End of Time: The Next Revolution in Physics*, Oxford: Oxford University Press.

Barbour, Julian (2003) 'Scale-Invariant Gravity: Particle Dynamics', *Classical and Quantum Gravity*, XX, 1543–70.

Barbour, Julian, Brendan Foster, and Niall Ó Murchadha (2002) 'Relativity Without Relativity', *Classical and Quantum Gravity*, XVIIII, 3217–48.

Barbour, Julian and Niall Ó Murchadha (1999) 'Classical and Quantum Gravity on Conformal Superspace', available online at: http://arxiv.org/abs/gr-qc/9911071

Belot, Gordon and John Earman (2001) 'Pre-Socratic Quantum Gravity', in C. Callender and N. Huggett (eds.), *Philosophy Meets Physics at the Planck Scale*, Cambridge: Cambridge University Press, pp. 213–55.

Brown, Harvey and Oliver Pooley (2001) 'The Origins of the Spacetime Metric: Bell's Lorentzian Pedagogy and its Significance in General Relativity', in Craig Callender and Nick Huggett (eds.), *Physics Meets Philosophy at the Planck Scale: Contemporary Theory in Quantum Gravity*, Cambridge: Cambridge University Press, pp. 270–71.

Butterfield, Jeremy (*forthcoming*) 'The End of Time?' *British Journal for the Philosophy of Science.*

Callender, Craig (2000) 'Shedding Light on Time', *Philosophy of Science* (Proceedings), LXVII, S587–S599.

Craig, William Lane (2001) *Time and the Metaphysics of Relativity*, Dordrecht: Kluwer Academic Publishers.

Kurt Gödel (1949) 'An Example of a New Type of Cosmological Solution of Einstein's Field Equations of Gravitation'. *Reviews of Modern Physics* 21, 447–50.

Godfrey-Smith, William (1979) 'Special Relativity and the Present', *Philosophical Studies*, XXXVI, 233–44.

Hinchliff, Mark (1996) 'The Puzzle of Change', in James Tomberlin (ed.), *Philosophical Perspectives10: Metaphysics*, Cambridge, MA: Blackwell, pp. 119–36.

Hinchliff, Mark (2000) 'A Defense of Presentism in a Relativistic Setting', *Philosophy of Science* (Proceedings) LXVII, S575–S586.

Maxwell, Nicholas (1985) 'Are Probabilism and Special Relativity Incompatible?' *Philosophy of Science*, LII, 23–43.

Monton, Bradley (*forthcoming*) 'Presentism and Quantum Gravity', in V. Petkov and A. Wayne (eds.), *The Ontology of Spacetime.*

Pooley, Oliver (2001) 'Relationalism Rehabilitated? II: Relativity', available online at: http://philsci-archive.pitt.edu/archive/00000221/00/rehab2ps.pdf

Putnam, Hillary (1967) 'Time and Physical Geometry', *The Journal of Philosophy* LXIV, 240–47.

Qadir, Asghar and John Archibald Wheeler (1985) 'York's Cosmic Time Versus Proper Time as Relevant to Changes in the Dimensionless 'Constants,' K-Meson

Decay, and the Unity of Black Hole and Big Crunch,' in Errol Gotsman and Gerald Tauber (eds.), *From SU(3) to Gravity: Festschrift in Honor of Yuval Ne'eman*, Cambridge: Cambridge University Press, pp. 383–94.

Rea, Michael C. (1998) 'Temporal Parts Unmotivated', *Philosophical Review* CVII, 225–60.

Rea, Michael C. (2003) 'Four-Dimensionalism', in Michael J. Loux and Dean W. Zimmerman (eds.), *The Oxford Handbook of Metaphysics*, Oxford: Oxford University Press, pp. 246–80.

Rietdijk, C. W. (1966) 'A Rigorous Proof of Determinism Derived from the Special Theory of Relativity', *Philosophy of Science*, XXXIII, 341–44.

Rietdijk, C. W. (1976) 'Special Relativity and Determinism', *Philosophy of Science* XLIII, 598–609.

Saunders, Simon (2002) 'How Relativity Contradicts Presentism', in Craig Callender (ed.), *Time, Reality and Experience*, Cambridge: Cambridge University Press, pp. 277–92.

Savitt, Steve (1994) 'The Replacement of Time', *Australasian Journal of Philosophy* LXXII, 463–74.

Savitt, Steve (2000) 'There's No Time Like the Present (in Minkowski Spacetime)', *Philosophy of Science* (Proceedings) LXVII, S563–S574.

Sider, Theodore (2001) *Four-Dimensionalism: An Ontology of Persistence and Time*, Oxford: Clarendon Press.

Sklar, Lawrence (1974) *Space, Time and Spacetime*, Berkeley, CA: University of California Press.

Sklar, Lawrence (1981) 'Time, Reality, and Relativity', in Richard Healy (ed.), *Reduction, Time, and Reality*, New York: Cambridge University Press, pp. 129–42.

Stein, Howard (1968) 'On Einstein-Minkowski Space-Time', *The Journal of Philosophy* LXV, 5–23.

Stein, Howard (1970) 'A Note on Time and Relativity Theory', *The Journal of Philosophy*, LXVII, 289–94.

Stein, Howard (1991) 'On Relativity Theory and Openness of the Future', *Philosophy of Science* LVIII, 147–67.

Tipler, Frank J. (1988) 'The Sensorium of God: Newton and Absolute Space', in G. V. Coyne, M. Heller, and J. Zyncinski (eds.), *Newton and the New Direction of Science*, Vatican City: Specola Vaticana, pp. 215–28.

Valentini, Antony (1996) 'Pilot-Wave Theory', in James T. Cushing, Arthur Fine, and Sheldon Goldstein (eds.), *Bohmian Mechanics and Quantum Theory: An Appraisal*, Dordrecht: Kluwer Academic Publishers, pp. 45–66.

Van Fraassen, Bas (1987) 'The Semantic Approach to Scientific Theories', in Nancy J. Nersessian (ed.), *The Process of Science: Contemporary Philosophical Approaches to Understanding Scientific Practice*, Dordrecht: Kluwer Academic Publishers, pp. 105–24.

11 The Special Theory and absolute simultaneity

John Lucas

Summary

The Special Theory has no place for absolute simultaneity. This seems to rule out the possibility of there being a God who knows everything that happens when it happens, and responds to the prayers of His children. Does this matter? Some theologians claim that it does not, because God is outside time. It is possible to argue that the Deity of the Greek philosophers is outside time, but implausible to maintain that the God of the Judaeo-Christian tradition is. Nevertheless that has been maintained by many theologians. Their arguments are not convincing. The only way a timeless Deity can be omniscient is for Its knowledge to be of timeless truths, and the only way for timeless truths to be ascribed to temporal propositions is to evacuate them, as in contemporary arguments against fatalism, of all substance. And if prayers are to result, as a consequence of the existence of a timeless Deity, in the boon asked for actually occurring, both must be the effect of the unchangeable structure of the universe. Not only must God's knowledge be insubstantial, but Man's freedom must be illusory.

The Special Theory is thought to downgrade time, but popular accounts are misleading. Although time is integrated with space in Minkowski space-time, Minkowski space-time has Lorentz signature (+++−), with time-like separations being fundamentally different from space-like separations. And although Minkowski space-time gives rise to the picture of a "block universe," the case is no different, logically speaking, from that of Newtonian space and time, which can similarly be regarded as a space-time formed by the Cartesian product of the two.

The Special Theory relativises simultaneity and seems to undercut the distinction between past, present and future, but the threat this poses is exaggerated. Many of the arguments adduced to subvert the distinction between past, present and future are fallacious, depending on an equivocation between different sorts of relative simultaneity. Moreover, the Special Theory is a theory of electromagnetism, but electromagnetism is not the whole of physics. Maxwell's equations, like Newton's laws of motion, are invariant under time reversal. But this does not prove that time does not have a direction. We do not feel impelled to jettison thermodynamics on the grounds that the Second Law of Thermodynamics is incompatible with the direction-indifferent time of mechanics and electromagnetism. Rather, we recognize that mechanics and electromagnetism do not tell the whole story, and "throw away advanced potentials," when we come to apply the theory to real situations.

Einstein formulated the principle that there are no privileged observers and elevated it from a truth about electromagnetism to a general metaphysical principle; whereas Newton, more cautiously, did not. And once we move from the Special Theory to the General Theory, privileged frames of reference are allowed: cosmic time flows generally and universally. Quantum mechanics likewise requires a real worldwide distinction between past, present and future. All over the universe wave functions collapse. There is a worldwide tide of co-presentness sweeping through the universe, as superposed wave functions, representing many alternative possibilities, collapse into one single eigen-function, representing what is, and remains, actually the case.

Modern physics vindicates our intuitive sense of the reality of time and does not require us to banish God to a timeless, insubstantial existence or to regard ourselves as mere worms crawling along a predetermined life-line. Time is real: and we are free agents, responsible for the decisions we make.

The challenge

Einstein's Special Theory of Relativity[1] seems to show that our common-sense view of time is mistaken. Time, we are told, is only relative, and doomed, like space, to fade away in a more scientific understanding of the universe. Or, again, time is merely something psychological, which we humans experience but with not objective significance in the real world. More seriously, it is argued that the Special Theory shows that different inertial frames of reference date distant events differently, so that the common-sense distinction between past, present and future no longer applies.

These claims bear weightily on our view of ourselves and our understanding of God. If time is unreal, the way we live now in the time of our mortal life is of no significance *sub specie aeternitatis*. All our notions of aspiration, achievement and responsibility are undercut, if there is no difference between the future and the past. God cannot know everything that is happening if there is no worldwide present tense, and cannot meaningfully hear our prayers or respond to them if the answers may equally well be given before the petitions as after them. The Special Theory threatens both the omniscience of God and any possibility of our having a real personal relation with Him. Modern physics, it seems, has dealt traditional theism a death blow.

Does this matter?

One strand of theological thought, going back to Plato and Parmenides, maintains that God is outside time, and that time is unreal, and hence that the findings of modern physics are nothing to be worried about. It is easy to say that God is outside time but difficult really to believe all its implications. The Judaeo-Christian account of God's dealings with His people is a temporal narrative. God led His people out of Egypt, parted the Red Sea for them, spake by the Prophets, gave His only son to be incarnate, to suffer under Pontius Pilate, to be crucified, and on the third day raised from the dead; and He continues to guide His church, and answer the prayers of the faithful. It is difficult to see how the timeless Deity contemplated by the philosophers in Athens could have anything to do with the temporal God known in Jerusalem.

Nevertheless efforts have been made, notably by Augustine, Boethius and Aquinas in the past, and in recent years by Eleonore Stump and Norman Kretzmann (1981); Paul Helm (1988); and Brian Leftow (1991) to explain how a timeless Deity could nonetheless be an agent in history and have personal relationships with human beings. But in spite of heroic efforts to give an intelligible account, the concessions that have to be made to timelessness destroy the personality which makes God not only an object of reverence but also the source of love. No convincing reconciliation has been achieved between the God of the Philosophers and the God of Abraham, Isaac and Jacob. The history of salvation in the Old Testament and the New Testament is through and through temporal. If Christ *did* not die, and *has* not *risen* again, then is our hope vain?

Misapprehensions of the Special Theory

Some of the consequences popularly supposed to be implied by Einstein's Special Theory are due to misunderstandings. The integration of time with space *is* a fundamental feature of the Special Theory. Whereas temporal duration was an invariant quantity under the Galilean transformations, it is not under the Lorentz transformation, which preserves only space-time separations, but not either spatial ones or temporal ones separately. In the excitement of discovery it was natural that exaggerated claims should be made – that all is relative, that time and space must fade away and be replaced by Minkowski space-time, that nothing can go faster than light. But despite its name, the Special Theory does not make out that all is relative: it stresses the absoluteness of space-time. Minkowski famously announced

> Henceforth space by itself, and time by itself, are doomed to fade away into mere shadows, and only a kind of union of the two will preserve an independent reality.
>
> (Minkowski 1923: 75)

But although Minkowski space-time can be described as a four-dimensional manifold, its metric (and hence also the topology based on it) is radically unlike that of a Euclidean 4-space. The topological difference between Minkowski space-time, with Lorentz signature (+++−), and a simple Euclidean 4-space, with Lorentz signature (++++), is profound. The Lorentz signature (+++−) marks an absolute distinction between time-like directions and time-like separations on the one hand and space-like directions and space-like separations on the other.

Mathematicians are sometimes tempted to discount this difference between time-like and space-like separations – they can transform Minkowski space-time into a simple 4-space by setting $t = -ix_4/c$, which would enable them to regard time as simply a fourth dimension of four-dimensional space, with a standard metric and topology. True, but that transformation and its converse, $x_4 = ict$ are not transformations of no consequence; transformations that convert simple exponential functions into periodic ones and *vice versa* are clearly highly significant ones, and, as we have seen, the metric and topology of Minkowski space-time are radically different from those of a Euclidean 4-space. Although there are important similarities between time and space, they fall far short of showing that the dimension of time was simply a fourth dimension of space multiplied by *ic*.[2]

Minkowski's space-time lends strong support to the picture of a "block universe" and suggests that we should reject our ordinary experience of time in favor of a much more abstract and mathematical concept, in which time is a fourth dimension, analogous to the three dimensions of space, with the future already in existence, just waiting to occur. We are to picture world-lines in Minkowski space-time as already timelessly existing and ourselves as going along a world-line encountering our pre-determined future as it takes place – in Weyl's words

> The objective world simply *is*, it does not happen. Only to the gaze of my consciousness, crawling upward along the lifeline of my body, does a section of this world come to life as a fleeting image in space which continually changes in time.
>
> (Weyl 1949: 116)

But this is only a picture. Although Minkowski space-time, a four-dimensional manifold with Lorentz signature, seems more integrated and solid than a Cartesian product of a three-dimensional space with a one-dimensional time, exactly the same picture can be imagined if we think of a world-line in Newtonian space-and-time – the Galilean transformation can be seen as just a special case of the Lorentz transformation in which the speed of light is infinite. Both are available, both misleading, both non-obligatory.

Frames of reference

The Special Theory teaches us that simultaneity is not, as we supposed, an absolute concept, but is always relative to some inertial frame of reference. If we change the inertial frame of reference, we change the events to be reckoned simultaneous. Putnam, Rietdijk and Lango have argued for a timeless view of time, because of inconsistencies in our ordinary concept of simultaneity, and hence of there being no absolute distinction between present, past and future (Rietdijk 1966: 341–44; 1976: 598–609; Lango 1969: 340–50; Putnam 1979: 198–205). If we consider two observers in two inertial frames of reference moving with a uniform velocity with respect to each other, they will have different lines of simultaneity, and an event that is future to one will be past with respect to the other. In Figure 11.1 the continuous line represents the line of simultaneity of, say, an astronaut, whose inertial frame of reference is his spaceship, and the line of dashes -- -- -- -- -- -- represents the line of simultaneity of someone in a laboratory, whose inertial frame of reference is the Earth.

The event E is both past and present, since it is in O's past, but is also simultaneous with an event F, say, which is itself simultaneous with the observer's "Now". So, it seems, we are forced to concede that there can be no inherent characteristic distinguishing the future from the present or the past, and that they are differentiated by nothing inherent but only their relation to particular observers. From this it is a natural step to think of space-time as a timeless entity, in which all events, present, past and future, already in some sense exist; and to think of the apparent passage of time as nothing real, but only our becoming aware of what is already there, waiting for its moment of disclosure to us, in the way that Weyl suggests.

There is clearly something wrong with this argument, plausible though it sometimes seems. It depends on a confusion between two-termed and three-termed relations. It assumes that simultaneity is a two-termed relation, whereas in point of fact in the Special Theory, simultaneity is a three-termed relation, and we always have to ask "simultaneous with respect to what inertial frame of reference?" Simultaneity with respect to a given inertial frame of reference *is* an equivalence relation, so that if one event is simultaneous with another and that other is simultaneous with a third, then the first is simultaneous with the third. But simultaneity with

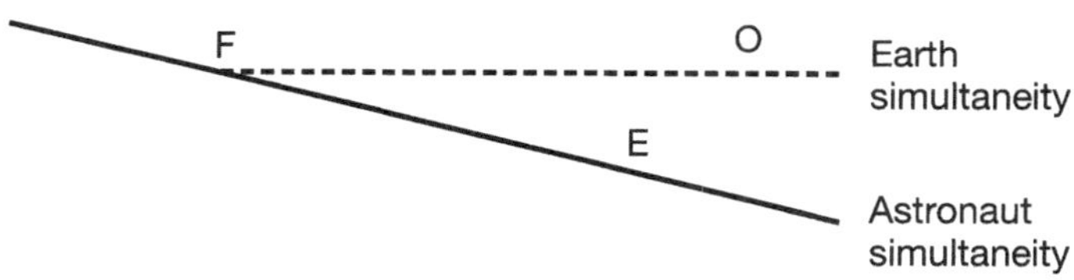

Figure 11.1 Event E is simultaneous with F which is itself simultaneous with O.

respect to *different* inertial frames of reference is not an equivalence relation at all.

Plato was the first to spot the flaw in this type of argument. He was thinking about the foundations of geometry and came to realize that two figures being the same could be taken to mean that they were the same size, or that they were the same shape. Two congruent triangles are the same size: two similar triangles are the same shape, but need not be the same size. It is in fact a defining characteristic of Euclidean geometry that it allows for figures of the same shape to be of different sizes (on the surface of a sphere, if we know that a spherical triangle has each of its angles a right angle, we know that the triangle is an octant, and that each of its sides is one quarter of the circumference of a great circle). Had it not been for the problem of irrational numbers, the Academy might well have based geometry on the independence of shape and size instead of on the axiom of parallels. As it was, the distinction was used by Plato and Aristotle only in political philosophy to show that equality in one respect need not imply equality in all respects.

The first argument of Putnam, Rietdijk and Lango fails: no event in the past of an observer can also be simultaneous with his "Now"; their argument depends on a simple equivocation in the use of the term "simultaneous". But still, it may be objected, there is an inconsistent ascription of dates, without appeal to lines of simultaneity. In Figure 11.2 the event G is both future and past, future with respect to observers on Earth, past with respect to the Astronaut.

The ascription of futurity, presentness, or pastness is, therefore purely relative, and indicates no real property of the event. But that argument also fails. The lines of simultaneity for a given inertial frame of reference do not determine what is currently going on at distant places, but only what dates should be ascribed to them in order to make electromagnetic phenomena coherent. The Special Theory is a theory of electromagnetic radiation and determines distances and durations, and hence positions and dates, by means of light signals. As far as electromagnetic phenomena are concerned, we have no other means of telling exactly when a distant event – an event in the "Absolutely Elsewhere and Anywhen" – takes place; but for any given inertial frame of reference, *if* we ascribe the same date to all events on a

Figure 11.2 Event G seems to be both future and past, future with respect to observers on Earth, and past with respect to the astronaut.

particular line of simultaneity, *then* the laws of electromagnetism apply neatly and yield harmonious results. So far as the Special Theory goes, simultaneity is a rather superficial and frame-dependent property, which we find useful for assigning dates to different events in different places, but which is not of fundamental importance in accounting for the propagation of causal influence by means of electromagnetic radiation. The ascription of presentness, pastness, or futurity, to events outside the light cone is nominal rather than real, and has no bearing on their ontological status.

Because the Special Theory speaks of space and time, it has been assumed that what it says about them is a metaphysical truth rather than a tenet of a particular physical theory. But electromagnetism is not the whole of physics, and physics is not the whole of knowledge. Other physical theories bear upon the interpretation of our theory of electromagnetism; so also our understanding of concepts we use in everyday life. Time in our ordinary experience has a direction. Even Weyl has the gaze of my consciousness crawling upwards and does not suggest that the lifeline of my body could reverse its direction in time, as it frequently does in space. It is similarly "anisotropic" in thermodynamics. The anisotropy of time has long been a problem for physicists, because it is difficult to reconcile with the fact that Newtonian mechanics, and likewise electromagnetism, are time-reversible: if we substitute $-t$ for t in the equations, the result is the same. Yet that is not how things happen – trees grow, plates break, fires radiate energy, milk mixes with tea, but not the reverse: we can tell at once if a film is being run backwards.

It was a great problem for classical physics how to re-insert "the arrow of time" into a physics that was apparently isotropic with respect to time. There were many attempts to explain the Second Law of Thermodynamics – that entropy increases – as a feature of boundary conditions. Causally antecedent conditions were more precisely specifiable than their effects, and so had higher negative entropy. If events were generally explicable, and if the sort of explanation that every event ought to have was a causal explanation, then just as the direction of explanation led to conditions of higher and higher negative entropy, so the passage of time led to conditions of higher and higher positive entropy.

But discomfort remains: the concept of explanation seems dangerously anthropocentric, and in the long-run entropy should decrease just as often as it increases. Twentieth-century physics, however, embeds the anisotropy of time more deeply in the structure of the universe. There could, so far as the theory of electromagnetism goes, be a reversal of the direction of time if there were also a change of parity, since the time-like dimension is integrated with a number of space-like dimensions that are all on a par. Professor Dummett could make warriors to have been brave and rain dances to have been performed, provided that their hearts were on the right and they danced their reels counter-clockwise (Dummett 1964: 338–59).[3] So there could be an instantiation of the laws of physics in which time was reversed, but it would be a looking-glass world that was, granted orientability, inaccessible from ours (Nerlich 1976, 1994: Chap. 2). Indeed, the invariance goes deeper, involving charge as well as time and parity: but the implication is the same; although it may be convenient on occasion to think of an electron moving backwards in time rather than a positron moving forwards, the difference between the two directions of time is as great as that between positive and negative charge (Gardner 1964, 1990). The theory of electromagnetism that allows time reversal does not give us the whole picture, and physicists doing simple electricity and magnetism were wise to distinguish advanced from retarded potentials, and to discard the former.

The directedness of time provides a further argument against assimilating time to space. For any space-like direction can be transformed by a continuous rotation into the opposite direction, whereas there is no continuous transformation of one direction of time into its opposite. The only transformation of one temporal direction into its opposite is a discontinuous reflection, and this is linked to the parity of the 3-space space-like sub-space of Minkowski space; and if, as we believe, that 3-space is orientable, its parity is a fundamental feature, not susceptible to change – a point reinforced by the concomitant connection with charge. So, again, however much it is convenient on occasion to treat time as a fourth dimension, it remains a significantly un-space-like one.

The implications of ontology

Difficulties remain. If we ascribe a difference of ontological status to what is present or past from what is future, it will have awkward consequences with our theory of electromagnetism. It will pick out some inertial frame of reference as the absolute one. Any God's-eye view of what is past and unalterable, as opposed to what is future and still open and indeterminate, must impose a privileged frame of reference that is contrary to Einstein's principle that there are no privileged observers. But Einstein propounded his Special Theory in the context of electromagnetism, and so long as we confine our attention to electromagnetism we have good reason for adopting the Lorentz transformation as our way of correlating positions and dates in different inertial frames of reference; but it does not follow that that is the way we must ascribe positions and dates to distant events, and that no other ascription could be correct (Capek 1961, 1968; Stein 1968: 5–23; 1970: 289–94; Sorabji 1980: 114–19; Torretti 1983: 249–51).

Newton also had difficulties with Absolute Space. Newtonian mechanics is relativistic. Newton's laws of motion come out the same in any uniformly moving frame, so that it is impossible to tell, within Newtonian mechanics, whether we are moving with respect to Absolute Space or not. Newton recognized this, but maintained, correctly, that Absolute Space was nevertheless a coherent notion, and opined that the center of the solar system might be at rest in it. Later, when electromagnetic theory was being developed, Michelson and Morley sought to determine the rest frame of the ether, and if they had succeeded, we should have undoubtedly taken that as being at rest in Absolute Space. Similarly now, if some superluminal velocity of transmission of causal influence were discovered, we should be able to distinguish inertial frames of reference and say which were at rest absolutely and which were moving. If, for instance, we were able to communicate telepathically with extra-terrestrial beings in some distant galaxy, or if God were to tell us what was going on in Betelgeuse now, then we should have no hesitation in restricting the Principle of Relativity to the phenomena of electromagnetism only. Hence it cannot be an absolute principle foreclosing absolutely any possibility of absolute time.

These are not mere possibilities. In many of the models that cosmologists use – solutions of the field equations of the General Theory – there is a worldwide cosmic time that flows, if not evenly and uniformly, at least generally and universally. There are thus also preferred hypersurfaces (not necessarily flat) of simultaneity constituting a worldwide present and separating a real unalterable past from a possible future not yet actualized. The Special Theory is countered not just by the General Theory, but more fundamentally by quantum mechanics. It is difficult to make sense of quantum mechanics, and the interpretation I adopt is rejected by many. But it is natural to try and interpret it realistically and to construe the collapse of the wave packet as a real event in which the many possible eigenvectors, with their associated eigenvalues, give way to a single eigenvector with one definite eigenvalue. If this is so, there is a definite moment of truth when possibilities become definitely true or definitely false. There is a fact of the matter, quite independent of whether we know it or not, and of how and when we know it. Knowledge may be unable to travel faster than the speed of light, but reality does not have to travel at all. Galaxies may be thousands of light-years away, and we shall be able to assign a date to a particular collapse only thousands of years after the event, and our assignment may well depend on an idiosyncratic choice of frame of reference, but nonetheless there will have been a definite moment at which the event occurred quite independent of any frame of reference. There is a worldwide tide of actualisation – collapse into "eigenness" – constituting a preferred foliation by hypersurfaces (not necessarily flat) of co-presentness sweeping through the universe: a tide which determines an absolute present. In this, quantum mechanics goes beyond anything that thermodynamics and cosmology suggest, which, although witnessing to there being a definite directedness in time, do not pick out any particular time as pre-eminently real – the moment of truth when possibilities become actual or else fade away. Quantum mechanics, however, does. Although we still do not understand it, and its interpretation is still much disputed, it now appears to be irremediably probabilistic, and not only insists on the arrow being kept in time, but distinguishes a present as the boundary between an alterable future and an unalterable past.

Reformulation of the Special Theory

The tension between the Special Theory and other parts of physics can be resolved by restricting the scope of the Special Theory. This is satisfactory to the working physicist. Different fields of research and their problems require different theories. We do not worry about the time reversibility of Newtonian mechanics when constructing a steam engine. It is quite reasonable to limit the application of the Special Theory to electromagnetism and simply say that when dealing with electromagnetic phenomena we should use "relativistic space-time" and "relativistic dates and durations" and "relativistic positions distances," thus ensuring that all our calculations are covariant under the Lorentz transformations.

In our metaphysical moods, however, we may feel uncomfortable. If, after all, there is an absolute frame of reference, defining absolute dates and durations and absolute positions and distances, then surely these are the ones we ought to use, as expressing what is really the case. We are not absolutely obliged to conform to this requirement, but it is reassuring to see that we can. In standard formulations of the Special Theory, we ascribe to the arrival of a radio pulse at a distant object a date that is exactly half way between the date of its emission and the date of the return of its reflection. But we are not obliged to. All that the principle of causality requires is that the date of the arrival of the radio pulse at a distant location should be after its emission and before the return of its reflection. It need not be half way between the two, but could be any fraction between 0 and 1. We can accommodate this by generalizing the Lorentz transformations to ε-Lorentz transformations, where the value of ε is not assumed to be one half, but can take any value between 0 and 1. The mathematics becomes more clumsy, but still works out. For most practical purposes we need not use it, for the same reasons as justify our using simple Newtonian mechanics to describe deck quoits on a ship steaming at a uniform velocity across the sea, or for describing terrestrial phenomena on the Earth which, we believe, is moving rapidly along its orbit round the sun. For practical purposes then the availability of ε-Lorentz transformations need occasion no worry.

A theoretical objection may still be mounted. The standard formulation of the Special Theory posits that the speed of light is the same in all directions, whereas the modified formulation does not. It assumes only that the round trip speed of light to a distant object and back again, is the same in all directions. And this latter assumption is intuitively less appealing. Arguments of symmetry can be adduced for the speed of light being the same in all directions; but if once it is allowed that light travels faster when outward bound than when coming back again (or *vice versa*), it no longer seems reasonable to postulate that the differences between the outward and return journey always cancel out. But they do. And if we are led to accept the theory on other grounds, we can accept that the Special Theory is in the same case as Newtonian mechanics. Neither postulates the existence of absolute space, but both are compatible with it. Although within each theory all inertial frames of reference are on a par, each theory is compatible with some being picked out on other grounds as being more privileged.[4]

Conclusion

The Special Theory led many philosophers to adopt an entirely tenseless view of time. They over-reified the Special Theory and extrapolated the Principle of Relativity beyond its proper sphere. So far as the Special Theory is concerned, distant events are distant. I cannot be there to know what happens when; I can only ascribe a position and a date on the strength of causal influences transmitted with a velocity no greater than the speed of light. If we ascribe positions and dates in accordance with the Lorentz transformation, then the laws of electromagnetism will come out the same whatever inertial frame of reference we adopt. But the laws of electromagnetism are not the only laws of nature, and other branches of physics give a different picture of time. Thermodynamics has time with a definite direction, cosmological versions of the General Theory postulate a cosmic time, and quantum mechanics is kind to the time of ordinary experience, regarding it mostly as an independent variable, not under pressure to be anything else, and as having a distinguished present and inherent directedness.

It is a good result. Far from undermining our intuitive sense of the reality of time, modern physics vindicates it. If we take science seriously, we are not required to banish God to a timeless insubstantial existence; nor are we required to regard ourselves as mere worms crawling along a predetermined lifeline. Time, as the concomitant of God being a person, is, as we have always known it to be, real: and we, as we always ought to acknowledge, are free agents, with minds of our own, able to decide what we are to do, and responsible for the decisions we make.

Notes

1 Throughout I shall talk of Einstein's Special Theory, which is a theory of electromagnetism, and his General Theory, which is a theory of gravitation, leaving out the words "of Relativity," which only darken counsel. Although relative in some respects, both theories are absolute in others and postulate a space-time that is invariant over all observers. It is, moreover, desirable to avoid possible confusion between the way Einstein differed from Newton, and the way that Leibniz and his relationist followers did.

2 For further arguments against simply construing t as an imaginary fourth dimension of space, see Misner *et al.* (1973: 51).

3 Reprinted in Dummett (1978: Chap. 19), and in Le Poidevin and M. MacBeath (1993: 117–33).

4 For a fuller exploration of an ε-Lorentz version of the Special Theory, see Tooley (1997: 337–73).

References

Capek, M. (1961) *The Philosophical Impact of Contemporary Physics*, Princeton University Press.

—— (1968) 'Time in Relativity Theory: Arguments for a Philosophy of Becoming', in J. T. Fraser (ed.), *The Voices of Time*, Penguin, pp. 434–54.

Dummett, M. A. E. (1964) 'Bringing About the Past', *Philosophical Review* 73, 338–59.

—— (1978) *Truth and Other Enigmas*, Duckworth.

Gardner, Martin (1964) *The Ambidextrous Universe*, Basic Books.

—— (1990) *The New Ambidextrous Universe*, W. H. Freeman.

Helm, Paul (1988) *Eternal God*, Oxford.

Lango, John W. (1969) 'The Logic of Simultaneity', *Journal of Philosophy* 66, 340–50.

Leftow, Brian (1991) *Time and Eternity*, Cornell University Press.

Le Poidevin, R. and M. MacBeath (1993) *The Philosophy of Time*, Oxford.

Minkowski, H. (1923) 'Space and Time', An Address delivered at the 80th Assembly of German Natural Scientists and Physicians, at Cologne, 21 September, 1908; reprinted in A. Sommerfeld (ed.), *The Principle of Relativity*, New York, 1923, p. 75.

Misner, C. W., K. S. Thorne and J. A. Wheeler (1973) *Gravitation*, W. H. Freeman.

Nerlich, G. (1976, 1994) *The Shape of Space*, Cambridge.

Putnam, H. (1979) 'Time and Physical Geometry', *Journal of Philosophy* (1967) 64, 240–47; reprinted in H. Putnam (ed.), *Mathematics, Matter and Method*, Philosophical Papers, I, Cambridge, pp. 198–205.

Rietdijk, C. W. (1966) 'A Rigorous Proof of Determinism Derived from the Special Theory of Relativity', *Philosophy of Science* 33, 341–44.

—— (1976) 'Special Relativity and Determinism', *Philosophy of Science* 43, 598–609.

Sorabji, R. (1980) *Necessity, Cause and Blame*, Duckworth.

Stein, Howard (1968) 'On Einstein-Minkowski Space-time', *Journal of Philosophy* 65, 5–23.

—— (1970) 'A note on Time and Relativity Theory', *Journal of Philosophy* 67, 289–94.

Stump, Eleonore and Norman Kretzmann (1981) 'Eternity', *Journal of Philosophy* 78, 429–58.

Tooley, M. (1997) *Time, Tense, and Causation*, Oxford.

Torretti, R. (1983) *Relativity and Geometry*, Oxford.

Weyl, H. (1949) *Philosophy of Mathematics and Natural Science*, Princeton.

Index

eBooks

eBooks – at www.eBookstore.tandf.co.uk

A library at your fingertips!

eBooks are electronic versions of printed books. You can store them on your PC/laptop or browse them online.

They have advantages for anyone needing rapid access to a wide variety of published, copyright information.

eBooks can help your research by enabling you to bookmark chapters, annotate text and use instant searches to find specific words or phrases. Several eBook files would fit on even a small laptop or PDA.

NEW: Save money by eSubscribing: cheap, online access to any eBook for as long as you need it.

Annual subscription packages

We now offer special low-cost bulk subscriptions to packages of eBooks in certain subject areas. These are available to libraries or to individuals.

For more information please contact webmaster.ebooks@tandf.co.uk

We're continually developing the eBook concept, so keep up to date by visiting the website.

www.eBookstore.tandf.co.uk

FSC
www.fsc.org
MIX
Papier | Fördert gute Waldnutzung
FSC® C083411

Kolibri 360 GmbH
Ferdinand-Jühlke-Straße 7
99095 Erfurt, Deutschland
produktsicherheit@kolibri360.de